PRODUCT ENGINEERING

Product Engineering

Eco-Design, Technologies and Green Energy

Edited by

DORU TALABĂ
University of Transilvania at Brasov, Brasov, Romania

and

THOMAS ROCHE
Research Group on Eco-Design, Galway Mayo Institute of Technology, Galway, Ireland

 Springer

A C.I.P. Catalogue record for this book is available from the Library of Congress.

ISBN 1-4020-2932-2 (HB)
ISBN 1-4020-2933-0 (e-book)

Published by Springer,
P.O. Box 17, 3300 AA Dordrecht, The Netherlands.

Sold and distributed in North, Central and South America
by Springer,
101 Philip Drive, Norwell, MA 02061, U.S.A.

In all other countries, sold and distributed
by Springer,
P.O. Box 322, 3300 AH Dordrecht, The Netherlands.

Printed on acid-free paper

TABLE OF CONTENTS

CONTRIBUTIONS

Part 4
ROBOTICS AND MANUFACTURING.............347

INVITED LECTURES

CONTRIBUTIONS

Part 5
GREEN ENERGY

INVITED LECTURES

CONTRIBUTIONS

PREFACE

This book contains an edited version of the lectures and selected contributions presented during the Advanced Summer Institute on "Product Engineering: Eco-Design, Technologies and Green Energy" organized at the Transilvania University of Brasov (Romania) in the period 14-21st of July 2004. The Advanced Summer Institute (ASI) was organized in the framework of the European FP5 funded project "ADEPT − Advanced computer aided Design of Ecological Products and Technologies integrating green energy sources" and was devoted to the Product Engineering field, with particular attention to the aspects related to the environmentally conscious design and green energy sources.

The objective of the ASI was to create the framework for meeting of leading scientists with PhD holders and advanced PhD students carrying out research in the field of Eco-Design, CAD, Simulation technologies, Robotics, Manufacturing and green energy sources. The aim was to create conditions for high level training through a series of 15 invited lectures presented by world reputed scientists, as well as to give possibilities for young researchers to present their achievements and to establish professional contacts. The ASI was seen also as an opportunity for academics, practitioners and consultants from Europe and elsewhere who are involved in the study, management, development and implementation of product engineering principles in the learning and teaching sectors, as well as professionals to come together and share ideas on projects and examples of best practice.

Out of the invited lectures, the ASI programme included a number of contributions from the other participants. In total, the event was attended by about 70 participants from 12 countries.

The topics covered areas of Product Engineering including new aspects related to the environmental issues, i.e.:

- ECO-Design,
- Computer Aided Design (CAD),
- Simulation technologies,
- Robotics and Manufacturing,
- Green Energy

x

Although usually these topics are addressed within distinct approaches, it was the idea of this ASI to bring together scientists from different areas of Product Engineering, such as to catalyze cross-fertilization and enable new ideas in an interdisciplinary framework.

The lectures included in the book have been presented as tutorials as well as state of the art papers in the respective areas of Product Engineering, providing thus a good overview of the current work in the field. Therefore it addresses a wide range of readers, from students to professors, from industrial experts to the researchers.

The publication of this book has been possible thank to the kind support from the European Commission in the framework of GROWTH programme of the Fifth Framework Programme for research and scientific development. For this reason the ASI Directors express hereby their full gratitude. The support from the Transilvania University and all the other partners in the project ADEPT is also acknowledged.

Braşov and Galway,
July 2004.

Doru Talabă and Thomas Roche

LIST OF PARTICIPANTS

Invited lecturers
Coiffet, Philippe - University of Versailles, France
Cuadrado, Javier - University of La Coruna, Spain
Dewulf, Wim - Katholieke Universitat of Leuven, Belgium
Duflou, Joost - Katholieke Universitat of Leuven, Belgium
Gogu, Grigore - French Institute of Advanced Mechanics, Clermont Ferrand, France
Horváth, Imre - Delft University of Technolgy, Netherlands
Kaplanis, Socrates - Technological Educational Institute of Patras, Greece
Mogan, Gheorghe Leonte - Transilvania University of Brasov, Romania
Nikravesh, Parviz - University of Arizona, USA
Rahnejat, Homer - Loughborough University, UK
Ray, Pascal - French Institute of Advanced Mechanics, Clermont Ferrand, France
Ritchie, James - Heriot Watt University of Edinburgh, UK
Roche, Thomas - Galway Mayo Institute of Technology, Ireland
Talabă, Doru - Transilvania University of Brasov, Romania
Tapia, Arantxa - University of Basque Country, San Sebastian, Spain

Participants
Alexandru, Petre - Transilvania University of Brasov, Romania
Andreica, Gabriel - Transilvania University of Brasov, Romania
Antonya, Csaba - Transilvania University of Brasov, Romania
Avram, Cătălin - Transilvania University of Brasov, Romania
Batog, Ionel - Transilvania University of Brasov, Romania
Bârsan, Lucian - Transilvania University of Brasov, Romania
Borca, Alina - Transilvania University of Brasov, Romania
Brănescu, Dumitru - Transilvania University of Brasov, Romania
Bucur, Camelia - Transilvania University of Brasov, Romania
Budală, Adrian - Transilvania University of Brasov, Romania
Butilă, Eugen Valentin - Transilvania University of Brasov, Romania
Butnariu, Silviu - FARTEC Comp. Brasov, Romania
Butnaru, Tiberiu - Transilvania University of Brasov, Romania
Cabezudo Maeso, Sara - University of Basque Country, San Sebastian, Spain

Canciu, Emil - Transilvania University of Brasov, Romania
Chişu, Emil - Transilvania University of Brasov, Romania
Ciofoaia, Vasile - Transilvania University of Brasov, Romania
Ciupercă Radu - Technical University of Moldova, Republic of Moldova
Creţan, Monica - Technical University of Iasi, Romania
Creţescu, Nadia - Transilvania University of Brasov, Romania
Daj, Ion - Transilvania University of Brasov, Romania
Dulgheru, Valeriu- Technical University of Moldova, Republic of Moldova
Eros, Izabela - Transilvania University of Brasov, Romania
Gavrilă, Cătălin - Transilvania University of Brasov, Romania
Girbacia, Florin - Transilvania University of Brasov, Romania
Gonzales, Manuel - University of La Coruna, Spain
Grama, Monica - Transilvania University of Brasov, Romania
Ionescu, Nicolae - "Politechnica" University of Bucharest, Romania
Jula, Aurel - Transilvania University of Brasov, Romania
Jaliu, Codruţa - Transilvania University of Brasov, Romania
Künzler, Urs - Martin Iseli, Berne University of Applied Sciences, Switzerland
Lateş, Mihai - Transilvania University of Brasov, Romania
Luca Moţoc, Dana - Transilvania University of Brasov, Romania
Neagoe, Mircea - Transilvania University of Brasov, Romania
Nedelcu, Anişor - Transilvania University of Brasov, Romania
Ochoa Laburu, Carlos - University of Basque Country, San Sebastian, Spain
Petra, Cosmin - Petru Maior University of Targu Mures, Romania
Pocola, Adrian - Technical University of Cluj Napoca, Romania
Podborschi, Valeriu - Technical University of Moldova, Republic of Moldova
Puiu, George - University of Bacau, Romania
Radu, Florin - Transilvania University of Brasov, Romania
Răşină, Cristina - Transilvania University of Brasov, Romania
Rusák, Zoltan - Delft University of Technology, Netherlands
Săvescu, Dan - Transilvania University of Brasov, Romania
Spyrogiannoulas, Antonis - Technological Educational Institute of Patras, Greece
Stareţu, Ionel - Transilvania University of Brasov, Romania
Şişcă, Sebastian - Transilvania University of Brasov, Romania
Teodorescu, Mircea - Loughborough University, UK
Tirziu, Florin - Transilvania University of Brasov, Romania
Vaculenco, Maxim - Technical University of Moldova, Republic of Moldova
Velicu, Radu - Transilvania University of Brasov, Romania

INTRODUCTION

In the last decades Product Engineering became more and more a multidisciplinary field including aspects from a wide range of scientific areas, still treated distinctly within the research institutions. Aspects related to the environment, aesthetic style, human factors and ergonomics are now critical issues for the success of a product on the market. The Advanced Summer Institute on "Product Engineering: Eco-Design, Technologies and Green Energy" focused on some of these topics with a particular attention paid to the aspects relevant for the environmental protection.

In this context, this book is structured on five chapters, covering the topics of Eco-Design, Computer Aided Design, Simulation Technologies, Robotics and Manufacturing and Green Energy.

Eco-Design is the main topic of the book. This methodology penetrates all aspects of design following the stream of waste and resource consumption across the whole product life cycle. This is the reason for which this subject was privileged including five lectures and three selected contributions that treated the eco-aspects from various perspectives: from the management and integration of the environmental impact information into the product life cycle and business environment, to the human aspects in the man-machine relation and aesthetics style, as part of the human natural environment.

The *Computer Aided Design* (CAD) topic is included with two lectures and two selected contribution focusing mainly on conceptual design aspects which are of crucial importance for the next generation of CAD systems. In the same idea new ways of interaction between the user and the CAD environment, e.g. the use of haptic immersion technologies have been presented.

Simulation technologies are frequently included in the Product Engineering textbooks, which usually present Finite Element Methodologies for analysis in various applications. For the Adavanced Summer Institute (ASI) and this book, the editors chosen to focus on a simulation technology with potential for assembling a wide range of simulation methods, including Finite Element Method ones: Multi-Body Systems (MBS) simulation, represented in this book by four invited lectures and four selected contributions was another privileged topic of the ASI. From the systematic and tutorial presentation of the MBS formulations and models to complex applications in multi-physics and real time simulation, a wide area was covered, illustrating the potential of this simulation technology.

Robotics and Manufacturing is another important field of the Product Engineering and was included in the book with two invited lectures and four selected contributions. Recent advances in this area are covered e.g. parallel robots and high speed machining.

Green Energy comes into the Product Engineering area of research in the context of the eco-design and generally the quest for alternative sources of renewable energy. This is demonstrated by the topics in the two lectures and three selected contributions, which present aspects on solar and wind energy technologies, two areas where Product Engineering is concerned both from the development of the systems themselves viewpoint and for the integration of this type of energy as much as possible into any other kind of product.

These apparently separate fields of research proved to belong to the same multidisciplinary mainstream and could not progress unless an integrated approach is adopted. New advances are now likely to produce changes in the entire product life cycle chain. This was illustrated within the ASI in many presentations from distinct sections illustrating similar methodologies used for different goals, e.g. the Virtual Reality techniques and Virtual Engineering in general, which are multi-purpose technologies or the particle model for CAD and simulation, which risen a special interest being subject of several lectures and contributions under different sections.

Taking into account the state of the art and the contemporary needs, this content is justifying the title of the book "Product Engineering: Eco-Design, Technologies and Green Energy", which addresses a wide audience in the engineering profession as the development engineers and practitioners, researchers, managers, academic staff, PhD and master students.

Doru Talabă and Thomas Roche

Part 1

ECO- DESIGN

1. INVITED LECTURES

2. CONTRIBUTIONS

THE DESIGN FOR ENVIRONMENTAL COMPLIANCE WORKBENCH TOOL

T. Roche
Galway Mayo Institute of Technology, Galway, Ireland

Abstract: The development of environmentally superior and compliant products and process is extremely important for electronic and vehicle manufacturers operating in, and servicing European markets. This is because of the existence of legislative drivers (Waste from Electronic and Electrical Devices, and End of Life Vehicle Directives), Standards (ISO 14000, EMAS) and increasing consumer pressure for the development of environmentally superior products. Design for the Environment represents an effective strategy for developing environmentally superior and compliant products (ESCP) and as an approach it needs to be implemented as early as possible in the design process. This paper describes a new framework for DFE methodology and tool development. A new CAD integrated DFE tool, called the Design for Environmental Compliance Workbench, which has been developed based on this new framework is also described.

Key words: design for environment, workbench tool.

1.   BACKGROUND

Global pressure, critically depleting natural resources and increasing market consciousness for the health of the environment has made the environmental superiority of products a critical competitive factor for manufacturers in the future. The Electronics and Vehicle Manufacturing sectors have come under particular pressure with the emergence of new European directives that is forcing them to become responsible for the safe disposal of their products at the end of life. This legislation is driven by two directives namely, the Waste from Electrical and Electronic Equipment (WEEE) for the electronics Sector and End of Life Vehicle Directive (ELV) for the automotive sector. According to the legislation OEMs are required to

3

D. Talabă and T. Roche (eds.), Product Engineering, 3–16.
© 2004 *Springer. Printed in the Netherlands.*

provide environmental information to life cycle stakeholders and legislative bodies regarding their products. (e.g. information includes the materials and composition in a supplied product, location of hazardous materials and their removal route, special handling concerns and dismantling instructions). Furthermore automobile manufacturers are forced to comply with environmental targets for example they are required to make a vehicle 95% recyclable (by weight) by the year 2015. Additionally Environmental Management standards (such as ISO14000 and EMAS) require OEMs to continuously improve the environmental properties of the products produced. Although voluntary, environmental management standard certification is crucial to organizations because of the increasing consciousness of the market and indeed trade barriers that can result.

For complex products (such as automobiles) the environmental legislation and standards creates two major problems, firstly *the generation, management and control of environmental information* and secondly *the implementation of methodologies to aid decision making for continuous improvement programs*. These problems are mainly because of the volume, dispersion and availability of knowledge and information. Clearly, with thousands of components in a vehicle, and the *diversity* of environmental information required for each (e.g. material constituents, fasteners, disassembly route) results in a *critically high volume* of information. *Availability* of information is also difficult because OEMs typically have hundreds of multi-tiered suppliers who are often unwilling to provide detail on the environmental characteristics of the products supplied (because of a perceived competitive threat). Fifty percent of original design work in the automotive industry is now done by suppliers, which greatly increases the needs for common standards. Also, this information is wildly *dispersed* both geographically and temporally (e.g. recyclers are at the end of life) and it is therefore difficult to synthesize to meet legislative compliance. Methodologies to aid decision-making with regard to continuous improvement of products are also compounded by these problems. These methodologies must be available to decision makers dispersed across the enterprise and must use standard and controlled decision criteria and information. Clearly there is a need to use software tools that aid OEMS to synthesize and manage appropriate information and to influence the continuous improvement process across the enterprise.

All product information is generated at the design stage and it is well known in research that over 90% of the life cycle costs (including environmental costs) are defined at this stage [39]. Design for the environment (DFE) represents an effective strategy for the development of environmentally superior and ELV/WEEE compliant products [28]. The author has worked on the development of design for environmental compliance tools for over six years. This work began by the development of design process models for design tools and continues today in the form of commercialisation of a CAD integrated tool called the Design for Environmental Compliance Workbench tool.

2. STRATEGIES FOR ENVIRONMENTALLY SUPERIOR PRODUCT DESIGN

There are many DFE strategies and each requires the inclusion of specific characteristics in the product. As the DFE field addresses the full product life cycle it is useful to address these approaches in the context of the life cycle model presented in fig 1.

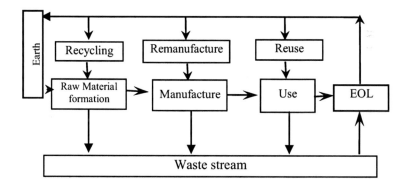

Figure 1. Product Life Cycle Model [28].

In this model the physical product passes through four generic phases in its lifetime, i.e. *raw material extraction, manufacture, use* and *end of life*. In each of these phases materials and energy are consumed either directly into the product or given off as waste streams.

When the product reaches the end of life a decision has to be made to reuse, remanufacture, recycle or dispose of it. Similar decisions have to be made regarding the materials and energies entering the waste stream. Four generic and interrelated strategies for the development of environmentally superior products can be derived from the model as follows [28]:

a) Select low impact materials and processes over all life cycle phases.
b) Reduce life cycle resource consumption (Materials and Energy)
c) Reduce life cycle waste streams (Materials and Energy).
d) Resource sustainment by facilitating first life extension and post first life extension, i.e. reuse, remanufacture and recycling.

Life Cycle Analysis (LCA) is the only method available to measure the *environmental impact* of products on the environment. The ISO14040 standard defines life cycle analysis as; *"a technique for assessing the environmental aspects and potential impacts associated with a product by: compiling an inventory of relevant inputs and outputs of a system; evaluating the potential environmental impacts associated with those inputs and outputs; interpreting the results of the inventory and impact phases in relation to the objectives of the study"* (ISO97). Life Cycle Assessment (LCA) is recognised as one of the most frequently used techniques for systematically evaluating environmental performance of a product throughout its life cycle [14, 38]. There are two main approaches to LCA, i.e. Full LCA and Abridged LCA. Full LCA is a rigorous quantitative method that systematically calculates and prioritises environmental impact of a product throughout its life cycle. Abridged LCA methods range from qualitative to quantitative methods that are less rigorous and data intensive and yield varying degrees of precision depending on the method applied. The important differences between the approaches, in the context of DFE, are represented in fig. 2.

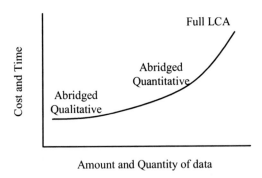

Figure 2. Different approaches to LCA, after [24].

There are many software tools developed to support the different types of LCA approaches, however many of the tools do not support the improvement phase and therefore need to be developed further for integration into the design process [28].

The reduction of *life cycle resource consumption* and life cycle waste streams requires resource minimisation solutions. Tools need to be developed and integrated into the design process to aid the designer to identify resource wastage directly and indirectly associated with the life cycle of the product. Some tools exist, however many are not integrated appropriately in the design process or indeed across the life cycle of the product. *Resource sustainment* is an extremely important and effective strategy for the development of ESCPs, particularly from the ELV and WEEE Directive implementation. First life extension may be achieved through designing for serviceability, maintainability, reliability and durability. Post first life extension strategies include policies to reuse, remanufacture, recycle and recover product at the end of life.

Reuse can be defined as the additional use of an item after it is retired from a clearly defined duty. Generic product characteristics that facilitate reusability have been synthesised from the literature as follows [4, 7, 19, 13, 16]: minimum number of components, serviceable, easy to clean, modular design, easy to disassemble, considers reduction of wear to components, considers corrosion protection, hazardous materials minimisation and the facilitation of part or subassembly removal.

Remanufacturing can be defined as a process that restores worn products to like new condition. Generic product characteristics that enhance its re-manufacturability include [4, 7, 19, 5, 13] : cleanability, modular design, and ease of disassembly, serviceable, testable subassemblies, and durable materials.

Recycling can be defined as a series of activities, including collection, separation and processing, by which products or other materials are recovered from the solid waste stream for use in the form of raw materials in the manufacture of new product other than fuel. Generic product characteristics that enhance recyclablility include [4, 19] : minimisation of material variety, minimisation of components, maximise material compatibility, minimise the use of hazardous materials, use recyclable materials, specify recycled content, label materials and facilitate ease of disassembly.

Clearly any holistic DFE methodology must be able to support the analysis, synthesis, evaluation and improvement of such characteristics.

As presented in the previous paragraphs the development of environmentally superior and compliant products is extremely complex. As much of a designer's work involves the use of CAD tools, there is an argument that new tools and methodologies need to be integrated in this type of environment.

3. LIFE CYCLE DESIGN FRAMEWORK

Traditional models of the design process have focused on the development of tools to improve the performance of a part of the life cycle of the product, e.g. design for manufacture or design for assembly. These tools can be described by the general term 'design for X tools', 'X' typically standing for assembly, disassembly or manufacture. The result is a proliferation of tools to aid the designer at individual life cycle stages with individual goals [17, 25]. As discussed in the previous section, new models must take a more holistic view, i.e. focus on the total life cycle system, to include raw material extraction, manufacture, use and end of life [20, 2, 3, 22, 36]. This is particularly true for DFE as a high degree of environmental coupling can occur across the life cycle stages [28, 15, 31, 6]. There is a need therefore to develop a life cycle design framework on which to build tools and methodologies to support DFE. By mapping the traditional design process model on to the product life cycle as shown in figure 3 a new design framework (called PAL) was derived.

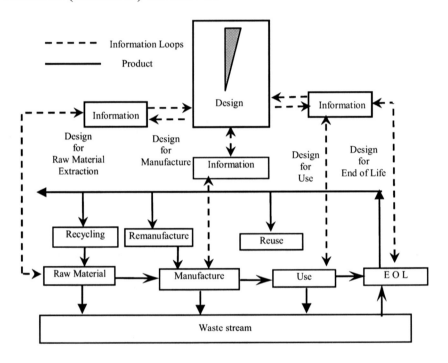

Figure 3. Mapping of life cycle phases to the design Process [28]

In the resulting model shown in figure 3 information is acquired through a set of life cycle design information loops, i.e. design for raw material extraction, design for manufacture, design for use and design for end of life. The design process transforms this information into product design characteristics, which are subsequently embedded in the product.

The resulting PAL framework for life cycle design is represented by a tri axial information transformation space, see fig. 4. The *vertical axis* consists of the degree of embodiment of a candidate design, i.e. the design traditional **Phase** (divided into requirements definition, functional design, concept design, and detailed design phases). On the *horizontal* axis is the phase dependent information transformation **Activity** that the designer is involved in at any particular design stage and is represented by the steps analyse, synthesis, evaluate in the model. In the context of DFE, environmental parameters and criteria have to be analysed, synthesized, evaluated and improved at every stage of the design process.

The life cycle information **Loops**, which define the third axis, represents the source of information for each life cycle phase of the product. These loops provide a focus for the type of information that has to be processed through the design framework. The activity and loop axis bound the life cycle problem-solving plane. This plane ensures the analysis, synthesis, and evaluation of life cycle information throughout each phase of the design process.

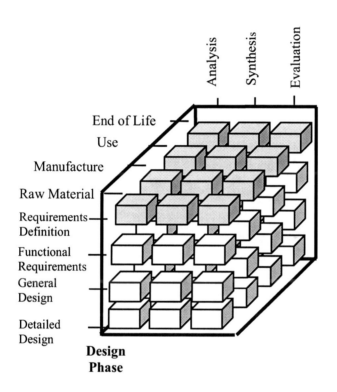

Figure 4. Phase Activity Loop (PAL) Life Cycle Design Framework [28].

The PAL Framework has been designed to support the development of methods, methodologies and tools to support life cycle design, particularly to cater to the coupling occurring between multiple product variables associated with DFE. It has formed the basis for the development of the DFEC Workbench described in this paper.

4.   DESCRIPTION OF DFE WORKBENCH TOOL

The design for environmental compliance workbench exists at three levels of abstraction to facilitate the deployment of such tools in a distributed design environment i.e. DFEC Workbench Desktop, DFEC Workbench Enterprise and finally DFEC Workbench Global. The development of the DFE Workbench methodology is focused on the analysis synthesis, evaluation and improvement of life cycle product *General* and *Detailed* design information (from the PAL Framework). The _DFEC Workbench Desktop_ exists in two forms, firstly the manual methodology, which is largely based on using special charts and reference information in a structured manner to evaluate and improve an emergent design. The second is a CAD integrated software tool, which effectively automates processes, associated with the manual methodology, as well as providing added functionality such as a WEB based report generator. The DFEC Workbench Desktop resides in a CAD Environment[1] and operates on virtual prototypes (VPs) created in that environment. The appropriate data is automatically synthesised from the virtual prototype and evaluated using different DFE tools. Each of the variables evaluated are prioritised and advice is given to the designer on alternative product or process characteristics that will enhance that variable. The designer optionally decides to accept the advice and makes the appropriate improvements in the CAD model. Data is then re-synthesised from the (new) model and the process begins again. This continuous improvement process continues until the best solution is found for that particular set of variables. The following fully integrated tools reside on the DFE Workbench.

- IAS Module
- SAM Module
- Advisor Agent
- Knowledge Agent
- Dynamic Report Generator

[1]The prototype DFEC Workbench has been developed in Solidworks 2000 and ProEngineer. Work is currently in progress to port the DFE C Workbench to CATIA V5.

4.1   Impact Assessment System (IAS)

The IAS is effectively a life cycle analysis tool that extracts the appropriate data from the virtual prototype. The user defines the life cycle processes and materials used by selecting options from the materials and processes databases. Improvements can be made at the *part* or *product system* levels.

4.2 Structure Assessment Method (SAM)

SAM focuses on the structure of the emergent virtual prototype in an attempt to enhance product structural characteristics in the DFEC context. SAM is a complex methodology, which quantitatively measures and records data such as follows:

1. Material Type and Variety
2. Material Intensity of Type/s (Mass)
3. Material Compatibility, (taking into account fasteners)
4. % Recycled Material Content
5. % Recyclable Material
6. % Hazardous Material
7. % Biodegradable
8. Number and Types of fasteners
9. Number and types of tools required for disassembly.
10. Total standard disassembly time
11. Standard part removal time and optimum route.

The coupling between all variables is managed and recorded by the DFEC Workbench Desktop. For example if an additional fastener is added to the virtual prototype then the number and variety of fasteners and disassembly times are recalculated for the product structure.

It should be noted that SAM is more focused on the product structure system therefore it can deal with individual parts and the relationships between them from an environmental viewpoint.

4.3 Advisor Agent

The advisor agent has two functions; firstly to prioritise variables generated by the IAS and SAM tools. Secondly the advisor agent actively gives advice to the designer on alternative structural characteristics to enhance either the environmental impact or structural characteristics of the emergent design. For example the advisor agent may suggest alternative materials or process to reduce the environmental impact of a product. The advisor agent uses a significant number of materials and processes stored in a propriety database, therefore making changes is very efficient. It should be noted that the advisor agent manages coupling between all variables in the product model. Hence, if the designer selects a new material then impact data and structural data is re-evaluated.

One important characteristic of the advisor agent is that it does not constrain the designer in any way. The designer is free to decide what he/she considers to be the optimal solution for the candidate design.

4.4   Knowledge Agent

The knowledge agent provides advice to the designer in a consultative mode. For example the designer can use the Knowledge Agent to find a material with specified mechanical properties and list them in increasing environmental impact. The designer can then use the selected material in the design.

4.5   Dynamic Report Generator

WEEE and ELV are likely to require reports to be created for compliance with directives. The report generator automatically generates reports on the product designed by the user. These reports are made available in two modes, i.e. as system reports that can be printed and viewed locally or as World Wide Web reports that can be made available via an Extranet model to people who need product data.

For example dismantlers may need to know the location of hazardous materials, the disassembly route and time for a specific product type. The report generator is designed so *that preferred dismantlers* can log on to the DFE Workbench site (of the associated manufacturer), type in a product descriptor code and get detailed product structural data directly for that *specific* product. If the designer makes a change in the product structure in the design process then the data is automatically updated on the web report server.

The *DFEC Workbench Enterprise* supports the synthesis, evaluation, analysis and prioritisation of both environmental impact and structural data associated with a full product system consisting of a number of subassemblies that have been previously analysed with the DFEC Workbench Desktop application. The enterprise version does not require a CAD environment to operate. The DFE Workbench enterprise identifies the highest environmental impact or structural problems associated with a particular subassembly/assembly and enables the system manager to notify the department developing the particular subassembly that modifications must be made in order to meet the desired environmental or structural criteria. Product Life Cycle Management tools are used to support this activity.

Both Desktop and Enterprise DFEC Workbench applications link directly with an oracle database server. The communication between the design team and system engineer/product manager is performed within an intranet network and allows instant access to the latest structural and environmental data from an emergent virtual prototype.

The DFEC Workbench Global has been developed as an intranet/internet application that allows easy communication and reporting of the environmental and structural data generated with the Desktop and Enterprise Workbench. It resides on a web server and is linked with the oracle databases, which allows customisation and control of data.

The DFE Workbench supports the development of environmentally superior products within distributed design environments. It allows easy collaboration between designers, as it is a centralised tool residing on oracle databases. The tool also allows quick access and customisation to meet the needs of various departments in a company.

The tool has been tested and validated with a large set of companies from the automobile and electronics industry sectors.

5. TESTING OF TOOL RESULTS AND CONCLUSIONS

The methodology was tested in three modes as follows:
a) In the analysis, synthesis, evaluation and improvement of an existing product using the manual method.
b) In the analysis, synthesis, evaluation and improvement of virtual prototypes in the design process, using the DFE Workbench software.
c) By performing consultancy for large multinationals in the electronics, automotive and electro-mechanical sectors.

Scientific tests on the DFE Workbench, using protocol analysis techniques were carried out with a large number of experienced engineers from both the electronics and electro-mechanical sectors. A summary of the results and conclusions established from the tests are as follows:

– The PAL Framework was a very powerful support for the development of tools and methodologies to support life cycle design activities.
– The proper application of the DFE Workbench methodology can result in the improvement of the design irrespective of the experience of the designer.
– It was established that the application of the DFE workbench took only 1.5% of the actual model creation time to use on a product, hence it can be concluded that the process of DFE did not have negative impacts on the process of design.
– The manual method takes a long time to complete. It is tedious to calculate all of the variables, particularly when having to iterate through a number of solution variants and having to recalculate every time. However the manual methodology was found to be a very useful, in fact essential tool for practical training on the principles of the DFE Workbench.

- The use of standardised criteria such as; standard times, labelling, and material compatibility's, is a very positive feature of the methodology particularly for benchmarking and design comparison.
- The strong and clear linkage between the global and local indices is identified as a very positive feature of the methodology.
- The prioritisation process was found to be very useful for the search and improvement activity.
- The inclusion of an advisor was seen as essential to the operation of the methodology.
- There are very distinct advantages for integrating the DFE Workbench in a CAD environment, not least the automation of data synthesis activity, the availability of quantitative data directly from the model, the manipulation of this data, the management of data interrelationships, and clearly the resulting improvement in a design before it is manufactured.
- The development of a web based dynamic report generator was viewed as a very positive feature of the DFE Workbench Software.
- A national award for eco design was won as a result of eco design work carried out by the DFEC Workbench for a local company.
- It is essential to have a company engaged with the development team in the future in order to focus development efforts on industry needs.

6.   FUTURE PLANS

The focus of work on the DFEC Workbench suite of tools currently is in the commercialisation of the tool for the electronics and automotive sectors. In the next year the DFEC Workbench is undergoing continuous development at the author's institution. Resources are now in place to port the tool to CATIA V5 for the automotive sector companies. Additionally and concurrently with this development we propose to deploy a PLM approach and implementation methodology for enterprises for rapid roll out to the associated tiers in the relevant sectors.

For additional information on the DFE Workbench please contact Dr. Thomas Roche at the email address: Tom.Roche@GMIT.IE

REFERENCES

1. B.R. ALLENBY, T. GRADEL, *Industrial Ecology* (Prentice Hall, 1995).
2. L. ALTING, *Life Cycle Design of Products: A New Opportunity for Manufacturing Enterprises* (Concurrent Engineering Automation Tools and Techniques, Wiley Press, 1993, pp 1- 17).
3. L. ALTING, H. WENZEL, M. HAUSCHILD, *Environmental Assessment of Products*

(Chapman Hall, 1997).
4. S. BENHRENDT, C. JASCH, M. PENEDA, H. WEENEN, *Life Cycle Design A Manual for Small and Medium Enterprises* (Springer, 1997).
5. K. BRADY, A. PAYNTER, *Evaluation of Life Cycle Assessment Tools* (Unpublished Report from Environment Canada, kbrady@synapse.net, 1996).
6. G. CADUFF, R. ZUST, *Increasing Environmental Performance via Integrated Enterprise Modelling* (Presentations at 3rd International Seminar on Life Cycle Engineering, 1996).
7. T. CLARKE, *Eco Design Checklists* (Center for Sustainable Design, Surrey Institute of Art and Design, May 1999).
8. N. CROSS, *Engineering Design Methods* (J. Wiley & Sons, 1994).
9. CSA, *Design For The Environment* (Canadian Standards Association, Z762-95, October 1995).
10. ANON, *Design for Environment Guidelines* (DFE Research Group, Manchester Metropolitan University, http://sun1.mpce.stu.mmu.ac.uk/pages/projects/dfe/pubs/pubs.html, 1997).
11. G. Q. HUANG, *Design for X - concurrent engineering imperatives* (Chapman Hall, 1996).
12. *Eco-Indicator 95, Final Report* (Mark Goedkoop. PRe Consultants).
13. J. FIKSEL, *Design For Environment Creating Eco-Efficient Products and Processes* (McGraw Hill, 1996).
14. C. FUSSLER, *Driving Eco-Innovation* (Pitman Publishing, 1996).
15. C. HENDRICKSON, N. CONWAY-SCHEMPF, L. LAVE, F. MCMICHAEL, *Introduction to Green Design* (Green Design Initiative, Carnegie Mellon University, Pittsburgh, PA).
16. ICER, *Design for Recycling Electronic and Electrical Equipment* (Industry Council for Electronic Equipment Recycling, 1997).
17. K. ISHII, L. HORNBERGER, *The Effective Use and Implementation of Computer Aids for Life Cycle Product Design* (in Advances in Design Automation, Volume 1, ASME , 1992).
18. International Standards Organisation, *Environmental Management - Life Cycle Assessment - Principles and Framework* (ISO14040, 1st edition, June 1997).
19. G.A. KEOLEIAN, D. MENEREY, *Life Cycle Design Guidance Manual* (EPA, 1993).
20. F. KIMURA, *Inverse Manufacturing: from Product to Services. Managing Enterprises-Stakeholders, Engineering, Logistics and Achievement* (First International Conference Proceedings. MEP, Ltd. London UK. 1997).
21. T. LAMVIK, O. MYKLEBUST, S. STOREN, *Nordlist LCA Project Final Report* (SINTEF report no. stf38 s97001, ISBN 82-595-9937-6, 1997).
22. D.E. LEE, *Issues in Product Life Cycle Engineering Analysis* (in Advances in Design Automation, V 65-1, ASME, 1993).
23. MAN, ELENA, *Development of a Design for the Environment Workbench Software Tool* (Master Thesis, June 2000).
24. H. LEWIS, *Data Quality for Life Cycle Assessment* (National Conference on LCA: Shaping Australia's Environmental Future, www.cfd.rmit.edu.au/DFE/lca1.html, 1996).
25. A. MOLINA, A.H. AL-ASHAAB, T.I. ELLIS, R. YOUNG, R. BELL, *A review of Computer-Aided Simultaneous Engineering Systems* (in Research in Engineering Design, V7, Springer-Verlag, 1995, p 38-63).
26. J. POYNER, *The Integration of Environmental Information with the Product Development Process Using and Expert System* (PhD Thesis, Manchester Metropolitan University, September 1997).
27. T. ROCHE, *A Green Approach to Product Development* (Proceedings of 1998 International Conference on Intelligent Manufacturing Systems", 1998, Lausanne Switzerland).

28. T. ROCHE, *The Development of a DFE Workbench* (Ph.D. Thesis, September 1999).
29. C. RYAN, *Life Cycle Analysis and Design - A Product Relationship?* (First National Conference on Life Cycle Assessment. Melbourne. Australia.1996).
30. S. SCHALTEGGER, *Eco Efficiency of LCA. The Necessity of a Site Specific Approach in Life Cycle Assessment (LCA) - Quo Vadis?* (Birkhauser Verlag, 1996).
31. J. FAVA, *A Technical Framework for Life Cycle Assessments* (SETAC, 1990).
32. M. SIMON, *Continuing Integration of the Ecodesign Tool the Product Development* (IEEE International Symposium for Electronics and the Environment, 1996).
33. V. TIPINIS, *Towards A Comprehensive Life Cycle Modeling for Innovative Strategy, Systems, Processes and Product/Services* (Life Cycle Modeling for Innovative Products and Processes, 1995, PP 43 – 55).
34. ANON, *US Field Study Aims to Further Reduce Amount of Vehicle Plastic Waste* (United States Council for Automotive Research, http://www/uscar.org/, Spring 1999).
35. M.B. WALDRON, K.J. Waldron, *Mechanical Design Theory and Methodology* (Springer Verlag, 1996).
36. G. WARNECKE, *A Co-Operation Model of Product Development and Recycling* (Proceedings of First International Seminar on Reuse, Eindhoven 1996).
37. *Random House Webster's Dictionary*
38. E.V. WEIZSACKER, A. LOVINS, L.H. LOVINS, *Factor Four Doubling Wealth Halving Resource Use* (Earthscan Press, 1998).
39. R. ZÜST, *Sustainable Products and Processes* (Presentations at 3rd International Seminar on Life Cycle Engineering, 1996).

ECO-IMPACT ANTICIPATION BY PARAMETRIC SCREENING OF MACHINE SYSTEM COMPONENTS
An Introduction to the EcoPaS Methodology

J. R. Duflou and W. Dewulf
Department of Mechanical Engineering, Katholieke Universiteit Leuven, Belgium

Abstract: The Eco-efficiency Parametric Screening (EcoPaS) methodology, described in this paper, offers a systematic approach to component selection based on environmental impact minimisation. Starting from functional systems requirements, which are known in a very early design stage and often form part of the task specification, designers can browse alternative solutions with the aim to translate functional block descriptions into specific system components. For this purpose different techniques are called upon, mapping functional parameters onto environmental cost defining physical parameters. These mapping techniques, inspired by cost estimating relationships (CER's,) offer opportunities to quickly screen system level design alternatives, resulting in early estimates for environmental performance indicators.

In this paper the different mapping techniques, used as underlying building blocks for the EcoPaS system library, are described. Practical examples offer better understanding of the concepts.

The functionality offered by the described methodology is illustrated by means of a comprehensive example of a machine system component.

Keywords: eco-design, conceptual design, parametric, environmental cost estimation relationship, EcoPaS.

1.   INTRODUCTION

It is a well-known fact that decisions taken in an early, conceptual design phase can influence the outcome of a design exercise more significantly than any optimisation step later on in the design process [1]. In an eco-design approach an early recognition of favourable system component solutions is

17

D. Talabă and T. Roche (eds.), Product Engineering, 17–30.

therefore of great importance. Generic eco-design guidelines form insufficient support for designers in this respect, while a detailed comparative study based on LCA techniques is too demanding in terms of required expertise and time consumption. Since material selection and exact dimensional specifications are typically determined in later design stages, building an LCA inventory only becomes feasible in an embodiment or detailed design phase (Figure 1). Even if appropriate competences would be available and time delay would not be an issue, the data requirements inherent to a conventional LCA study make the technique unsuitable as a support tool for conceptual design decision-making.

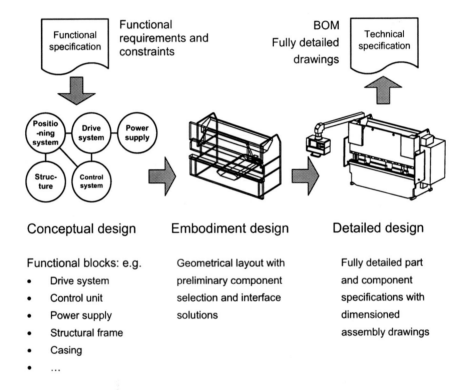

Figure 1. Decision scope and available information for eco-design support in different design phases

The specific nature of machine design offers opportunities to overcome this status quo. In a systems approach, design of machine tools largely consists of the identification of appropriate system components that can fulfil predefined functional requirements and constraints in an optimised way. The selection of such components leaves open a large number of possible configurations, since for each functional block in a conceptual design scheme a range of solutions is normally available. Design catalogues illustrate the alternative options available for elementary functions. Project specific constraints typically limit the range of choices in the solution space for every functional block in a design. Where multiple alternatives, however, can meet the requirements and constraints imposed by the design specification, a series of combinations remains open for optimisation based on selected criteria, such as eco-impact minimisation.

A systematic screening of functional design alternatives on a sub-system level, based on well-automated eco-impact prediction methods, could support the designer in making strategic choices leading to a minimal ecological footprint for the system being designed.

In this paper a number of techniques, developed for the purpose of systematic screening of functional alternatives, based on parametric input of functional requirements and constraints, are presented. The Eco-efficiency Parametric Screening (EcoPaS) system forms the basis for an extension of the existing lightweight LCA methods from materials and process oriented analysis techniques to sub-system level selection tools suitable for pro-active conceptual design support.

2.   METHODOLOGY DESCRIPTION

The basis of the methodology is the assumption that only functional requirements and constraints are known at the outset of an early, conceptual design stage. Although there may be high uncertainty about the nature of the system to be designed, requirements and constraints are normally well documented and are often available as specific, quantitative data.

It is the case when, for example, designing a lighting system, the intensity of use and the expected service life of the system may be documented or can easily be estimated. The type of environment in which the system has to function will be given, often documented with sketches or drawings of the construction. The nature of the activity that the system has to support (office environment, living quarters, sports facilities, …) will be known from the outset. This data can be treated as functional parameters that, for every type of lighting system under consideration, allow the performance of a dimensioning exercise according to some well-known procedures.

The aim of the methodology described in this paper is to link the functional parameters directly to environmental impact indicators for a range of design alternatives, thus allowing a quick screening of these alternatives to support early design decisions. In this framework eco-impact can be expressed in monetary units, such as external costs [2] or willingness to pay [3], or by a commonly used environmental performance indicator, such as, for example, the Eco-Indicator 99 [4].

For this purpose different methods can be called upon, as documented in the following sections. These methods are inspired by cost estimation relationships (CERs) that allow early estimation of product costs based on perceived strong correlations between cost as a dependent variable and a number of independent cost driving variables [5]. Therefore these methods are further referred to as Eco-Cost Estimating Relationships (E-CERs).

2.1   Empirically Derived E-CERs

Using techniques like regression analysis, dominant eco-impact drivers can be determined for many system components, starting from more detailed LCA output. The purpose of this approach is to identify one or more functional parameters that are sufficiently strongly correlated with the environmental impact created by a category of components in their production and/or utilisation phase. An example is summarized in Figures 2 to 4, illustrating that for 3-phase electromotors the nominal power can be used as an independent variable to estimate the eco-impact caused during both the production of the motor and its utilisation phase.

Figure 2 demonstrates a strong correlation between the nominal power and the amount of copper required to construct a 3-phase induction motor.

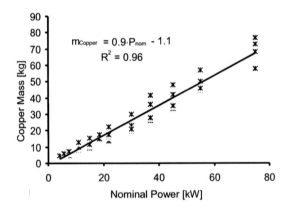

Figure 2. Amount of copper used in the construction of 3-phase induction motors in function of nominal power (based on a single supplier survey [6])

From Figure 3 it can be concluded that for the same type of motors the nominal power is also strongly correlated to the total cradle-to-grave Eco-Indicator score by combining the large amount of copper, as a dominant material in the inventory of an electromotor, with the relatively high eco-impact per kg of this material, the strong resemblance between Figures 2 and 3 can be anticipated.

When evaluating the environmental performance of an induction motor as an operational drive system, distinction should be made between the energy that is effectively passed on to other sub-systems and the energy that is dissipated in the motor itself. Only the latter should be taken into account as an environmental impact attributed to the efficiency of the system under evaluation. Figure 4 illustrates the strong correlation between the nominal power of an electromotor and the Eco-Indicator 99 score corresponding to the dissipated energy per time unit on condition that the motor is loaded according to its nominal output power.

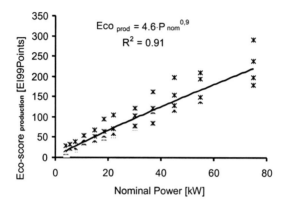

Figure 3. Cradle-to-grave Eco-Indicator 99 score for 3-phase induction motors (only taking into account material production impact)

For a given utilisation scenario, specifying the anticipated intensity of use and the projected service life duration, and an estimated required nominal power, a total environmental impact can thus be calculated using these empirically derived relationships.

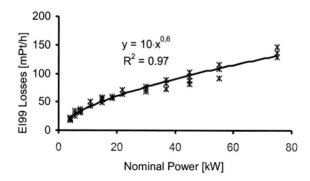

Figure 4. Eco-Indicator 99 score for energy losses in 3-phase electric motors as a function of nominal power

With similar E-CERs available for other drive system alternatives, a quick comparison of the environmental efficiency of the solutions under consideration becomes feasible for a given application scenario (required nominal power, intensity of use and expected service life).

2.2   E-CERs Based on Underlying Models

Theoretical model development is a technique that can be used, among others, for developing E-CER's for structural components.

Different types of load bearing structures can be distinguished and modelled using graphostatic relations between independent load variables, dimensional parameters and properties of the applied construction materials. Functional constraints, such as maximum allowed deformations, can be taken into account to determine the relevant relationship model. Inspired by the ratios determined by Ashby for minimal weight design [7], ratios can be derived that allow a rapid comparison between the environmental impact created by different construction material alternatives.

For a uniformly loaded rectangular plate-like construction with a maximum allowed deformation as the main focus, the graphostatic relationship would, for example, be:

$$y_{max} = \frac{\alpha \, q \, b^4}{E \, t^3}, \tag{1}$$

with α as a parameter depending on the plate proportions and the boundary conditions, q a uniform load, b the plate width, t the thickness and E the Young modulus of the plate material.

For ξ_i representing the environmental impact score per unit weight for a given material i (for example expressed as an Eco-Indicator 99 value), for a the length of the plate and ρ the specific weight of the material, the total impact of the plate $\xi_{plate \, i}$ can then be written as:

$$\xi_{plate,i} = V \, \rho_i \, \xi_i = t \, a \, b \, \rho_i \, \xi_i = \sqrt[3]{\frac{\alpha \, q \, b^4}{E_i \, y_{max}}} \, a \, b \, \rho_i \, \xi_i, \tag{2}$$

which is equivalent to:

$$\xi_{plate,i} = \left(\sqrt[3]{\frac{\alpha \, q \, b^4}{y_{max}}} \, a \, b \right) \left(\frac{\rho_i \, \xi_i}{\sqrt[3]{E_i}} \right). \tag{3}$$

Since only the second factor in this formula can be influenced by the material selection, the screening of different material alternatives can be limited to a maximisation of the following ratio:

$$\frac{\sqrt[3]{E_i}}{\rho_i\,\xi_i}. \tag{4}$$

This only contains structure independent parameters which can be obtained from a material database.

Similar ratios can be defined for other construction and functional constraint types (a few number examples are listed in Table 1).

More refined theoretical models can be derived for structural components to be built into dynamic systems. For different types of transport, for example, the average environmental impact per unit weight can be taken into account with the estimated transport distance as an independent parameter.

Table 1. Coefficients representing the adequacy of four materials in given construction situations: beams and plates dimensioned for respectively optimal stiffness and optimal strength. Scores are based on average European cradle-to-grave scores, and only take into account impact of material production steps.

	Constr. Steel (80% virgin, 20% recycled)	Stainless Steel (100% virgin)	Aluminium (100% virgin)	Aluminium (100% recycled)
ρ, kg/m^3	7800	7800	2700	2700
ξ, mPt	86	910	780	60
E, MPa	200,000	200,000	70,000	70,000
σ_f, MPa	250-400	250-800	40-400	40-400
Stiff beam, $\sqrt{E}/\rho.\xi$	0.67	0.06	0.13	1.63
Stiff plate $\sqrt[3]{E}/\rho.\xi$	0.009	0.001	0.002	0.025
Strong beam, $\sqrt[3]{\sigma_f^2}/\rho.\xi$	$6\text{-}11\cdot10^6$	$6\text{-}32\cdot10^5$	$1\text{-}28\cdot10^5$	$2\text{-}49\cdot10^6$
Strong plate, $\sqrt{\sigma_f}/\rho.\xi$	0.02-0.03	0.002-0.004	0.003-0.009	0.04-0.12

For machine components dynamic behaviour can be modelled in a similar way, taking into account, for example, independent functional parameters such as the type of movement (translational, rotational), acceleration/ de-acceleration patterns, and the average duration of operation.

2.3   Dealing with uncertainties

When compiling materials databases for LCA support, it is a known problem that the production methods applied by different material producers will correspond to different eco-impact scores per unit weight of a specific material. Similarly non-uniform operational practices and different machine configurations can lead to rather large variations in the eco-impact that can be allocated to specific manufacturing processes. When using E-CER's as a method to screen design alternatives, different sources of uncertainties can also influence the outcome of the analysis. Both the variation of the underlying data points, in case of an empirically derived E-CER function, and the early estimations used as values for the independent parameters can be a cause of uncertainty.

Just like a breakdown of LCA support databases into more detailed sub-categories may provide a way out for high variability in collected data, distinguishing different component variants can allow a higher coefficient of determination when working with empirically derived E-CER's. In the case of the electromotor, that could, for example, mean making distinction between asynchronous and synchronous motors.

The approximate knowledge of the boundary conditions for systems components can create uncertainty about the appropriate values to select for the independent input parameters. Since the use of E-CER's to estimate the environmental impact provides an immediate response for a given set of input data, conducting a sensitivity analysis, however, requires a very limited effort. From such an analysis it can be easily concluded whether variation of the important input parameters can significantly influence the outcome of the parametric screening exercise. In case the outcome of a comparative study proves to be sensitive for limited parameter variations, determining a more exact input value for the involved parameters may require extra attention.

3.   ECOPAS SYSTEM ARCHITECTURE

It is an explicit objective of the EcoPaS methodology to confront the designer only with input requirements that are feasible to determine in the early conceptual design stage. Therefore, besides the actual eco-efficiency assessment module, eco-design support systems based on the EcoPaS methodology contain imbedded parametric dimensioning functions for the different component types supported by the system. The integration of such a dimensioning module into the system architecture is illustrated in Figure 5.

The system procedure reflected in this flowchart should be interpreted as follows. Starting from a function selection, a pre-selection of component alternatives is obtained from the design catalogue. At this point a designer can steer the system by eliminating unfavourable alternatives. The pre-selection will determine the specifications of the functional component requirements and constraints to be entered interactively.

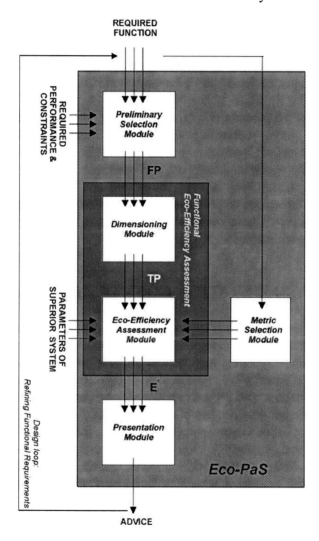

Figure 5. Modular overview of the EcoPaS system architecture with FP representing functional parameters and TP intermediate technical parameters

In the next step these requirements and constraints are processed in the fully automated dimensioning module to determine the dominant technical characteristics for a given set of component alternatives. Although this dimensioning module is represented in the scheme as a single functional block, separate modules can be defined for the different component types. Additional modules can thus be added when the library of supported functions and component types is enhanced.

In principle some inputs for the system could be obtained as outputs of other system components. Although such links may affect the simplicity of the procedure for the EcoPaS system end-user, the modular system architecture can support such modelling scenarios.

Using the E-CER's, an estimated environmental impact can then be generated for the respective design alternatives. The preferred metric for this output can be preset based on company standards or can be selected interactively.

The output of the parametric screening exercise needs to be presented in a comprehensive way, supporting a component selection procedure. A graphical presentation module is preferred in this respect. Where other decision making criteria, such as for example cost, reliability, or maintainability, could play a role, based on the selected metrics the environmental impact can be provided as an absolute or relative score for the different alternatives. These values can then serve as input for multi-criteria decision-making techniques.

4.   IMPLEMENTATION CONSIDERATIONS

In order to serve as a generic eco-design support tool, a rather complete catalogue of functions and corresponding component types would need to be available in a robust software implementation. It is obvious that building the corresponding dimensioning modules and collecting the data for the underlying parametric E-CER's form considerable tasks that require input from many different parties. Just like building LCA material and process databases did not happen overnight, an exhaustive catalogue of functions and components types can only emerge through a systematic and continued effort. Since, however, the evolution of machine component concepts can be perceived as fairly gradual, a systematic build up of the envisaged catalogue forms a slowly moving and thus realistic target.

While working towards a generic EcoPaS implementation, the individual dimensioning and parametric impact assessment modules can serve their purpose as part of the services offered to machine builders by individual component suppliers. As a complementary source of information, such modules can be added to on-line catalogues, creating opportunities to optimise component selection from a life cycle engineering perspective within the offering of an individual company.

As far as platform requirements for an EcoPaS implementation are concerned, once the E-CER's and parametric models have been established the computational complexity is low and memory requirements are limited. An entry level PC platform can therefore provide all required support.

Special considerations are required for the development of an ergonomic graphical user interface. Defining a user-friendly access to a comprehensive and non-ambiguous taxonomy of functions forms the biggest challenge in this respect. Once a specific function has been selected, offering a simple input-output interface is feasible, as can be witnessed from the case study in the next section.

5.   CASE STUDY

To illustrate the EcoPaS methodology, a simple case of a commonly known category of components is included here.

Most machines contain dynamic sub-systems, the relative movement of which needs to be facilitated with a minimum loss of energy. For this example rotational movements are considered. The related function for which suitable components need to be identified can be described as "isolation of rotational relative movement between machine sub-systems transferring forces to each other". This functionality can be provided by different types of bearings.

Depending on the independent parameters imposed on the system as functional requirements, some type of bearings may be excluded from the solution space. However, typically several types can fulfil all functional requirements and a selection can be made based on life cycle engineering considerations such as minimal total life cycle cost or minimal environmental impact.

The bearing type categories used in this example are friction bearings, ball/roller bearings and air bearings.

A detailed study of the dimensioning methods proposed by different manufacturers of these bearing types pointed out that the dominant technical parameters can be determined starting from the following functional specifications: the maximum force on the bearing (F), the internal diameter (di) and the rotation speed (n) [8]. The anticipated duration of the active use phase (L_h) is required as an additional input parameter to estimate the total energy dissipation during the component's service life and the number of times a component may need to be replaced to cover the complete projected life span of the machine tool.

Starting from this data, the technical feasibility of the respective alternatives can be verified, the technical dimensions can be generated and the friction characteristics of the different bearings can be estimated. Full details of the underlying dimensioning modules can be found in reference [8].

The graph in Figure 6 illustrates the dependency of the relative ranking of the different bearing types on the functional input variables. The figure shows the bearing type with the lowest estimated environmental impact for a number of parameter combinations.

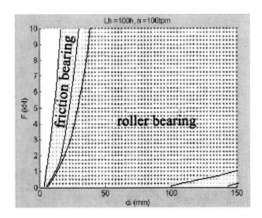

Figure 6. Graphical representation of the evaluation results in function of independent functional parameters: F, radial force; d_i, internal diameter for fixed values of L_h, operational life time and n, rotation speed (rpm)

Taking into consideration the uncertainty on the underlying E-CER and the possible variance on the estimated values for the independent variables, the preference zones corresponding to the two relevant types of bearings can not be considered to be defined with a strict boundary. For the depicted combination of Lh and n this is indicated by means of a transition zone in which the difference in estimated total environmental impact for the compared systems does not exceed a preset uncertainty threshold level.

As this example illustrates, the preference for a certain component type clearly depends on the functional parameter combination. In most cases summarizing the parametric analysis results as a limited number of generic guidelines is therefore not evident.

Figure 7 illustrates the simplicity of the user interface that can be offered for the considered component types. For a given set of functional input parameters, the technical parameters are automatically generated in the underlying dimensioning module. The depicted environmental impacts, expressed as Eco-Indicator 99 values for both the production and utilisation phase of the respective alternatives, allow to draw a clear conclusion concerning the preferred solution from a life cycle engineering point of view. For the depicted combination of independent functional parameters, the air bearing system does not offer a technically feasible alternative. While the environmental impact linked to the production phase of both remaining bearing types is of the same order of magnitude (lower part of bars), the energy dissipation during the utilisation phase (upper part of bars) is clearly less favourable for the friction bearing solution. Other combinations may lead to different conclusions. High rotational speeds, for example, can lead to situations in which the energy consumption of compressed air supply to an air bearing is clearly compensated by reduced friction losses, resulting in a clear preference for this type of components.

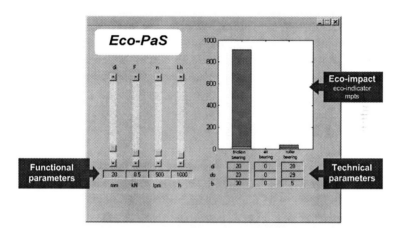

Figure 7. User interface with input fields for functional parameters (left) and graphical output (bar chart right). Technical parameters, as generated by the dimensioning modules, are offered as a secondary output.

6.   DISCUSSION AND CONCLUSIONS

The described Eco-efficiency Parametric Screening methodology can form the basis for a series of pro-active tools supporting eco-design of machine tools in an early design phase. The method allows designers to make a quick approximate environmental impact comparison for different components that form potential alternative solutions to fulfil preset functional requirements. The required data input for this purpose is limited to functional parameters that are typically already known in an early conceptual design stage.

Implementation of the EcoPaS methodology for a number of component categories has allowed to validate the effectiveness of the method as an eco-design support tool. Additional components are being added to the list of covered functionalities in cooperation with a number of component suppliers.

Ideally the EcoPaS methodology would need to be implemented covering an exhaustive catalogue of functionalities and corresponding machine components. The development of the required automatic parametric dimensioning and impact estimation modules will require systematic and coordinated efforts by many parties active in this domain.

While working towards this target, the implementation of the EcoPaS concept on an individual company level can already demonstrate a company's compliance with the emerging European legislation concerning eco-design requirements for machine tools (e.g. [9]).

It should be noted that the applicability of the described method for comparative screening of alternative system components is not necessarily limited to eco-impact minimisation support. The method can also contribute to a systematic life cycle cost minimisation strategy. By using monetary cost estimation relationships, a quick life cycle cost estimation can be conducted for a range of design alternatives with minimal time delay as a consequence.

ACKNOWLEDGEMENTS

The authors would like to recognise the support of the Flemish government through IWT (Institute for the Promotion of Innovation by Science and Technology in Flanders) and of the Belgian federal government (Belgian Science Policy agency), respectively in the framework of the postdoctoral research program and the CHASM project (TAP program).

REFERENCES

1. M. M. ANDREASEN, L. HEIN, *Integrated Product Development* (IFS Publications, Bedford, 1987).
2. P. FASELLA (Ed.), *ExernE - Exernalities of Energy.* (Vol. 1: Summary, European Commission, Brussel, 1995).
3. B. STEEN, *A Systematic Approach to Environmental Priority Strategies in Product Development.* ((EPS), Version 2000 - General System Characteristics, CPM report 1999:4, CPM Gothenborg, 1999).
4. M. GOEDKOOP, S. EFFTING, M. COLLIGNON, *The Eco-Indicator 99 - Manual for Designers".* (Pré Consultants, Amersfoort, 2000).
5. B. BRUNDICK, *Parametric Cost Estimation Handbook.* (US Navy, Arlington 1996).
6. Catalogue of A1 Standard three phase motors, ATB, 1995.
7. M. ASHBY, *Materials Selection in Mechanical Design.* (Butterworth Heinemann, Oxford, 1992).
8. C. VERMEIREN, *Ondersteuning van eco-efficiënt ontwerpen op basis van parametrische analyse.*(Master thesis 01EP15, Department of Mechanical Engineering, K.U.Leuven, 2001).
9. NN, On Establishing a Framework for the Setting of Eco-Design Requirements for the Energy-Using Products and Amdending Council Directive 92/42/EEC (Proposal for a Directive of the European Parliament and of the Council, Brussels 01-08-2003 COM(2003)453.

MAN MACHINE INTERACTION AND MAN MACHINE INTERFACE - CRITICAL ISSUES FOR THE HUMAN ENVIRONMENT

Ph. Coiffet

Laboratoire de Robotique de Versailles, France

Abstract: Machines invade our daily and professional life. That is a fact. The invasion is peculiarly sensitive to anyone for the last decades with expansion of computers and telecommunications never seen before. Consequently, the phenomenon generates questions and fears for people. It is usual to hear on questions about: where are we going to with machines? Are machines going to totally replace us? Are they in contradiction with our way of life, our private life etc.? All questions appear legitimate as far as there is no structured and adapted information accessible to everyone about machines developments prepared by specialists. The paper wants to show that the central core of the relational problem of man with machine is located into the friendly-using matters (section 1). Then, in spite of technological opportunities (section 2), three points must draw attention (section 3): the "natural" evolution of MMI can bring out some dangers, the "natural" evolution could not meet some important needs in matter of MMI, at last, it is imperative to provide new education to designer and users of man-machine systems. In section 4, some recommendations are proposed to stay on the good way based on humanism.

Key words: man-machine interaction, man-machine interface, man-machine systems, robotics, automation.

1. INTRODUCTION

Machines exist since the period of Antiquity and man did not cease to improve them in terms of utility, efficiency and safety. This process has accelerated in the last century and particularly rapid for some decades. A

31

D. Talabă and T. Roche (eds.), Product Engineering, 31–54.

consequence, at present time, is exhibited through the significant **complexity** and some **autonomy** capacity of machines.

These new features have consequences on humans: some **distance,** sometimes a true separation between the designer and the user, and a **confusion** of user in the presence of machine.

Attitudes of users in difficulty with machines can be divided into two groups: a first one shows a **defiance**, a fear, a kind of blocking position or even sometimes a refusal to use machines; a second one has an opposite behaviour: people exhibit a **total trust** in machines and neglect potential risks. From another viewpoint, facing the functioning complexity of machines, sometime humans show an attitude connected to magic, using some **rites** in order to get results.

Nevertheless, lucky users are also satisfied with their machines. And technological means exist to solve a large part of problems if they are correctly implemented.

Designer's attitude is of course different, but his objective is more focused on increasing machine **safety** and **capabilities** than on a better adaptation to human user. However, for some years, designer is aware of man's role in relation with machine, but this problem is not yet sufficiently mastered. At present time, endeavours in that way are limited to tentatively separate, decentralize, and set a **hierarchy of functionalities** of man-machine interfaces in order that the user can have a better understanding.

Finally, machine design obliges to solve very complex, non-linear problems with numerous parameters, as well as pluridisciplinary problems. So, a full understanding of machine is impossible to a single person. That alights the man-machine links difficulties.

An immediate consequence deals with the necessity for the designer to consider machine as a **man-machine system.** Taking into account relations between these two last entities becomes so important in system design to design of machine entity *stricto-sensu.*

The foregoing considerations lead to put **questions** whether to have inside reasonable delays machines correctly adapted to potential user requirement. Some main questions are the following:

a) about some **prospective** aspects: will only market laws lead to machines in conformity with a humanistic vision of society? How to fit a humanistic approach into a market approach?

b) about needs and demands: in a humanistic vision, some needs are expressed and do not meet offer, whereas demand, in case it generates profitability, always meets an offer. How to take into account true needs neglected by markets?

c) about risks: what dangers are coming from machines and their user-interfaces? What juridical consequences are in case of an accident?

d) about **scientific progresses**: can man-machine interaction be mastered with present or near future scientific and technological knowledge? That point evocates the necessary symbiosis of so called "tough" or "hard" sciences with other sciences including more suggestive aspects.

e) about machines adaptation: how to take into account the various categories of human beings: young and old, professional and amateurs, disabled persons and healthy persons, single operator and grouped operators…considering the large variety of machine applications?

f) about communication: how to explain to public actualities of things concerning machines and to give it a positive and critical judgment about technological progress?

g) about education: how to educate the future designers and users of machines and eventually to make attractive professional activities in relation with technology?

The paper tries now to describe the present situation in matter of MMI (section 1), to review technological possibilities in order to make more performing MMI in the future (section 2) and to consider the possible evolution of MMI inside a liberal policy (section 3). Some recommendations are proposed in section 4 to preserve humanistic values.

2. PRESENT STATE OF MAN-MACHINE INTERACTION PROBLEMATICS

2.1 Man-Machine Interaction Problem Description

2.1.1 Machines classification

To comment man-machine interaction it is required to define what is named "machine" and how to classify them. Numerous possibilities exist. Here, due to the subject, it is simple to consider a classification based upon the machine situation relative to the environment. So, there are:

- machines that act; they are used either as transportation tool or they physically modify their environment such as manufacturing tools,
- machines that perceive; they gather information from their environment to transform it according to treatments or shaping,
- machines that reason; they have to understand information or to shape it in order to be treated.

Numerous machines at present time own the three previous features or at least two because the presence of computer and sensors is now generalised.

Considering the way they are used and some features of MMI (interfaces), machines can be divided into **machines for general public** and **machines for professionals.** In this last category, the cars (although addressing general public) will be included as well as **machines for personal assistance** (e.g. to overcome physical handicaps, to get reeducation or training) because safety problems are identical to those dealing with professional machines and therefore inducing the same type of MMI.

2.1.2   Machines and decision making

Functioning of any machine passes through compulsory steps of decision making. The basic question concerning the control modalities of any machine is which or who will take a decision in the functioning process?

The answer is very complex and depends on instant situation of machine relative to the physical and mental world. The situation is under deep influence of very different contexts, taking into account present, past and future time (the goal).

According to applications, know-how, opportunities offered by science and technology and a lot of other criteria, decision making for a machine will come either from **man** (pilot) or from the **machine** itself. For any machine, there is a **sharing** of decisional tasks, with two extreme cases: the **fully automatic** machine (after starting human order) and the **fully manual** machine.

The natural trend of engineer or researcher is to go toward fully automatic machines with autonomous decision making. That is possible only accepting strong limitations in the tasks. In fact numerous mental and physical tasks needed or useful in daily life, very easily performed by man, are not performable with a fully automated machine. That is the reason why an association of man with machine is often chosen to make the sequence of decisions imposed by the desired work performance.

2.1.3   Decision making and time responses

Success of exchanges of man with machine is very sensitive to the temporal or **time aspect.** In fact, when man sends an instruction to machine.(gesture or speech or other type of instruction...) he expects to receive an answer or response inside a delay that will be almost the same as he could get from another person or from his natural environment. For instance if he represents a virtual dog on his computer he wants the dog to move with the same speed as a true dog moves. Operator wishes a so called **real time response** (such it would happen in his daily environment).

Desire of real time comes from features specific to man. His mental and physical performances are bound to his transmission-reception system relative to his environment. Man owns **specific frequencies** depending on components of his sensory or control system to which machine has to be adapted in order to get friendly-using exchanges. There are **proper times to man.** Three such times can be considered. One is linked to the global performances of his body associated to his reflex nervous system with a maximal frequency of about 10 hertz. Another one is bound to his aware brain; it can be slower because it needs some duration for reasoning. At last, a kind of virtual time is felt *a posteriori* when he focuses his attention on something becoming unaware of all other things. Frequency of that last proper time is very variable and can be very low according to the concentration level of man.

Machine, at present time, as composed with a physical body and a control computer, owns two proper times that can be compared, from one hand, to the proper time of man linked to his transmission-reception system and to the inertial features of the moving parts of body, and from another hand, to the proper time of the aware brain. The first one seems to have a similar nature of what happens with human body but can reach frequencies of several hundreds hertz. The second one cannot be compared with what human brain exhibits: computer basic frequencies reach several megahertz generating computations speeds thousands or millions times larger than these ones performed by brain. However, algorithms implemented in computer are not the same as these ones used by human brain. Finally, taking into account **intelligence** to be used to reach a goal, to solve a problem, the time and quality advantages of result can be provided by machine or by man. It is known machines intelligence and machines awareness are very ill mastered (in spite of numerous research works devoted to that subject). So, as soon as some control has to call for anticipations, inventiveness, feelings, for suggestive aspects in general and intuition, machine is far behind man.

Proper frequencies for computer and man can be enlarged via **learning/teaching.** That phenomenon always exists for man facing a machine he wants to use (even when it is believed the machine has been sufficiently well designed to not use that phase). Familiarization within a new tool is compulsory. It can take few minutes or demand a very long time. When designing a MMI, an important objective deals with reduction of the learning/teaching step. Such a phenomenon results an improving of performances and is also searched to be implemented on computers. Learning/teaching methods are numerous and is inspired by what has been observed with man, but remain limited in performances. They constitute a promising way to qualify machines and their MMI.

The foregoing five proper frequencies are basic parameters of man-machine relations that put difficult problems, but they have to be considered when looking for improving such relations.

Consequently, the man-machine link problem, the man-machine interaction problem and the creation of the corresponding **interfaces**, appear as very complex problems and have to be solved in urgency.

2.2   Difficulties in MMI Design

2.2.1   General goal

In MMI design, the objective copes with two properties:
1. **safety,** commented further on with the risks matters;
2. **friendly-using** that can be summarized with machine using easiness, or man-machine symbiosis, or the fact that man has a good feeling with machine as its operator.

2.2.2   Friendly-using

Although man has the best adaptation facilities due to his brain properties, he shows efficiency and readiness to accept his machine operator role only when his physical, sensory and mental ergonomy are respected as well as possible.

So, in MMI design, it is tried to adapt large parts of machine to that ergonomy, or the adaptation means are connected to machine inputs-outputs. Generally, that connection does not correctly work excepted if is forecasted during the machine design process.

In these adaptations two parts have to be considered:
a) one can be considered as **generic,** but it remains a **model.** It deals with knowledge of (global) human ergonomics, including sensory models, (representation modes of information and its rhythm) to be used to transmit to operator pertinent information from machine and to allow him to easily and efficiently react,
b) another one is the implementation of these models on physical matters and they are **depending on each application**, on each operator and on each machine.

A list of main difficulties in MMI design and implementation can be presented:
− spreading of machines applications (therefore spreading of needs in man-machine relations) does not allow to think to a totally universal or generic MMI. In fact, under assumption such an interface could be

technologically made, it would be, from one side, complex and **costly**, and from another side, unuseful and **ill-adapted** to number of applications.

– For a chosen application and imposed information channels for control, the application goal could not be reached (for instance, in computer scene description understanding via speech, an additional visual image understanding system could be of interest to reach the goal)

– For a chosen application, identification of interface pertinent features can be difficult (a typical instance is teleoperation to remotely handle and assemble parts). Some people believe that visual feedbacks are sufficient to correctly teleoperate (any other problem being solved). Other people believe that haptic feedbacks are imperative to be added to visual feedbacks in order to take advantage of good conditions in teleoperation).

– In fact, to make a perfectly ergonomically interface including the three physical, sensory and cognitive levels, it is necessary to know good models of man. Man anatomy and physiology are well known. Sensory interpretation modes by brain are less known as well as all things concerned by substitution and cooperation of senses during task execution. Psychological and mental motivations of man during a machine piloting are also not known.

– In man-machine system, machine as well as man can make decisions. The splitting line for decision making attribution is a fundamental problem. It is not solved because it is linked for each action to its goal and to its safety level. For instance, in cognitive sciences, an a priori was precisely that that line did not exist. At the opposite, a fact such as the Three Miles Island nuclear plant accident proved that the accident was a consequence of a succession of human errors. If the automation process had been designed differently, the accident did not probably occur. So, the risk notion (associated to machine using) is a very influent parameter for the responsibilities repartition to man and to machine. That parameter is not sufficiently known and modelled.

– Technologies do not yet allow implementing really ergonomical interfaces. From one side, the "true" ergonomics is not known. The system "transparency" is an important factor to be studied. From another side, techniques respecting detected ergonomical constraints are often inadequate. For instance they result in an operations which are unpleasant and tiring, computers are not still fast enough to allow complex visual analysis and there are a lot of other technological limitations.

In front of design and implementation difficulties of MMI, a clarification can be done classifying them into user's categories, machines categories and applications categories (i.e. environments bound to application).

2.3   Actual MMI weaknesses

Anyone is a witness of actual MMI defects when using a computer or a home appliance or driving a car...

2.3.1   Main weaknesses

2.3.1.1 About machines for general public

Looking at machines for general public that are the most numerous (mobile phone, fax, TV set, personal computer, etc...), their MMI has always the same nature. Command is performed pushing or touching buttons and generating or not a written text. Control is visual via reading of texts or symbols displayed on variable size screens. Sometimes there is sound and more scarcely speech. This relation mode with machine is basically acceptable even if it is improvable. However, MMI includes two different aspects when used. One is the **exchange mode** or the manner with which operator exchanges information with MMI (one is pushing a button, one is reading a screen etc. - that refers to physical and sensory ergonomics). The other one is the manner with which operator generates the **sequence of operations** or information exchanges to be performed in order to reach a specific goal. That generation should be very easy, that is to say it should be immediate to understand "how it works". That is the cognitive ergonomics problem. But a lot of progresses have to be done about that point as explained now:

1. too **long directions** to be read, ill organized, with bad translations whether the system is not made in the user's country,
2. using of a too specialized or customized vocabulary (e.g. of professional computer science or a manufacturer vocabulary) is difficult to be understood,
3. designers believe to meet user needs making multi-functional systems. It results in additional length to directions, in making difficult to remind the instructions and finally in leading to using errors. It can generate a full refusal to use machine. Obviously, none is able to know and use all functions of his computer or his mobile phone etc. without spending a very long time. And time to do that does not exist... Designers are aware of this problem but, up to now, proposed solutions are not satisfactory. And to provide appliances oversized in their functionalities has a cost which is supported by customers who does not take advantage of the corresponding service.

2.3.1.2   About professional machines

Professional machines MMI have other features. First, **learning/teaching** is considered as a normal process (cars, planes, manufacturing systems etc.). But man-machine system has to allow a good performance of two functionalities: a **correct execution** of the desired work (depending on machine quality but also on control mode, therefore on MMI), and keeping of a **high level of safety** (any mobile machine or machine modifying its environment is potentially dangerous for man).

Numerous MMI present **unsuitable aspects about** physical **ergonomy** (for instance ill situated displays relative to pilot), about sensory ergonomy (for instance difficulty to read a message) and about cognitive ergonomy (for instance ambiguous message). All that can lead (and has already led) to accidents.

Cognitive ergonomy is delicate to master because it is basically a **communication** problem with man that has to respect human characteristics. From one side, language ambiguity allows to transmit a lot of things about what is not said and allows great flexibility. From another side, when the communication result generates decision making to physically act on a machine, error is not permitted and obliges to delete any ambiguity, which is at the opposite of a basic feature in human communication.

2.3.1.3   About assistance and training machines

MMI plays an important role in assistance machines devoted to overcome physical handicaps (prosthesis, orthotic devices, telethesis, assistance to blind persons etc.) as well as in training machines or machines for reeducation.

The first ones put the problem of machine design, of its **function.** For instance, considering an arm, a hand or a leg, what functions owned by the living link will be chosen to be reproduced knowing it is impossible to make an identical system? For instance, about a telethesis, what tasks useful to patient will be chosen to be performed among the infinity of tasks performed daily by a healthy man? etc. Generation of criteria and choices has not yet found its ideal solution.

Then, these machines put the problem of **interfacing with man.** That is a peculiar problem because the ideal solution should result in a feeling for the user that this machine is a part of himself (artificial hand for instance). And, it is important to note that the machine user did not choose to be in his physical situation. Numerous **psychological problems** are added influencing acceptance or rejection of machine by users selected by the designer.

All assistance systems in that category provide performances very far from desired results, considering machine itself or its MMI. Nevertheless, services they bring out have not to be minimized.

Training machines can be considered as true professional machines. The only **safety** problem can be considered as different. To intrinsic safety (to not generate accident during machine using), learning/teaching **quality** or reeducation quality has to be added. In fact, the machine goal deals with (positive) modifications of physical or mental performances of the user who is the machine operator. A bad reeducation of a wounded leg can result in a permanent handicap. An incorrect training can decrease performances instead of increasing them. So, the MMI machine is determining for success. These MMI have to be improved. For instance, with the last training systems in virtual environment a new question occurs: what role has the **virtual environments** in **competences acquiring**?

2.3.2   Some causes of MMI weaknesses

In all fields of application of machines, progresses in matter of MMI are necessary. Improvements seem possible with present knowledge. Other improvements demand long-term researches. A condition to progress is a close collaboration of physical sciences with life sciences and social sciences. But the inheritance of past in the matter makes difficult these types of collaboration. From another side progresses demand investments in men and equipments.

2.4   Physical and mental risks using man-machine systems

2.4.1   Presence of risks

Care of protection of physical integrity of man driving a machine was developed in parallel whith the machine's complexity, their potential danger and the awareness of human life importance. At present time, **person safety** is the first priority followed by **equipments safety.**

Thanks to automation, great progresses were done in these two fields. Handling and transportation systems are very secured through automation. Hardware is well protected in case of faults or errors thanks to associated electronic systems reacting more fast than man. Software are highly secured for a lot of potential problems. In case of accident issued from unforeseen situations, man has urgency stoppers usable in dark or smoky environment…

The first care about machine being safety, it also shows *a contrario* that using machines is not without **risks**. Those risks can concern the machine operator (individual accident), groups of persons without any link with the operator (plane accident), whole populations (massive destruction weapons or plant explosion).

Danger presents different forms. It can concern **physical integrity** of persons or their **mental stability.** It can also concern a kind of **slavery** of persons or their **manipulation** without they are aware of the phenomenon (general computerization of society).

It is necessary to consider the following as a theorem: **zero risk does not exist** when using machines or tools. A perfect safety is impossible because it is impossible to foreseen neither **human creativity** nor **human unawareness** in the matter.

The general public does not understand the idea that zero risk does not exist. If the industry dispatches the truth it could result in a lot of problems and over reactions. An illustration takes place in USA with numerous haulings into court bound to an unforecast using of products or machines.

2.4.2 Risks evaluation

Risk assessment becomes very difficult when introducing all human factors issued from the previous section. Risk calculations are generally based on mastered parameters. It assumes a "correct" machine operator, well knowing the system, and absence of phenomena or individuals external to the system under consideration. Mastered parameters are limited to technical data such as components reliability (that can be computed).

However, **human factors**, even limited to the adequate operator behaviour, appear as the **first source** in case of accident (tiring, clumsiness, interpretation errors, voluntary risky situation…). These factors are more and more studied but remain limited for an efficient application.

MMI study is a quite important way to understand and **decrease risks bound to human factors**. In fact, to control a situation passes through knowledge and understanding of this situation (relative to the desired situation). Knowledge and understanding are initiated by sensory excitations transmitted to the operator. These **sensory feedbacks** have to be in conformity with the triggering of adequate mental processes. But this also depends on the **operator state** (physical and mental) that has to be appreciated. So, MMI studies is important to assess risks.

2.4.3 Risk feeling

As zero risk does not exist, as science is unable to completely assess risk, as communication with general public is not adequate, all that results in a separation of real risk with felt risk. Finally, risk feeling is a psychological affair and media have a big impact on the feeling that can be at the opposite of the actual situation.

2.4.4   Responsibility in case of accident

Imperative presence of risk results in certainty of accident, even its probability is very low. Accidents bound to man-machine systems do exist. So, the accident responsibility problem occurs as well as damages repairing and all consequences referring to laws.

Laws about machines and responsibility are complex and were elaborated around "classical" machines without any possibility of "self-decision". So, all things deal with the operator or the machine manufacturer whether there is no external intervention. In man-machine system, as explained, machine can have some capacity in matter of decisional and behaviour autonomy. Machine can take "initiatives". In case of accident sharing of responsibilities become less clear because a lot of persons can have contributed to the implementation of such properties in the machine under consideration. Consequently, man-machine systems have to be seriously studied from the responsibility point of view and new texts of laws have to be carefully prepared.

3.   CONTRIBUTION OF TECHNOLOGICAL PROGRESS TO MMI IMPROVEMENT

What is requested in MMI is maximal **safety** in system using and **friendly-using**, making attractive machine.

3.1   In matter of safety

Studies relative to safety are proposed for a long time and technology offers numerous solutions. Safety means implemented in machine for a normal functioning, including faults and perturbations bound to technical causes, are efficient and only demand some improvements in every specific case. But safety means involving human factors as origin of accidents are not available and studies are undertaken about the point, peculiarly for plane and car transportation applications.

3.2   In matter of friendly-using

This problem was relatively neglected in the past by designers. Cause relies on simplicity of ancient machines without integrated computer. Using such type of machine does not need complex adapted- to- man equipment and it does not draw attention on the problem. Priority given to function and efficiency of machine by designers contributes also to not pay attention to the friendly-using problem.

Consequently these two facts have lasted the approach of man-machine relation problem. Now that work is under consideration but it is a very complex problem. It demands ideas on sharing of decision making between man and machine and also demands numerous technological new components to get a fluent communication of man with machine (because man receives information via all his senses and can transmit information via any part of his body under voluntary control). Oral communication was considered as the ideal solution for communication. At present time, it exhibits weaknesses in application cases and it has to be considered as one component among others in a MMI system.

Progresses in matter of friendly-using can be considered from two ways. One is the software way, relying on artificial intelligence techniques or cognitive computer science. It considers machine as an infra-human partner of operator (search of autonomous reflective behaviours). The other way deals with virtual reality techniques and is more sensitive to a material aspect. Bases of Virtual Reality consist in making interaction of man with the real world he wants to transform through an artificial world generated by computer. The real world is connected to its virtual representation, receives command orders from it and sends information to it. The same thing is true for the human operator.

Man-machine system transposition into a simulation brings out interesting **advantages** to control machine. First it allows a lot of choices to represent man, machine and environment. These choices are used to make **functional representations** that can be evolving and adapted to various phases of the task to be performed. These functional representations facilitate understanding of situations by the operator and contribute to improve cognitive ergonomy. Its also allows to disconnect simulation from real system in order to work on strategic explorations "off line". At last, it allows man to be artificially **immerged** into simulation, which can solve a part of physical and sensory ergonomical problem of command/control.

Connection of machine to virtual scene generates "ordinary" technological difficulties dealing with linkage of computer with machine. But interface to be implemented between man and his virtual representation is more difficult. Man "immersion" in virtual world is desired. The basic reasoning consists in saying the following: immersion illusion will be perfect if the virtual world can excite the operator sensory system exactly as the equivalent real world should do. But a real world excites the whole human sensory system and every sense individually. So, the suitable artificial excitation is not easy to create. Priority is given to 3D coloured vision (vision interfaces are numerous: HMD, CRT cameras, stereo spectacles etc.), haptic forces (datagloves, joysticks, trackballs and instrumented mice are currently used) and 3D sound. Works on smelling are now very active.

Immersion into virtual scene via adequate sensory excitations does not allow to act but only to look at. In order to act and interact other means are needed. First, specialized sensors allow knowing where man is into the scene. Then it is necessary to move, and that implies all command interfaces that can be or not integrated into control interfaces (giving back sensations).

VR (virtual reality) techniques allow to progress on the way of friendly-using interfaces, but main defects deal with technological problems. Sensory feedbacks are provided by not ideal systems and the operator risks to delete advantages provided by VR on other plans. Researches cope with sensors that have no contact with man. Although VR exhibits results below expectations, it is a promising approach to solve a lot of problems connected to interfaces creation.

The two main issues to get satisfactory MMI concern, from one side, a better **knowledge of man** (at physical, sensory and mental levels), and, from another side an improvement of **computers performances.** They remain two slow for good real time complex image analysis, but expectations of solutions in a close future are credible.

4.   PREVISION OF MMI EVOLVING. RISKS AND CONSEQUENCES

4.1   The only market-oriented evolving can prevent dangers

Looking at the past it is clear that way of life and length of human life have permanently progressed in parallel with progresses in matter of scientific and technological developments. This phenomenon is more sensitive looking at the last century, so the parallel motions could go on.

Nevertheless, new factors such as globalisation of trade and financial exchanges, the important influence of markets and invasion of computerization in professional and daily life question products designers in terms of cost criteria. MMI offer in matter of new technologies as well as in terms of products can be important thanks to technological works on computers and interfaces, and research on human factors, but the problem is the using of these progresses to users who claim needs and demands. For instance, is there a risk of absence of new products that can be made and that should respond to a "real" need (a "real" need is there defined as one corresponding to existence or know-how of not manufactured or not sold products that would allow a sure improvement of life conditions to potential users)? For instance, is there a risk of too many products that could be without true utility or used to goals in opposition to human values?

Such risks cannot be discarded because MMI development will track the ordinary laws of offer and demand on markets. Machines devoted to general public and the market engine will be, in several cases, the customer attraction and they not to satisfy a true need. About machines for handicapped persons, as another instance, market can go on ignoring the need if it is not profitable whereas the humanitarian need is present.

Another type of risk can be the following. Could using of sophisticated future machines become obligatory (via regulations) or **imperative to survive** and participate to the society life (for instance, let us assume that there is no access to information and official papers without using a computer)? Obviously, the risk is to exclude de facto parts of populations that cannot use computers for various reasons.

Under these conditions, market criteria could push computerization to limits for which user, obliged to use machines and obliged to pay at every using, would be dependant on manufactures or sellers in **monopolistic position** that could be dangerous. Among dangers, absence of pressure on designers to improve machines and their interaction with users has to be noted.

At last, market pressure can lead MMI designers to handle the problem of physical risk for operator from an **insurance** point of view and not as endeavours to be done in order to progress in matter of **risk prevention.** That attitude would drive to take few initiatives about phenomena understanding and MMI improvement.

We have only exhibited some data linked to the nature of the development logics in which we are immersed and that could brake the desired development of MMI. Solutions have to be proposed in order to prevent those types of shifting trends, but it is also necessary to be aware of the contribution of investors to technological progress. Although we are inside a development logics based upon profitability, this last feature is sensitive to the notion of time or delay: profitability at short, mean or long term. Governments have in hands the general public interest and stand in position, under some conditions, to influence markets, making profitable companies dedicated to manufacturing of products selected from their social utility. So, these opportunities can be exploited to suitably take into account human desires and needs in man-machine systems and MMI.

4.2 The market-oriented evolution could not suitably fit to some important needs

Demand or need in matter of man-machine systems concerns two levels: demand or need of machines offering various **functions,** and demand or need of **friendly-using** and **safe** machines. Friendly-using property comes out from MMI. Financial or economical aspects are not considered here because their impact on demand is the same as for any product.

Quality of friendly-using property influences demand and leaves need intact. Demand can be generated by a lot of motivations. Offer in new products depends on demand but also on technological progresses among which a choice must be done (key technologies). Below, needs and demands are classified into two groups: according to machine categories and according to user categories.

4.2.1 Needs and demand according to machines categories.

4.2.1.1 MMI of man-machine system for general public

Little demand concerns **new functions** for machines. General public does not know what can be added to a phone or to a washing machine and does not know potentialities of the own personal computer.

What is strongly felt about demand and needs deals with **friendly-using improvement** of machines, that is to say, improvement of MMI. Demand and needs can present a **contradictory aspect.** A machine that does "all things" can be requested but people want that machine remaining very simple in its control. To meet the demand big works on MMI ergonomy are necessary.

Besides utilitarian machines, a part of general public is attracted by entertainment machines. More and more technology-based games offer not only visual feedbacks but also force feedbacks. And these games constitute a good target to improve MMI. Beyond the impact of offer and demand on development of entertainment machines, the question of orientation to give to them, and therefore this one of a possible influence on demand, can be put. At present time simulations of scenes of violence are in majority. Demand of useful or educational systems that could be successful if they are attractive could be encouraged. For instance simulators to learn skiing, to drive a car, to learn dance, to develop sport and so on are good target to start their development. That is an example of influence on demand easy to implement by designers if they want to show interest to the problem.

4.2.1.2 MMI of professional machines

With machines dedicated to **individual using** the general purposes are efficiency of operator within a shortest time through teaching/learning, possibility to replace the operator with another one with a minimal cost, understanding by operator of machine reactions and safety in machine using. These goals support demand in matter of MMI.

About collectively controlled systems or cooperative systems (for instance production systems), efficiency is more related to personal quality of the team in charge of piloting than to MMI sophistication. Nevertheless MMI play an important role in communication of team members. Needs are less to create new MMI than to improve existing ones and integrating them in order to get a better organization of work and communication.

Medical assistance

When mobility recovering becomes medically impossible, it is necessary to call for machines in order to link them or to integrate them to the patient (active **prosthetic** or **orthotic devices**). Needs are extremely important. Technological problems about machine and its MMI are very difficult to be solved. Each case is specific from one side, and the concerned population is numerically limited, from another side. So there are very few investors. However, the raised problem is very generic and its solution could benefit to design of a lot of MMI usable for other applications.

In matter of assistance to surgical activities, development of adapted MMI is a positive contribution to safety of delicate operations. Needs are very important.

About re-education after repairing of a motor function (fracture for instance), systems assisted with virtual reality techniques give spectacular results in matter of recuperation shortening. These good results seem to be linked to the presence of a customized re-education programme well tracked by the patient because he can control by himself his performances and he is invited to compete with the machine. Furthermore, exercises are recorded and medical doctors can adjust the programme development according to the actual state of patient. Needs are related to development of new MMI according to a range of pathologies and to improvement of existent devices.

Assistance to training

Technologies are not far from those ones used for general public. Need for a weighted teaching/learning relative to sport activity is obvious. And there are a lot of different sports every one demanding a specific MMI.

Simulators for machines piloting (plane for instance) are probably the most advanced systems in matter of MMI. They are used additionally to test new MMI in order to improve them. They allow interesting machines behaviours simulations relative to extreme situations. These simulations are very useful to generate automatic strategies in dangerous situations allowing to keep pilots in life. Needs in matter of such simulators are growing.

4.2.2   Needs and demand according to users categories

In MMI design it is aimed to respect physical, sensory and mental features of user(s) (ergonomy). But all men are not identical relative to those three classes of characteristics.

MMI designers, almost always, work with in mind the idea of **standard human user(s)** or of **mean user** relative to each of the three previous classes. But standard means conceptual standard man who has no material actuality. No one is totally relevant to standard assumption relative to the mind or body features. Consequently, a standard type of MMI will be perfectly adapted to nobody. As an instance we can quote a force feedback data glove used to interact with virtual worlds. It cannot be used by persons with too small or too large hands or by hand-leftist.

Every machine, through its desired purpose, aims at a **category of populations.** For instance, a professional machine is barely devoted to retired persons.

Every machine has MMI belonging to one of the main three groups: adapted to physical ergonomy, adapted to sensory ergonomy, adapted to mental or cognitive ergonomy. Machine vocation informs in general about the group inside which its MMI will be positioned. The MMI group being identified, the dedicated population has to be divided into sub-populations with common features or, at least, homogeneous enough relative to the selected MMI group (for instance: tall, means and small persons, tall and thin and tall and fat etc. for an MMI belonging to the physical ergonomy group). This group therefore will be dispatched into a number of MMI types. Every type, then, can have tuning parameters in order to be adapted to every peculiar person.

That simple approach largely remains without implementation. In fact, it seems that identification of populations and of their features in relation with MMI design has never been undertaken. Reasons are probably, from one side, related to the fact that awareness of MMI importance has just started, and, from another side, that solutions are difficult to find, and, finally, adaptation is expensive because, often, it is not a simple modification of an existing MMI but it demands deep remake.

However, imperative needs are present and they should be taken into account:

– population of old persons (that grows) who are disoriented in front of general public machines in daily life (phone, TV, washing machines, cameras). Age brings out decreasing physical performances (about mobility, transportable load) and sensory performances (vision, audition).

Memory can also show some weaknesses. These evolving in ergonomical characteristics do not reach a pathological level for a large part of the population. So, it seems not too much difficult to design MMI adapted to the situation generated by age, but few things are done.

– population of functionally disabled persons has been previously evocated. It puts more difficult problems for it is necessary to go down up to the individual handicap level in order to adapt any MMI.

– The current general public faces difficulties in front of machines scarcely adapted to mental ergonomy.

– Population of pupils and students also raises the same problem at a highest level.

Population of teams involved into piloting professional machines (manufacturing systems for instance) should need making of multi-users MMI to improve communication between collaborative workers.

All these populations, of course, do not call for the same urgency level but they are essentially composed with persons with too low financial resources to have in impact on industrials. So, the institutional sector has to pay attention to these problems.

4.3   Designers and users education is necessary

Some societal risks were swept and it is expected they will be overcome thanks to adapted institutional interventions. Nevertheless, MMI market, at present time, is just starting, and it is important to think about the consequences of its development in order to reach a sound and safe introduction of MMI ad man-machine systems into public. Two points about that seem crucial: education of MMI designers and users, and information quality dispatched to the general public.

4.3.1   MMI designers and users education

New technological domains occur and public institutions have difficulties to acknowledge the fact.

In fact, in 19th century, Auguste Comte proposed a classification of sciences that was adopted by French public authorities, and then by other countries. The general topics are still respected in France. But that classification is ignoring the technology and the crossed scientific domains, both generating at present time the main part of innovation. Some fields having in the past generated disputes and problems can be quoted as examples: computer science, then automatic control are often still identified as a part of mathematics, and, more recently, robotics considered as a subset of automatic control, microelectronics, mechatronics and the same approach can be noticed about man-machine interaction.

Without minimizing difficulties for authorities neither to propose a new classification of knowledge nor to take in charges its consequences, the absence of display of technologies that have a strong impact onto society evolution, from on side, generates discomfort and claims or delays in promotion or departure for foreign countries of the best researchers in the concerned field, from another side, brakes the public research-industrial research cooperation, and, finally, introduces a lag in implementation of educational programmes to be set for benefit of all economical and sociological components of the country. Nevertheless, in France, there are some initiatives. For instance, at CNRS: creation of a department devoted to new information and communication technologies and sciences gathering traditionally separated scientific domains. Other instances can be quoted in universities and schools of engineers. Nevertheless, disputes are permanent and illustrate difficulties that are met to create the necessary dynamics toward technologies that revolution our way of life.

MMI field is intrinsically dependant on crossing of several classical domains because the matter is concerned with making of a technical system well suited for man. So, knowledge in computer science, mechanics, automatic control, anatomy and physiology, cognitive ergonomy and psychology…are requested.

4.3.1.1   About designers education

A good MMI designer is probably an experienced engineer or researcher. His initial education was provided by university or school of engineers. Initial formation is supposed to educate so called "generalist" engineers although teaching is often specialized enough. They have know-how to contribute to design of machines that demand mechanical, electrical, electronical and computer constraints. However, constraints due to user adaptation are often neglected, considered as of second urgency, and not taken into account in design starting. They generate, only in a final state, some MMI more or less satisfactory.

Initiatives could be taken to improve the machine design process:

a) to create really "generalist" streams of education i.e. with teaching integrating several types of knowledge, belonging from one side to classical "physical" domains, but also, from another side, to life and human sciences etc.
b) to propose complementary courses specialized on MMI,
c) to implement really multi-disciplinary research teams,
d) to implement permanent adapted courses all along the professional life…

4.3.1.2   About users education

As far as general public is concerned it cannot be a classical education. The basic problem deals with adaptation and acceptance of machines. Either market will provide intuitive and friendly-using machines generating a vast acceptance, or machines will stay difficult to be used and education could adopt the form of explanations or of psychological assistance to user in order to overcome his fears.

About professionals, a machine is never an entertainment… Results to be inserted into economical sectors are expected from it. Probably, machine will keep a complex driving in order to draw the best from it. It will stay dangerous (physically) and generating penalties (economically) in case of hazardous actions. MMI will help to correct using. A peculiar teaching will stay necessary for user. But education and learning will be easier with good MMI than with bad ones…

About persons assisted or aided with machines, friendly-using is a necessity not yet implemented in order to improve the daily life of these persons. They will not request an education but rather a psychological assistance and an aid to learning. It is expected future MMI will satisfy demands and needs of this category of population.

4.3.1.3   And at school?

Children seem to own gift to easily integrate handling of technological new systems, even when their friendly – usage is imperfect. They know how to use them without learning how they work. Advantages (for learning/teaching) and inconveniences (forget of elementary mathematics and of writing) are known. MMI could succeed in bringing out an almost full transparency between child and his machine. Nevertheless there is a risk of confusion for young children about a virtual object or person and its corresponding reality (Super Mario is alive and he is in the computer…).

So, studies are necessary about the role of machines at school or at scholar age as well as about implementation of ad hoc information actions in direction of parents.

4.3.2   Communication with general public

Education is a long way. Its successful achievement also depends on information continuously transmitted during the educational process toward general public. This has a big influence on mind state of people. Public must be informed with realistic data. It has to be aided in case of unfortunate events involving machines and MMI. But, general public does not permanently exhibit an adequate level of realism. Almost any false idea can be accepted mainly when media push in that direction. Beyond the awareness, actions should be undertaken to introduce sound ideas in general public minds.

5.   SOME RECOMMENDATIONS

After the foregoing overview on situation and probable evolving of MMI, the main lesson copes with awareness of difficulties for a wished development. Nevertheless these difficulties being foreseeable they could be overcome through adapted initiatives of authorities. Among them we push the following ideas.

5.1   About getting performing, safe and friendly-using MMI

Machine design as soon as its first stage should integrate interfaces presence and all things concerned by man-machine interaction.

Priority in research subjects in order to progress towards safe and friendly-using machines deals with:

a) identification of human being, of his sensory system working and of man behaviour while driving machine;

b) understanding of human factors impact upon human behaviour, peculiarly in unforecast situations.

Qualified MMI generation demands multi-disciplinary research works that are few at present time. Creation of such hybrid research teams is recommended.

Creation of educational sets providing flows of engineers submitted to balanced teaching about "hard" sciences, life sciences and human sciences would allow education of designers well adapted to work on any technological problem involving man and peculiarly on MMI.

5.2   About MMI users

Most of the time, MMI are devoted to a standard human who is mentally mature and in good physical conditions. An important part of population does not meet these assumptions. A peculiar attention must be dedicated to these persons, mainly old persons, disabled persons, children who are subject to mental risks dealing with machine addiction or mental merging of real and virtual objects or living beings.

5.3 About responsibilities

Current regulations in matter of responsibility in case of accident occurrence during machine using refers to ancient machines without any behaviour autonomy capacities i.e. without capacity in making decision by themselves. Introduction of that opportunity into modern machines makes greatly complex the attribution of responsibilities in case of accident. Studies must be undertaken in order to adapt laws and regulations to real machine properties.

The MMI development engine will come from offer-demand market process. But market could neglect not profitable sectors whereas humanitarian demand and needs are there at high level. Authorities must watch phenomena and propose aids to solve problems.

5.4 About information of general public in matter of machines and technology

General public does not hold a sound and realistic judgment on technology and machines. Public is ill informed whereas the way of life depends on technological successes. It is necessary to dispatch information preventing the general public from fears and inviting to trust in technology and machines.

6. CONCLUSIONS

Definitively, the deep reason to design and make machines is man service. It has to contribute to his personal blossoming, to develop his abilities and cleverness, to allow him exhibition of his faculties and free him from all weighting things in life. Performing MMI making machine friendly-using bring out a very large added-value to the expression of human vocation. Any initiative meeting that direction must be encouraged.

Beyond this humanistic aspect, let us notice that a fast MMI development is **possible** thanks to actual evolving of technology. Such a development is also **wished** considering it is an important source of machines safety improvements and considering it can positively transform the way of life of categories of population. But, furthermore, it is **economically required** to stay in international competition in technological matter.

ACKNOWLEDGEMENTS

The author thanks for their kind contribution: F. de Charentenay, J. Dhers, F. Ewald, P. Fillet, J. P. Marec, R. Masse, M. Pélegrin, P. Perrier, E. Spitz, all from the National Academy of Technologies of France, F. Kaplan (Sony France), J.P. Papin (LRV), P. Rabischong (Faculty of Medicine of Montpellier) and G. Sabah (CNRS).

REFERENCES

1. G. C. BURDEA, P. COIFFET, *Virtual Reality Technology*. (Second Edition, Wiley Interscience, 2003).

INTEGRATING ECO-DESIGN INTO BUSINESS ENVIRONMENTS
A multi-level approach

W. Dewulf and J. R. Duflou
Department of Mechanical Engineering, Katholieke Universiteit Leuven, Belgium

Abstract: A successful implementation of eco-design not only requires the availability of appropriate tools, but also a thorough integration of eco-design within the business operations. This paper discusses this integration on different levels. The first level comprises individual design projects, where links with more traditionally used tools and procedures need to be made. On the second level the focus is widened to an eco-design involvement of the entire company, which can be supported by a product-oriented environmental management system. The third level reflects the life cycle perspective. The paper discusses how sector-wide initiatives can provide incentives for a wider application of eco-design in today's companies. Finally the discussions are synthesized into a conceptual framework for integrating eco-design into business practice based on a multi-level approach.

Key words: eco-design, product oriented environmental management, sector, environmental performance indicator.

1.   INTRODUCTION

Despite of legislative actions and intensive research programs, the wider industrial community has still not adopted eco-design as an evident part of business practice [1]. One reason is the unbalance between the numerous research efforts concentrating on eco-design tools and the limited attention paid to the integration of eco-design thinking in the overall functioning of designers, companies and sectors [2]. However, other Design for X programs have proven the importance of a thorough integration ([3, 4, 5]).

This paper discusses the envisaged integration of eco-design from

D. Talabă and T. Roche (eds.), Product Engineering, 55–76.

various viewpoints. First, the integration on the level of individual design projects is introduced. Links to both the overall design toolbox (Section 2) and the traditional design procedures (Section 3) need to be made. Section 4 stresses the importance of involving the entire company in the eco-design process. The advantage of even larger cooperative actions covering industry sectors is introduced in Section 5. The observations are then synthesized into a conceptual framework for integrating eco-design into business practice (Section 6).

2.   INTEGRATION OF ECO-DESIGN TOOLS IN THE OVERALL DESIGN TOOLBOX

A wide variety of requirements are imposed on the design process, including cost, quality, manufacturability and functional requirements. Consequently, designers make use of a wide range of tools. Eco-design tools need to be integrated with these tools to the largest possible extent in order to ensure adoption by the designers. Both sharing of input data, to avoid redundant input efforts, as well as integration of output data are necessary.

2.1   Integration of tool input data

Many of the tools presented in the previous section require an extensive amount of input data. Integration of eco-design into business practice calls for easy retrieval of the required data, comprising, on the one hand, data on life cycle processes such as LCI data and, on the other hand, product attributes such as product geometry and materials inventory.

Information on life cycle processes can be recovered from previous generation products through direct data exchange with other actors in the life cycle, thus creating a link between different actors of the product life cycle. Both systems with distributed databases [6] or with a central database [7] have been proposed. By lack of representative historical data, virtual prototyping and simulations can be used ([8, 9]).

Product attributes can, at least partially, be recovered from non-eco-design oriented product development tools, such as CAD/CAM or PDM systems. Especially the field of Design for Disassembly has taken advantage of the developments in the CAD/CAM field with respect to storing assembly information, and recovers similar information for disassemblability evaluation (e.g. [10]).

The cradle-to-grave environmental performance can be estimated, combining geometrical and feature-based CAD/CAM data on one hand, and life cycle inventory data related to materials and processes on the other hand

[11, 12]. Both the impact of different designs and of different manufacturing conditions can thus be assessed. Life cycle modelers that calculate the effect of different use, wear-out, and end-of-life scenarios can be added to cover the full product life cycle [10, 13].

When a substantial amount of product components is bought, material inventory data needs to be captured from the supplier. This calls for standardized communication formats [14] and life cycle wide PDM applications [15].

2.2 Integration of tool output data

A product design that will realistically compete on the market, should not only be environmentally friendly but also needs to meet other customer requirements, such as a competitive pricing and a competitive overall life cycle cost. Consequently, methods and tools are needed to compare the environmental performance of designs with their economic costs and benefits.

The approaches proposed in literature can be classified in three categories:

a) *presentational integration*, restraining from quantitative aggregation of the criteria, but taking care of presenting environmental aspects next to technical or economic performance information;

b) *multi-criteria decision making (MCDM)*, supporting the derivation of weighting factors or weighting functions allowing for aggregation of the criteria using a weighted sum;

c) *eco-efficiency*, aggregating typically an economic value-indicator and an environmental impact indicator by dividing the former by the latter.

2.2.1 Presentational integration

In the area of decision theory, two competing opinions can be distinguished. Some believe that formulating decisions quantitatively by use of a mathematical language is crucial to thinking clearly and rigorously. Others believe that most decision problems are too complex to be realistically formulated mathematically [16]. Discussions with industry representatives have shown that the latter opinion is shared by a considerable fraction of current business decision-makers. Consequently, many tools do not aggregate environmental scores with economic or technical performance indicators. However, they contribute to eco-design integration by offering both economic and technical as well as environmental metrics within the same working environment. For example, material selection tools such as IDEMAT [17] or EuroMat [18] support material selection based on simultaneously considering technical, ecological and economic criteria, however without integrating these aspects into one single score. On a product planning level, simple two-dimensional eco-portfolio matrices are commonly used [19, 20, 21]. They are an extension of the well-known portfolio matrix concept, and represent products simultaneously according to an economic and an environmental indicator (Figure 1). While products in quadrant I could be called 'eco-efficient', products in quadrant II require eco-design attention, products in quadrant III need 'traditional reengineering', and products in quadrant IV might as well be abandoned.

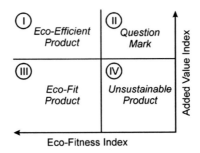

Figure 1. Eco-Portfolio matrix of Dow Chemicals [19]

2.2.2   Multi-criteria decision making

Multi-criteria decision making (MCDM) is a formal approach to assist decision makers in selecting the optimal alternative when facing conflicting criteria by deriving weighting factors or weighting functions. A large number of MCDM techniques have been developed, and are currently also applied in eco-design applications.

Kara et al. [22] present a simple MCDM application, deriving weighting factors for ecological performance, product performance, product cost, design speed, and design expenses based on paired comparison. Youngchai [23] and Azzoni et al. [24] use the Analytic Hierarchy Process (AHP) technique for integrating environmental and economic criteria. Multi-attribute Value Analysis Theory has been used by Mueller et al. [25] for aggregating environmentally related product attributes, such as recyclability, energy consumption and toxicity into a single score using value functions representing customer preferences. The technique has been implemented into a Green Design Advisor tool for a consumer electronics manufacturer.

The previously mentioned techniques fully allow complete compensation between attributes: a large gain in a lesser attribute will eventually compensate for a small loss in a more important attribute. Since this is not always considered opportune, Geldermann et al. [26] propose to use outranking MCDM techniques.

Major advantage of MCDM methods is the resulting transparency of the decision-making process. Knowing the set of weighting factors, potential suppliers can in theory optimize a product in order to meet customer requirements in an optimal way. Opponents of MCDM, however, question the ability of mathematical formulae to express a decision-making process.

2.2.3   Eco-efficiency

The concept of eco-efficiency, which can be generally defined as a combined economic and ecological efficiency, was established during the last decade as the business-oriented part of sustainable development. The concept was widely distributed via the organization Business Council for Sustainable Development [27]. The following definition is now widely referenced: "Eco-efficiency is reached by the delivery of competitively priced goods and services that satisfy human needs and bring quality of life, while progressively reducing impacts and resource intensity throughout the life-cycle, to a level at least in line with the earth's carrying capacity" [28].

This conceptual definition has been interpreted and concretized many times. Schaltegger [29] proposes to concretize eco-efficiency as the relationship between a measure of economic output (such as e.g. added value) and a measure of added environmental impact. Accordingly, a common approach to measure eco-efficiency needs to consist out of a ratio between added value and an indicator that measures the environmental impact [30]. Both WBCSD [30] and the Global Reporting Initiative [31] propose a number of nominators and denominators for eco-efficiency formulas, primarily focused on company environmental or sustainability reporting applications. Examples of eco-efficiency metrics are Net sales [EUR] / Material Consumption [kg] and Number of products sold / Amount of greenhouse gas emissions [kg CO_2-equivalents].

Boks et al. [32] propose using the inverse of the common eco-efficiency definition, i.e. environmental impact avoided divided by the financial effect. On a product level, an LCA result could also be considered to be the inverse of an eco-efficiency indicator, since it represents the environmental impact per functional unit (i.e. the added value).

Characteristic for the eco-efficiency approach is the ratio: an indicator to be maximized (being a desired output) is divided by an indicator to be minimized (being an undesired input). When widening the interpretation of "value" from merely economic value to functional value, it is possible to propose more technically oriented eco-efficiency indicators based on the ratio of a technical performance indicator by an environmental performance indicator.

Indicators belonging to this category of technically oriented eco-efficiency indicators have been proposed by Ashby [33] and Wegst [34]. Basis for the approach are the material indices introduced by Ashby [33] as a way to integrate criteria on two different material properties in view of performance optimization. It can, for example, be shown that the material index $E/q.\rho$ (with E being the Young modulus [kg/m^2], q the cradle-to-factory-grave energy consumption for the material [J/kg], and ρ the density of the material [kg/m^3]) is an adequate performance measure for the material selection of a stiff and low-energy tie. This means that, from all ties with similar geometry and energy content, the tie with highest $E/q.\rho$ ratio will be the stiffest, or, vice versa, from all ties with similar geometry and stiffness, the tie with highest $E/q.\rho$ ratio will have least energy content. For beams and plates, the material indices are respectively $E^{1/2}/q.\rho$ and $E^{1/3}/q.\rho$. Considering the stiffness of the structure as the desired added value and the cradle-to-grave energy content of the construction as the undesired environmental input, Ashby's materials indices can thus be considered a technical implementation of the eco-efficiency concept.

3.   INTEGRATION OF ECO-DESIGN IN THE COMPANY'S DESIGN PROCEDURES

Although product development procedures significantly vary between different companies and product types, design research has developed generic design process models based on high-level parallels. Similarly, detailed prescriptive eco-design procedures for the integration of environmental aspects throughout the product development process need to be tailored to the specific company situation, though high-level models can be proposed based on the generic design process models. Well-known examples include the PROMISE Ecodesign Manual developed in co-operation with the UNEP [21], the US EPA Life Cycle Design Manual [35], the eco-design consensus model of Bakker [36], and the recently published ISO TR 14062 report on 'Integrating environmental aspects into product design and development' [37]. Despite using different terminology, these eco-design procedures show a lot of similarity. Figure 2 presents an example of a generic procedure for how to approach eco-design and when to use specific tools throughout a product development process as presented by ISO TR 14062. Concrete implementations of this procedure can be found at different large manufacturers, as depicted in Figure 3.

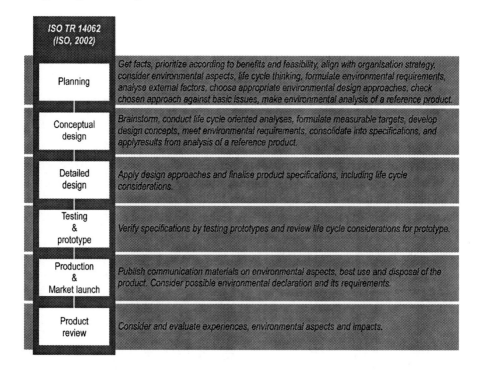

Figure 2. Example of a generic procedure of integrating environmental aspects into the product design and development process presented in ISO TR 14062 [37]

4. INTEGRATION OF ECO-DESIGN WITH OTHER COMPANY PROCESSES

Until now, eco-design has often been regarded as a technical task to be handled by industrial or engineering designers. However, the importance of integrating eco-design in all other business processes has recently been stressed [37], [41]. It is, for example, evident that the purchase department needs to be involved in order to select suppliers that offer components complying with the envisaged environmental performance of the end use equipment. Moreover, sales and marketing need to be trained to pay attention to the green market segment. Meanwhile, the production department is required to operate according to clean production principles in order to avoid unnecessary spills of lubricants, cooling liquids,...

But most important is undoubtedly the involvement of the company management. Indeed many decisions with respect to eco-design need to be taken at a level well beyond the decision sphere of the individual designer. The environmentally most beneficial improvement options require typically a change of technology or business strategy, for which authorization needs to be granted higher up the decision hierarchy. For example, a thorough life cycle assessment of a refrigerator indicated the need for a change of cooling and foaming agent, which evidently touched the core competences of the manufacturer [42]. Moreover, major copier manufacturers have shifted from selling products to selling services: customers no longer buy a copier, but pay for the service of getting copies. This implies that copiers are no longer owned by the customer, but by the manufacturer, who can take them back at any appropriate time for upgrading or refurbishment. Moreover, the manufacturer can recover components from end-of-life machines to be reused in 'new' copiers (e.g. [43, 44]). Similar strategy shifts have been taken in other sectors, e.g. for single use cameras [45]. It is evident that these strategy changes, which required important redesign of both the product and the logistic chains, required the full involvement of the company management.

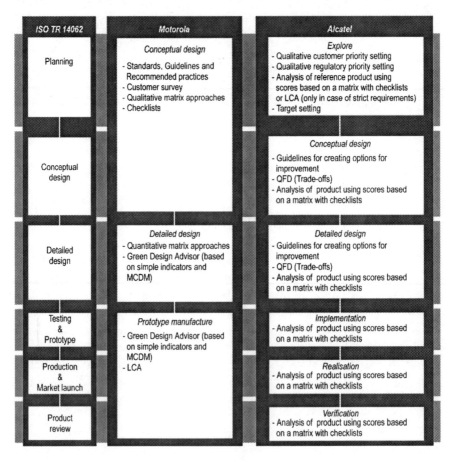

Figure 3. Selection of eco-design procedures used in industry [37, 38, 39, 40]

A sustained management involvement is however often lacking. This missing link between the strategic level (policy) and the daily green design activities is recognized as the major obstacle to a sustained practicing of eco-design in today's pro-active companies [46, 47]. Due to the current lack of company-wide integration, efforts spent on eco-design pilot projects remain without long-lasting effects: companies tend to return to business-as-usual once the pilot study has been finished [48].

These observations call for a comprehensive, company-wide eco-design framework. Basis for such framework is provided by product oriented environmental management, which mutually links the stated business processes. This product-oriented environmental management in turn needs to be supported by a product-oriented company environmental strategy.

Product-oriented environmental management systems (POEMS) have received increasing attention over the last few years. A POEMS follows the structure of a 'traditional' environmental management system (EMS), which has proven very appealing to industry: in December 2003, nearly 4000 companies were registered under the European Union Environmental Management and Auditing Scheme EMAS, and more than 45.000 companies had received ISO 14001 certification [49]. The recent revision of the EMAS scheme (EMAS II) now requires the EMS to include a product-oriented section [50].

Basis for an EMS - and consequently also for a POEMS - is the well-known Deming circle "Plan-Do-Check-Act", starting from a company's environmental strategy. A proposal for a generic POEMS model has been presented by Rocha et al. [48].

5. INTEGRATION OF ECO-DESIGN IN AN INDUSTRY SECTOR

Eco-design is, by definition, a life cycle encompassing activity. The environmental performance of the product is not limited to its existence in between the factory gates, but includes the potential impacts caused by suppliers and sub-suppliers up to the product's cradle during material mining, as well as by distributors and users down to the product's grave at the final waste management companies. Adequate eco-design requires an intensive exchange of data and information between these actors. Material inventories or even cradle-to-grave inventories form the basis for life cycle assessment studies performed during many eco-design projects.

However, the time investment for data collection required from suppliers is far from evident. Although suppliers that are heavily dependent on a single customer will be prepared to invest the necessary manpower, suppliers of off-the-shelf components will often consider the overall cost-benefit balance negative when dealing with such request from a single customer. On the other hand, many actors meet each other in the life cycles of different products. Suppliers of machine components deliver goods for multiple machine types, trademarks and manufacturers. Consequently, when a number of customers join forces in formulating data requirements and exchange formats, business opportunities increase, and the overall return on the invested time might well be beneficial. Efficient, standardized information exchange, which simultaneously takes into account proprietary and confidentiality considerations, is therefore a must throughout the product life cycle. A number of sector-wide co-operation and standardization efforts have been initiated over the last few years.

Within the rail sector, the Scandinavian railway operators have joined forces in developing uniform requirements for Design for Environment in the Nordic Manual for Rolling Stock Material" [51]. Moreover, the RAVEL project, described further in this paper, laid down the basis for a sector-wide co-operation based on Environmental Performance Indicators [52].

Within the electronics sector, the Electronic Industries Alliance (EIA) developed a Material Declaration Guide [53], proposing a uniform set of materials and threshold levels to be included by manufacturers in material declaration questionnaires. This set is based on, amongst others, existing and expected legislation as well as voluntary industry commitments. Moreover, the European Computer Manufacturers Association has issued the Technical Report TR/70 on "product-related environmental attributes" advising on the contents of supplier declarations [54].

Within the automotive industry, the ten largest European car manufacturers co-operate in the establishment of a common raw materials database and the development of methodological recommendations in view of LCA [55]. Furthermore, an International Material Data System [56] keeps archives of all materials used in the sector. Based on this system, material inventories are required from suppliers and lists of restricted or prohibited substances can be provided. The International Dismantling and Info System IDIS [57] is an industry-wide system for providing dismantling centers with relevant information, including parts lists, information about contained polymers, service handbooks and 3D-drawings.

These examples show two major, interconnected subjects of sector-wide standardization activities: a measurement system and a communication system.

6. A 3-LAYERED FRAMEWORK FOR ECO-DESIGN

In this section, the above discussions on the need for eco-design integration in the whole of business processes are conceptualized into an overall framework consisting of three levels: the individual design project, the company and the sector level. Since the ability to measure lies at the basis of all improvement potential, a measurement system using product oriented environmental performance indicators forms the core of the framework.

6.1 Environmental Performance Indicators

The ISO 14031 standard on Environmental Performance Evaluation [58] very generically introduces Environmental Performance Indicators (EPIs) as *specific expressions that provide information about an organization's environmental performance*. Selection and definition of EPIs is, according to both ISO 14031 and ISO 14001, a management process that needs to take into account the significance of the environmental aspects, the influence the organization has over the environmental aspects, the organization's environmental policy, the environmental legislation, and the views of other stakeholders. This definition of EPIs thus fits into the vision of the ISO 14001 standard on Environmental Management Systems, allowing companies to define and select their proper indicators and targets, and subsequently to focus on continual improvement. The absence of predefined minimum performance levels is, on the one hand, a point of strong criticism, but guarantees, on the other hand, a low threshold level: companies can start the EMS with easily-understood EPIs and limited but achievable targets, and subsequently evolve towards environmental excellence by continual improvement.

Being a part of a company's activities, products and their life cycles can also be subject of EPIs. A wide range of potential indicators can be found on different levels of the cause-effect chain, as depicted in Figure 4.

Since a simple EPI often defines only one environmental dimension of an object of study, it is generally necessary to use a set of multiple EPIs to describe a product's environmental profile.

A number of requirements apply for selecting and defining EPIs for eco-design applications [58, 59]:

a) An EPI should, according to the current state of understanding, drive or represent a significant environmental aspect of the product, while taking into account the environmental priorities of relevant stakeholders. Moreover, the set of EPIs should, together, cover all significant environmental aspects over the product life cycle.

b) An EPI should be measurable. This criterion comprises, on the one hand, the availability of unambiguous procedures and mathematical formulae to gather, measure, or calculate the EPI and, on the other hand, the availability of data needed to calculate the EPI score for both supplier and customer (controllability and transparency);

c) An EPI is only useful to a business situation if it can be influenced by the organization;

d) It should be possible to unambiguously accumulate EPI scores of subsystems to an overall product EPI score.

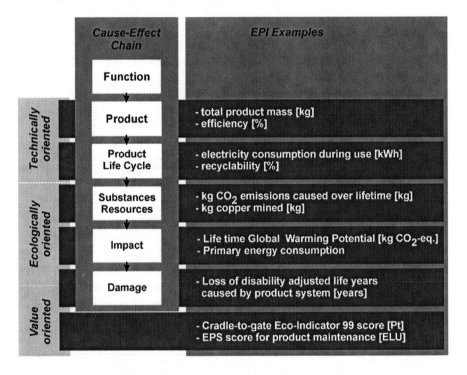

Figure 4. Examples of Product Oriented Environmental Performance Indicators and their link to the technological-environmental cause-effect chain

While using the defined EPIs in eco-design projects, it is important to clearly state the functional unit for which the EPI is calculated, as well as the life cycle model.

Within the product life cycle environmental assessment domain, there is an eternal discussion on whether environmental performance assessments should be built on holistic and scientifically sound considerations, aiming at exactly quantifying an organization's environmental impact, or whether it should be driven by practical considerations such as usability and simplicity [60]. The EPI concept allows to start with readily available product attribute

EPIs and evolve towards more complex, ecologically oriented EPIs once a higher level of experience has been reached.

6.2 The concept of the 3-layered eco-design framework

The concept for the 3-layered eco-design framework is depicted in Figure 5. It is a combination of Deming circles, interconnected through the use of environmental performance indicators.

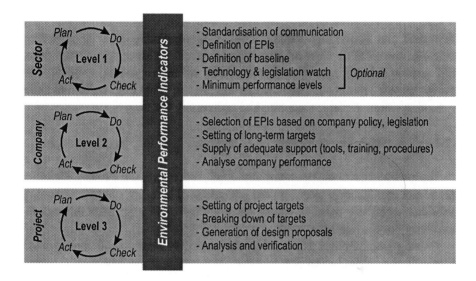

Figure 5. Concept of the 3-layered framework for eco-design

6.2.1 Level 1: the sector

Tasks within the sector layer concentrate on standardization of both the measurement system and the communication system. The former comprises the development and unequivocal definition of EPIs, including e.g. setting of system boundaries, providing lists of restricted materials, supplying basic environmental data (e.g. EPI scores for individual materials), and guiding the life cycle modeling. The latter supports efficient and transparent data exchange. Moreover, the sector can decide to increase co-operation by calculating EPIs scores for a baseline reference, thus providing input to target setting tasks. Furthermore, the installation of a legislation and technology watch, providing up-to-date insights in state-of-the-art and future eco-design conditions is in many cases beneficial on a sector level. Finally, the most far-reaching cooperation entails the development of a set of minimum performance levels, expressed in terms of minimum EPI scores.

The current draft proposal for a European Directive on establishing a framework for eco-design for End Use Equipment [61] comprises a "presumption of conformity" for products that have been developed in accordance with non-compulsory technical specifications adopted by a recognized standards body. The sector level of the proposed 3-layered eco-design framework could serve as a development platform for such technical specifications on the basis of the identified set of EPIs, thus creating a tangible business advantage in terms of cost reductions through facilitated regulatory compliance.

Daily maintenance of the provided eco-design support in terms of e.g. basic data and knowledge support should be supplemented with periodical review of the EPIs in view of new legislative and technical developments as well as new scientific insights in the effects of environmental aspects. This review should moreover include an update of the EPI communication vocabulary, the reference baseline, and potentially the targets.

6.2.2   Level 2: the company

The second layer, on the company level, represents the Product-Oriented Environmental Management System (POEMS). The efforts required from an individual company have, however, been significantly alleviated through the availability of the sector layer.

Starting point of the POEMS, as for every environmental management system, is the setting of targets based on the company environmental policy, and with the aim of continual improvement. Based on its proper environmental policy, the company selects a number of EPIs from the set proposed by the sector. For all selected EPIs, minimum performance levels are defined which should be reached either on average or by each product developed or operated by the company. The EPIs and EPI target levels form the basis of design procedures, research and development projects, supplier selection procedures, development and acquisition of adequate tools, allocation of resources, etc.

Next to the POEMS audits and the follow-up of the actual product performance of company products on the market, a regular management level review of the POEMS and of the product-oriented environmental policy of the company is necessary.

6.2.3   Level 3: the project

The third layer is situated on the level of individual product development projects. This level is consequently closest to most current eco-design implementation projects.

The planning phase includes organizational issues, such as the set-up of a project team, as well as the setting of targets. For a purchaser in a business-to-business relationship, this implies the selection of EPIs and the setting of quantitative EPI targets based on company decisions (Level 2) as well as on project specificities. For a supplier, this implies combining customer targets with the proper company EPIs. It should be emphasised that, in practice, the eventual environmental performance of the product is decided upon at this stage, i.e. before the actual creative design tasks start.

The project level targets are set at the highest level of the product structure. However, in many sectors a product is developed by a number of design teams and designers, each individually responsible for a subsystem of the final product: improvements of the environmental performance must, consequently, take place at a subsystem level. It is therefore necessary to break down the overall targets into design targets on the level of individual designers. This breakdown of targets is current practice in e.g. the rail vehicle development sector with respect to system mass.

Design targets are the basis for generating ideas and design proposals, as well as for assessing the performance of the subsystem designs. Moreover, regular review is needed to assess the overall product performance and the suitability of the initial EPI selection and target levels in order to adjust the breakdown of targets used.

7. CASE STUDY: THE RAVEL SYSTEM SUPPORTING GREEN DESIGN AND SUPPLY CHAIN MANAGEMENT FOR THE RAIL VEHICLE SECTOR

This section briefly presents a green design and supply chain management system for the rail sector from the viewpoint of the above mentioned EPI centered eco-design framework. This system was developed within the Brite-EuRam project RAVEL (RAil VEhicLe eco-efficient design) [52] as a joint effort of railway operators, a major rail vehicle manufacturer, a rail vehicle subsystem supplier, eco-design consultants, and universities.

The main driver of the project was the awareness that the environmental requirements, imposed by authorities, railway operators and - indirectly - the passengers, drastically increased over a few years' time span, while no satisfying support solutions were found on the market. At the time of starting the project, environmental requirements imposed on rail vehicle manufacturers by railway operators included maximum levels for energy consumption and exhaust emissions, recyclability scores for the vehicle and its systems, the compliance with lists of forbidden or restricted materials, obligatory marking of polymers, and the provision of full material inventories in support of a LCA screening by the customer.

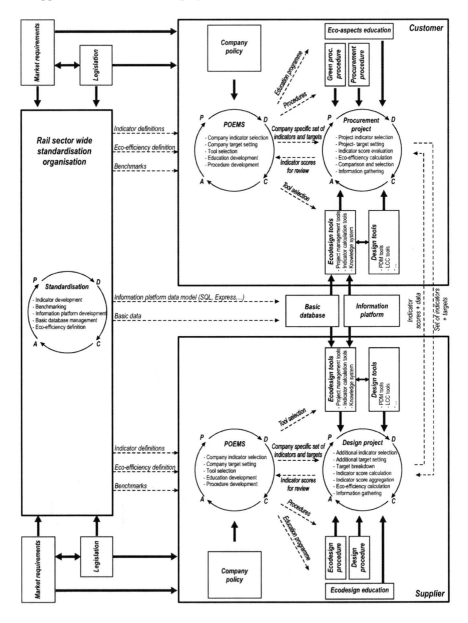

Figure 6. Schematic overview of the RAVEL system, interpreted according to the 3-layered eco-design framework [62]

Figure 6 depicts the RAVEL system interpreted according to the generic 3-layered model presented in the previous sections. Due to the large amount of actors involved in a rail vehicle development process, unequivocal communication was recognised to be a crucial factor early in the project.

Therefore, the development of EPIs was largely withdrawn from the company specific POEMS, and centralised within the framework of what should become a sectorial standardisation body. This standardisation body is planned as a co-operation between the International Union of Railway Operators (UIC) and the European Union of Rail Vehicle Manufacturers (UNIFE), and is meant to provide:

a) standardized definitions of EPIs;
b) standardized basic data needed for EPI calculations. This includes both a standardized list of materials to be used when describing products, and basic material properties required for EPI calculation;
c) a standardized eco-efficiency definition to allow combined economical and environmental performance evaluations;
d) a standardized data model for improved communication. This data model (information platform) has been developed both as a relational database structure and in EXPRESS format [63];
e) a calculated baseline reference for benchmarking.

8. DISCUSSION

The presented 3-layered eco-design framework has a number of advantages. First, the communication between all actors in the product development process is improved through the availability of standardised EPI definitions as well as a common information platform structure. The EPI definitions allow an early integration of eco-design in the design process, i.e. during the setting of requirements. Moreover, the availability of the standardized information platform, that unequivocally defines the way products and their environmental properties need to be defined, allows the development of adequate software connections between an eco-design platform and other design tools. Due to the sector-wide use of this information platform, the development of this connecting software can become profitable. Moreover, EPIs can be used in MCDM techniques or eco-efficiency formulae in order to integrate eco-design criteria with other design criteria.

Another asset of the framework is the compatibility with the successful EMS philosophy with respect to both its structure as well its focus on continual improvement: sectors and companies can start with low-threshold level EPIs before moving to more ecologically oriented EPIs such as life cycle impact category indicators. Through allowing the latter type of indicators, life cycle assessment is fully supported. Nevertheless, EPIs allow for including environmental aspects which are not always visible from LCA results, such as recyclability or small accident risks with important consequences, as well as for focussing on items specifically emphasised in the company policy in view of greening the company image, e.g. a ban on PVC. While an LCA in first instance calculates the full life cycle impact, EPIs are focused on the life cycle aspects, which can be influenced by the company.

The organizational requirements form the major disadvantage of the framework. Especially in view of the lack of strict borders between sectors, suppliers will sometimes be forced to follow more than one sector-wide system.

9.   CONCLUSIONS

The integration of eco-design into business has been discussed from various viewpoints. At the designer's desk, the integration of eco-design tools with his familiar toolbox is necessary. This encompasses both the sharing of input data as well as an integrated presentation of analysis results. Moreover, eco-design activities need to be embedded into the company's design procedures. This not only holds true for the designer's work, but for all employees of the company. Finally, sector-wide initiatives can be stimuli for suppliers to get involved into the eco-design activities. These observations are synthesized into a conceptual framework for integrating eco-design into business practice based on a multi-level approach. Characteristics of the system are the sector-widely coordinated effort towards standardization as well as the core function of environmental performance indicators.

REFERENCES

1. A. TUKKER, E. HAAG, P. EDER, *Eco-Design: European State of the Art - Part I: Comparative analysis and conclusions*. (IPTS Publication, JRC, Sevilla, 2000).
2. H. BAUMANN, F. BOONS, A. BRAGD, *Mapping the green product development field: engineering, policy and business perspectives* (Journal of Cleaner Production 10 (2002), pp.409-425).
3. B. WATSON, D. RADCLIFFE, *Structuring Design for X Tool Use for Improved Utilization.* (Journal of Engineering Design, Vol. 9, No. 3, 1998, pp.211-223).
4. M. MØRUP, *Design for Quality*. (PhD dissertation, Institute for Engineering Design, DTU, Lyngby, 1994).
5. C. GRÜNER, J. LAGERSTEDT, *Organisational Implementation of DfE in Industry.*

(Proceedings of 7th CIRP International Seminar on Life Cycle Engineering, Tokyo, November 27-29, 2000, pp.233-240).

6. S. MIAYMOTO, J. FUJIMOTO, *Development of Environmental Information Systems with a Distributed Database.* (Proceedings of Ecodesign'99, 1-3 February, 1999, pp.148-153).

7. T. OHASHI, T. ISHIDA, Y. HIROSHIGE, T. MIDORIKAWA, T. URAKAMI, M. UNO, *Ecological Information System - Data Exchange System Between Manufacturers and Recyclers.* (Proceedings of Ecodesign'99, 1-3 February, 1999, pp.143-147).

8. Z. SIDDIQUE, D. ROSEN, *A virtual prototyping approach to product disassembly reasoning.* (Computer-Aided Design, Vol. 29/12, pp.847-860).

9. F. KIMURA, *A Methodology for Design and Management of Product Life Cycle Adapted to Product Usage Modes.* (33rd CIRP International Seminar on Manufacturing Systems, 5-7 June 2000, Stockholm).

10. T. ROCHE, E. MAN, J. BROWNE, *Development of a CAD Integrated DfE Workbench Tool.* (Proceedings of 2001 IEEE International Symposium on Electronics and the Environment, Denver, May 7-9th, 2001).

11. S. NAWATA, T. AOYAMA, *Development of Life-Cycle Design System for Machine Tools - Linkage of LCI data to CAD/CAM data.* (ECP Newsletter 19, JEMAI, Tokyo, 2001).

12. G. SCHLOTHEIM, P. GROCHE, D. SCHMOECKEL, A. WANSEL, *Object-Oriented Modelling of Select Metal Forming Processes.* (Helsinki Symposium on Industrial Ecology and Material Flows, 30th August - 3rd September 2000, Heslsinki. Available from: http://www.cc.jyu.fi/helsie/).

13. O. WEGER, P. HENSELER, L. BONGULIELMI, H. BIRKHOFER, M. MEIER, *The interchange of environmentally related information - a prerequisite for the next generation of LCA-tools.* (Proceedings of Environmental Informatics 2001, 15th International Symposium on Informatics for Environmental Protection, October 10th - 12th, 2001, ETH Zürich, Switzerland).

14. K. KURAKAWA, T. KIRIYAMA, Y. BABA, Y. UMEDA, H. KOBAYASHI, *The Green Browser: An Internet-based information sharing tool for product life cycle design.* (Proceedings of 4th CIRP International Seminar on Life Cycle Engineering, Berlin, 26-27 June, 1997, pp.359-370).

15. M. ABRAMOVICI, D. GERHARD, L. LANGENBERG, *Application of PDM technology for Product Life Cycle Management.* (Proceedings of 4th CIRP International Seminar on Life Cycle Engineering, 26-27 June, 1997, Berlin).

16. J. MULLEN, B. ROTH, *Decision-Making: Its Logic and Practice.* (Rowan and Littlefield, Savage, 1991. Cited in: FEATHER T., HARRINGTON K., CAPAN D., Trade-off Analysis for Environmental Projects: An Annotated Bibliography, Planning and Management Consultants, Carbondale, 1995).

17. N., IDEMAT Software, T.U.Delft Design for Sustainability research group, Delft, 1998.

18. G. FLEISCHER, G. REBITZER, U. SCHILLER, W. P. SCHMIDT, *euroMat'97 - Tool for Environmental Life Cycle Design and Life Cycle Costing.* (in: KRAUSE F.-L., SELIGER G., Life Cycle Networks, Kluwer Academic Publishers, Boston, 1997).

19. OECD, *Eco-Efficiency in Transport: Workshop Report and Background Paper.* (ENV/EPOC/PPC/T(98)5, OECD - ENVIRONMENT DIRECTORATE - ENVIRONMENT POLICY COMMITTEE, Paris, 1997).

20. P. EAGAN, J. KONING, G. HAWK, *Application Principles for the Use of DfE Tools* (Proceedings of IEEE International Symposium on Electronics and the Environment, Orlando, 1-3 May 1995, pp.110-112).

21. H. BREZET, C. VAN HEMEL, *Ecodesign - A Promising Approach to Sustainable Production and Consumption.* (UNEP, Paris, 1997).

22.S. KARA, M. SUN, H. KAEBERNICK, *A Trade-off Model for Sustainable Product Development*. (Proceedings of 9th CIRP International Seminar on Life Cycle Engineering, Erlangen, 9-10 October, 2002, pp.103-111).

23.H. YOUNCHAI, *Methodology for Prioritizing DfE Strategies Based on LCA and AHP*. (M.S.Thesis, Ajou University, Korea, 2001).

24.G. AZZONE, G. NOCI, *Measuring the environmental performance of new products: an integrated approach*. (International Journal of Production Research, 34(11), 1996, pp.3055-3078).

25.K. MUELLER, W. HOFFMAN III, *Design for Environment - Methodology, Implementation and Industrial Experience - Part 2: Environmentally Preferred Products - How to evaluate, improve our products and report to our customers?*. (Proceedings of Electronics Goes Green 2000+, Vol1., September 11-13, Berlin, 2000, pp. 237-241).

26.J. GELDERMANN, T. SPENGLER, O. RENTZ, *Fuzzy outranking for environmental assessment. Case study: iron and steel making industry*. (Fuzzy sets and systems, 155(2000), pp.45-65).

27.S. SCHMIDHEINY, *Changing Course*. (Business Council for Sustainable Development, MIT press, 1992).

28.WORLD BUSINESS COUNCIL FOR SUSTAINABLE DEVELOMENT, Eco-efficient leadership, WBCSD, Geneva, 1996.

29.S. SCHALTEGGER, R. BURRITT, *Contemporary Environmental Accounting*. (Greanleaf Publishing, 2000).

30.H. VERFAILLIE, R. BIDWELL, *Eco-efficiency measuring - a guide to reporting company performance*. (World Business Council on Sustainable Development, 2000).

31.J. HENDERSON (Ed.), *Sustainability Reporting Guidelines 2002*. (Global Reporting Initiative, Boston, 2002).

32.C. BOKS, J. HUISMAN, A. STEVELS, *Eco-efficiency explored from the basis: Applications in an end-of-life context*. (Proceedings of the 9th CIRP International Seminar on Life Cycle Engineering, Erlangen, 9-10 April, 2002, pp.79-86).

33.M. ASHBY, *Materials Selection in Mechanical Design*. (Butterworth Heinemann, Oxford, 1992).

34.U. WEGST, M. ASHBY, *Environmentally-Conscious Design and Materials Selection*. (Proceedings of 2nd International Conference on Integrated Design and Manufacturing in Mechanical Engineering, IDMME'98, Compiègne, May 27-29, 1998, pp.913-920).

35.G. KEOLEIAN, D. MENERY, *Life Cycle Design Guidance Manual – Environmental Requirements and the Product System*. (EPA Publication No. EPA/600/R-92/226, United States Environmental Protection Agency, Office of Research and Development, Washington, 1993).

36.C. BAKKER, *Environmental Information for Industrial Designers*. (PhD Dissertation, T.U.Delft, 1995).

37.ISO TR 14062:2002, Environmental management - Integrating environmental aspects into product design and development, Technical Report, ISO, Geneva, 2002.

38.W. F. HOFFMAN III, A. LOCASCIO, *Design for Environment Development at Motorola*. (1997 IEEE International Symposium on Electronics and the Environment, IEEE, Piscataway, 1997, pp.210-214).

39.K. MUELLER, W. HOFFMAN III, *Design for Environment - Methodology, Implementation and Industrial Experience - Part 2: Environmentally Preferred Products - How to evaluate, improve our products and report to our customers?*. (Proceedings of Electronics Goes Green 2000+, Vol1., September 11-13, Berlin, 2000, pp. 237-241).

40.T. BERVOETS, F. CHRISTIAENS, S. CRIEL, D. CEUTERICK, B. JANSEN, *DfE Strategies for Telecom Products: Follow-up Eco-Performance throughout the Design*

Process flow. (Proceedings of Electronics Goes Green 2000+, Vol.1, Berlin, September 11-13, 2000, pp.177-182).

41. A. STEVELS, *Integration of EcoDesign into Business, A New Challenge.* (Proceedings of Ecodesign'99, 1-3 February, 1999, pp. 27-32).

42. H. WENZEL, L. ALTING, *Danish Experience with the EDIP Tool for Environmental Design of Industrial products.* (Proceedings of Ecodesign'99, 1-3 February, 1999, pp. 370-379).

43. P. WOUTERS, *Design for Reuse of copiers and Printers, Handouts of Symposium on Synergien durch die integrierte Betrachtung von Montage und Demontage.* (Fraunhofer-Institut IPA, Stuttgart, 7 October, 1998).

44. T. WATANABE, *Closed Loop Manufacturing System for Copiers Based on Parts/Module Reuse.* (Proceedings of 7th CIRP International Seminar on Life Cycle Engineering, Tokyo, November 27-29, 2000, pp. 258-263).

45. H. TANAKA, *Fully Automatic Inverse Factory for Re-Using Single-Use Cameras.* (Proceedings of 7th CIRP International Seminar on Life Cycle Engineering, Tokyo, November 27-29, 2000, pp. 256-257).

46. H. BAUMANN, F. BOONS, A. BRAGD, *Mapping the green product development field: engineering, policy and business perspectives.* (Journal of Cleaner Production 10 (2002), pp.409-425).

47. S. LARSSON, M. BERGENDORFF, G. GLIVBERG, *Requirements for a Rail Vehicle Specific Design for Environment System.* (in: DEWULF W., DUFLOU J., ANDER A., *Integration Eco-Efficiency in Rail Vehicle Design*, Leuven University Press, Leuven, 2001).

48. C. ROCHA, H BREZET, C. PENEDA, *The Development of Product-Oriented Environmental Management Systems (POEMS): The Dutch Experience and a Case Study.* (6th European Roundtable on Cleaner Production ERCP'99, Budapest, September 1999).

49. K. TSUJII, *The number of ISO14001/EMAS registration of the world.* (ISO World, see: http://www.ecology.or.jp/isoworld/english/analy14k.htm, March 11th, 2003).

50. European Union Environmental Management and Auditing Scheme. Homepage: http://europa.eu.int/comm/environment/emas/

51. N., *Nordic Environmental Manual.* (VR/NSB/SJ/DSB, 1999).

52. W. DEWULF, J. DUFLOU, Å. ANDER (Eds.), *Integrating Eco-Efficiency in Rail Vehicle Design.* (Leuven University Press, Leuven, 2001).

53. N., *EIA Material Declaration Guide.* (EIA, Arlington, 2001).

54. N., product-related environmental attributes, ECMA Technical Report TR/70, 2nd Edition, ECMA, Geneva, 1999.

55. See http://www.acea.be/eucar

56. See http://www.mdsystem.com/html/en/home_en.htm

57. See http://www.idis2.com

58. ISO 14031:1999, *Environmental management - Environmental performance evaluation - Guidelines.* (Internal Standard, ISO, Geneva, 1999).

59. Å. ANDER, M. BERGENDORFF, V. MANNHEIM, *Overview of the RAVEL Project.* (in: DEWULF W., DUFLOU J., ANDER Å. (Eds.), *Integrating Eco-Efficiency in Rail Vehicle Design*, Leuven University Press, Leuven, 2001, pp.53-67).

60. T. JACKSON, P. ROBERTS, *A Review of Indicators of Sustainable Development: A Report for Scottish Enterprise Tayside.* (School of Town and Regional Planning, Dundee, 2000).

61. N., *Draft Proposal for a Directive of the European Parliament and of the Council on establishing a framework for Eco-design of End Use Equipment.* (European Commission DG Enterprise, Brussels, 2002).

62. W. DEWULF, *A Pro-Active Approach to Ecodesign: Framework and Tools.* (PhD dissertation, K.U.Leuven, Leuven, 2003).

63. ISO 10303, *Industrial automation systems and integration – Product data representation and exchange.* (International Organisation for Standardisation, Geneva, 1994).

VIRTUAL ENVIRONMENTS - THE ECO-FRIENDLY APPROACH TO PRODUCT DESIGN?

J. Ritchie
School of Engineering and Physical Sciences, Heriot-Watt University, Edinburgh, UK

Abstract: The use of virtual environments and rapid prototyping in product engineering is expanding rapidly, with research now being carried out over a wide range of product areas. This paper will focus on some of the virtual technologies and applications that have been applied and are currently being used in both research laboratories and industry. It will also highlight the key issues that the author believes should be researched to enable the effective design of environmentally-friendly products.

Key words: virtual reality, eco-design, product development, prototyping.

1.   INTRODUCTION

The traditional company has been driven by the idea that industrial investment and innovation drove economic growth and satisfied the demands of the customer. However, in the modern world, it has become clear that industry is a major contributor to environmental destruction through the products and processes it uses and the materials it consumes. The disposal of this waste causes major environmental problems in the developed world with a high proportion of, even recyclable, waste being disposed of in a manner that goes contrary to good environmentally sound practice. In the UK, for example, landfill accounts for the largest proportion of the 30.5 million tonnes of annually disposed household waste (Figure 1) with another 70 million tonnes of industrial waste dumped as well. Efforts to control this through a landfill levy in 1996 have not slowed down the use of landfill and new permit legislation was issued to address this in 2003.

D. Talabă and T. Roche (eds.), Product Engineering, 77–88.

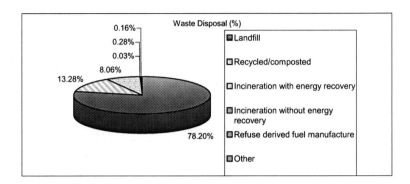

Figure 1. UK Waste Disposal Distribution

(Source: Department of the Environment and Rural Affairs, HM Government, UK, 2002, [1])

It is estimated that around two thirds of this waste is biodegradable and recommendations have been made with regard to using the waste gases to produce more energy. However, this means that around 33 million tonnes are not biodegradable and this highlights the recycling issue and the need for products that can be effectively disassembled and reused. Over 2.6 million tonnes of such waste appears to be associated with recyclable categories such as office machinery, electronic goods, motor vehicles and other forms of equipment.

Developments in business thinking have begun to change with more emphasis on how products should be designed for the whole life cycle. A typical example of this is the view produced by the Centre for Human Ecology entitled the "Triple Bottom Line" [2] which suggests that growth can no longer be the sole objective for business. Growing public awareness of global issues such as ozone depletion has forced industry, with the help of legislative measures, to include environmental impact within the business equation. One such example is the "Waste Electrical and Electronic Equipment (WEEE) Directive (2002/96/EC)" [3] which has been on the EU statute books for some years but is due to become law in the UK in late 2004. The WEEE Directive encourages and sets criteria for the collection, treatment, recycling and recovery of WEEE. It makes producers responsible for financing most of these activities ('producer responsibility'). The categories covered relate to a wide range of household appliances, computers, telecommunications equipment, televisions, videos and other similar products. Ron and Panev highlight some of the issues associated with disposal of such equipment [4].

In the automobile sector, although a high proportion of car parts are recycled in some way, there is an even greater emphasis through legislation for cars to be 95% recyclable, for example, in Japan over the next 10 years or so [5].

It is now recognised that technology alone cannot solve the environmental problems of industry itself or cure past wrongs; however, with some imagination, it can substantially support it. Industrial attitudes are moving towards the concept that production and consumption are part of a greater equation where business growth, the environment and a general social responsibility must be balanced. Companies are now moving from being simply product developers and sellers to product service developers.

Hawken [6] argued that industry in the 1990s was on the verge of a transformation that could make ecology part of the solution contending that business may be the only establishment powerful enough to facilitate the far-reaching changes required to reverse environmental destruction.

In this way economic interests could be part of a greater system of balancing forces in modern society. This will include the need for companies to adopt approaches in product development from the cradle to the grave rather than products which will be developed, sold, forgotten about and then disposed of as someone else's problem. The new strategy is that products should be designed to be recyclable where reuse of components and materials is paramount and companies have the responsibility of disposing of, in an ecological fashion, any materials that they do not reuse. Techniques such as life cycle analysis/life cycle assessment (LCA) and life cycle costing (LCC) will become more important at the product design phase as both consumer demand for ecological products – although this is a debatable issue- and legislation begin to impact on product creation, manufacture, use and disposal. Many studies have been carried out in this area [7, 8, 9]; however, the challenge for engineers is to come up with consistent models to consider these factors during the design phase of a product to assess its eventual environmental impact. Techniques such as DFX have long been mooted as supporting the design of ecologically friendly products but there is a need for cost models, such as those mentioned previously, to be standardised and built into design support systems to facilitate interactive evaluations in real time.

Another problem is that of prototyping new products. It is estimated that over 3.5 million rapid prototypes were produced by various rapid prototyping (RP) techniques in 2001 [10]. This makes no mention of functional pre-production prototypes and eco-design does not appear to be addressed adequately. For engineers, prototypes must be developed to quickly evaluate products and allow aesthetics and functionality to be determined before actual product release takes place. However, there appears to be very little mention of how prototypes can be used to develop environmentally friendly products, particularly at the end-of-life stage. Most applications appear to focus on aesthetics and form although it has been found that new methods have the potential for producing real functional alternatives; this has to be investigated further [11].

It would appear that, rather than focussing on physical prototypes, there is considerable potential to investigate other prototyping techniques that involve digital tools beyond those of traditional CAD, particularly to help evaluate product end-of-life issues. More intuitive, ergonomic tools could potentially support the sustainable aspects of product design more effectively. New virtual tools are becoming available to support the ergonomic analysis of products, particularly in the area of assembly and disassembly – the latter being central to product disposal. However, it is not only this aspect that is important to recycling: being able to evaluate all of the processes used during manufacture, assembly and disposal as a design incrementally evolves would be a more robust method of determining how ecologically sound is a particular concept. One major technology that will impact on this area and requires further investigation with respect to concurrent eco-design is that of virtual reality (VR). The remainder of this paper will highlight a number of technologies and applications associated with VR and how, it is felt, the future design environment will evolve to support LCA for contemporary product development.

2.   VIRTUAL REALITY

Virtual reality (VR) is a technology which, in the immediate future, will change the way in which products are engineered as well as the way in which the engineer of the future will generate their product ideas. This form of interface will take many forms and the solution chosen at a particular stage in the product design cycle will greatly depend on its fitness for purpose in relation to the type of task being carried out. However, the intuitive nature of an interface can greatly enhance the user experience and ability to carry out a specific task and it is this that will revolutionise the way in which engineers work. For more detailed information on the history of VR and associated technologies see the following texts [12, 13].

The design, manufacture and disposal of mechanically engineered products of all scales are ripe for the development of application solutions using virtual technologies. This section focuses on the general uses of VR, or virtual environments (VEs), in the product engineering domain and attempts to project where the development paths currently being followed will come together for the benefit of the mechanical engineer. It also attempts to address the impact it will potentially have on the eco-design of products.

There is an ongoing debate as to whether VR is only really experienced if the user is physically immersed in the computer generated environment. Others say physical immersion is not necessary as long as the user *senses* that they are part of the overall experience and are unaware of outside factors. This paper assumes that VR involves that latter experience, however this must take place in real time. It is the time-phased element that makes it different from CAD and other 'simulated' computerised experiences.

Another aspect is that of reality itself and how the experience visually 'looks' to the user. How real must the visual cues be to the user for it to be a virtually real experience? It is the author's contention that as long as the user feels they are psychologically involved in – what appears to be – an immersed experience then 'realism' as such is not an issue. Other human factors work actually requires to be done to investigate this further but, with no definitive answer, this reasoning will suffice for the purposes of this paper.

2.1   Virtual Reality Enabling Technologies

A wide range of VR technologies is available and many of these have been used as part of investigations into their practical use for engineering products. In this paper they have been divided into three main areas, namely: immersive/semi-immersive systems, desktop systems and augmented reality.

Within an immersive system of any kind the real time interaction is carried out whilst the user is, in some way, surrounded by the data and is able to interact with it via some form of peripheral device or devices. This computer generated world must be supported by the stimulation of as many senses as possible, the main one being sight but others such as touch and hearing are also possible. Within immersive and semi-immersive virtual environments (VEs) there are a number of different devices and systems that can be used.

The most common forms of semi-immersive and immersive visual displays are the head mounted display (HMD, the counter-balanced stereoscopic viewer or BOOM™ [14], a CAVE® (Cave Automatic Virtual Environment) [15] - variations include single wall displays and dome-based technologies, virtual workbenches [16] and augmented reality (AR) which combines real world images superimposed with computer generated pictures or data using a special HMD [17, 18].

Desktop virtual environments are simply flat screen computer monitors supported by special graphics generation software based on a normal PC or workstation with a time-phased element of interaction which is fundamental to its definition and application [19]. Desktop VR requires more to stimulate the senses then than just a computer screen since a flat screen and a mouse only really offer two or three degrees of freedom [20].

Traditionally, positional monitoring normally uses electromagnetic tracking or gyros to monitor the orientation of the user's head and hand or any other part of the body to which it is attached as well as mechanical trackers [14].

Hand operated gloves are a common way of manipulating or interacting with objects or interfaces, e.g. the Pinch Glove . A 3D mouse, spaceball or 3D wand may also be used for interaction via the hand in 3D space.

Other developments have been in the field of haptics [21], which enable the user to actually feel like they are grasping an object or touching a surface, e.g. the Phantom   a 3-dimensional pen-style haptic interface and the CyberGrasp , an exoskeleton glove-style haptic interface that allows users to 'touch' computer-generated objects and experience realistic force (kinesthetic) feedback.

3.   VR APPLICATIONS IN PRODUCT ENGINEERING

3.1   Design

Virtual reality can be applied in a wide range of engineering environments in both design and manufacture. This type of technology will considerably change the way in which engineers design products and evaluate downstream manufacturing, assembly and sustainable processes. Applications have advanced rapidly in the last ten or so years especially with regard to interactive applications and its potential for downstream data reuse.

In the field of virtual design, Bao et al [22] developed an immersive virtual product development (IVPD) environment utilising large screen and BOOM™ technology. This was used as a non-interactive digital mock-up tool throughout the whole design cycle. Data modifications were made through CAD systems and viewed in VR. Virtual design for assembly was investigated but no information is apparent on the actual checking of assembleability. Using a C2 CAVE environment, Cruz-Niera [23] investigated the domain of architectural design and automobile products to look at models at full scale. Interactive element placement and modification was possible as is recording of the process and the storage of design intent. Collaboration is seen as a big advantage with the main drawbacks being the lack of data standardisation, data structures and networking technologies. The use of VR by Ford, Chrysler, Caterpillar, Volkswagen and Damlier Benz was alluded to by Gomes de Sa and Zachmann [24] who see a role for immersive VR throughout the whole product development cycle. They integrate CAD, VR and PDM. They feel that human factors research should be a priority in this area whilst studying the engineering process. They feel that for immersive VR, "It must be at least as easy as designing with a CAD system." Ng et al [25, 26] and Ritchie et al [27] address such a comparison. They compared immersive VR and CAD showed significant advantages for the former; the first example of an immersive interactive industrial design task being carried out and compared with traditional design tools and methods. A number of proprietary CAD systems were compared with immersive VR for the routing of cable harnesses involving expert cable harness designers from a number of companies. Users were able to move around in an immersive 3D environment laying out cables, creating bundles and changing their properties via a 3D mouse. However, it was not clear what aspects of the virtual environment were giving these advantages; the next stage of this research is to investigate these human factors issues. Weyrich and Drews [16] researched a virtual workbench approach for designing and constructing virtual prototypes. Interaction was via space mouse, dataglove and stylus. This table-top approach suited the cognitive abilities of engineering designers. Although not demonstrating an actual design task, they also emphasise the need for human factors studies in this area. COVIRDS (COnceptual VIRtual Design System) [28] is a hand tracking, voice input interfaced immersive VR-based CAD environment where rapid concepts are modelled through free-form shape creation, an important advance in interactive virtual design. Kan et al [29] developed a desktop virtual reality collaborative environment (VRCE) and compared this to the dVISE, PIVOTAL and Deneb virtual collaborative engineering environments. Users can view and comment on designs and can load a model and share their ideas and experiences via logged text-based communications. Multi-user environments, when feasible, will revolutionise the way in which engineering design and manufacturing is carried out. This kind of work supports the supposition put forward by Ritchie et al [30] that VR will have an important role to play in future collaborative product engineering. TWR [31] have used large screen VR to carry out team-based design evaluation as well as visualising crash test data to determine car structural performance reducing the need for physical prototypes as well as helping to circumvent manufacturing problems and focuses on analysis. Since one of the earliest published examples of immersive virtual engineering analysis, using a BOOM [14], other demonstrations have been publicised involving automobile structural and vibration analyses at BMW (BOOM) [32], stress analysis at John Deere (CAVE) [33] which enhances collaborative design and allows visualising at a scale of 1:1, computational fluid dynamics (CFD) flows (surround screen) [34] and vibration analysis (CAVE) [35]. As systems become more powerful, having the ability to interactively analyse whilst simultaneously designing will greatly enhance the creative process and designer intuitiveness. New 3D interfaces, such as that provided by Digital Artforms [36], allow the creation, manipulation and editing of 3D objects, give a glimpse of what the interface of the future will be like for engineers. Also, the capability to log users in such a creative environment can potentially provide knowledge elicitation opportunities [37].

3.2   Manufacture

There are also many VR research applications that have investigated the simulation and/or generation of manufacturing and assembly information, demonstrating its potential for both downstream use and its application in concurrent engineering design throughout the whole life cycle. Using an immersive VR assembly system, Heger et al [38] enabled process planners to immerse themselves in a model and choose assembly equipment to snap objects into place ergonomically. An immersive VR planning system was also studied for generation of shop floor assembly plans [39]. Logging the user's movements and model joints, positions and precedence enabled the generation of usable shop floor process plans. Expert knowledge was also elicited [37]. Gardiner and Ritchie [40] have also suggested the use of VR as a tool for project planning. The effective checking of project planning before project commencement can potentially reduce energy and waste during the build process. The Virtual Assembly Design Environment (VADE) at WSU [41] allows engineers to combine assembly design and build using a haptic interface using VR and ProEngineer™ environment. Gomes de Sa and Zachman [24] outline the potential for virtual reality as a rapid prototyping tool in BMW through its function as a digital mock-up tool, particularly using immersive VR. They monitored ergonomic position and reach data using supplementary tools and cues to replace the lack of haptic feedback. They also interfaced with a PDM system. Early desktop machining VR work carried out by Bowyer et al [42] where virtual machine tool modelling and machining were demonstrated, now typified by products such as those at DENEB [43]. However, the ergonomic design and assembly of set-ups is not tackled effectively by modern CAM systems. However, the use of 'experiential' virtual material removal process planning will revolutionise the process planning activity and provide engineers with a 'feel' for processes [44] and designs. Work at Cardiff University in the 1990s [45], showed how the ergonomic assessment of an operator carrying out a laser welding process can be greatly enhanced using virtual technology. Virtual training tools were developed by Chuan-Jun et al [46] for CNC machine tool training and flexible assembly. Although focussing mainly on collision detection algorithms, the main lesson from this paper's point of view is that it appears possible to develop realistic machining and assembly operations using VR systems. Duffy and Salvendy [47] examined an Internet-based VR system for the simulation of a factory machining process and demonstrate the capabilities of this technology for collaborative work. A further study of virtual machining is discussed by Lin et al [48] comparing virtual and real machine operators during training where learning was enhanced more by the virtual tool.

4.   CONCLUSIONS

There are many VR applications being research and developed in the field of product engineering. The time is now ripe for exploiting this technology within the eco-design domain. The projections for computer power [49] imply an exponential growth of capability equivalent to that of the human brain by the 2020s. Through the evolution of proprietary systems and interfaces, VR will become the norm for engineering design, planning and analysis and provide an ideal, intuitive, ergonomic environment for the investigation of pre-prototype sustainable product development.

Access to globalised interaction and support databases and expertise will show that globalised team-based product development will become possible. However, as individuals, engineers will become multi-skilled virtual artisan [50]; designing new products and participating in globally team-based activities if necessary.

Support systems and human factors [51] should be researched in the area of sustainable product design to enable engineers to determine the ecological impact of their concept thinking throughout all stages of the life cycle process. How does VR help people communicate better when developing solutions? How does it enhance the production of more cost effective, sustainable engineering solutions? What should the interface of the future look like? How can this support ecologically effective product design? How can LCC or LCA be built in to support the product design process. As companies begin to implement this technology then it will be interesting to evaluate the economic costs of products pre and post use.

The use of virtual prototypes and virtual concurrent engineering practices will reduce the need for physical prototypes and allow the evaluation and checking of product life cycle costs. Future research should focus on the technological, information support and human factors issues associated with this.

REFERENCES

1. http://www.environment-agency.gov.uk/yourenv/?lang=_e Environment Agency, UK Government, 2002.
2. ***, *The triple bottom line: the challenge of the new industrial revolution* (Centre for Human Ecology, UK, 1998).
3. ***, *Waste electrical and electronic equipment (WEEE) directive.*(EU Directive 2002/96/EC., 2003).
4. A. DE RON, K. PENEV, *Disassembly and recycling of electronic consumer products: an overview* (Technovation, v15, n6, 1995, pp.363-374).
5. H. YAMAMURA, T. ISHIURA, KOBAYACHI, *Emerging situation and MMC's policy on automobile recycling – new technical issues and future activities* (Technical Review, Japanese Mitsubishi Motor Corporation, n15, 2003, pp. 6-13).
6. P. HAWKEN, *Business and the environment* (Edited by. Welford and Strarkley, Earthscan Publications, 1996, pp. 5-16).
7. C. J. ANDREWS, M. SWAIN, *Institutional factors affecting life-cycle impacts of microcomputers* (Resources, Conservation and Recycling, v31, 2003, pp. 171-188).
8. P. H. NIELSEN, H. WENZEL, *Integration of environmental aspects in product development: a stepwise procedure based on quantitative life cycle assessment* (Cleaner Production, v10, 2002, pp. 247-257).
9. M. D. BOVEA, R. VIDAL, *Increasing product value by integrating environmental impact, costs and customer valuation* (Resources, Conservation and recycling, v41, 2004, pp. 133-145).
10. T. WOHLERS, *Rapid prototyping and tooling state of the industry* (Annual Worldwide Progress Report, Wohlers Associates, 2002).
11. T. LIM, J. R. CORNEY, J. M. RITCHIE, J. B. C. DAVIES, *RPBloX - A novel approach towards rapid prototyping"* (3rd Nationl Conference on Rapid Prototyping, Tooling, and Manufacturing, CRDM, Editors Rennie, A.W.E., Jacobson, D.M. and Bocking, C.E., 2002, pp.1-8).
12. R. KALAWASKY, *The science of virtual reality and virtual environments* (Addison-Wesley, 1993).
13. G. BURDEA, P. COIFFET, *Virtual reality technology* (2nd Edition, John Wiley and Sons, 2003).
14. S. BRYSON, C. LEVIT, *The virtual wind tunnel* (IEEE Computer Graphics and Applications, 1992, pp. 25-34).
15. http://cave.ncsa.uiuc.edu/ (2003) National Center for Supercomputing Applications.
16. M. WEYRICH, P. DREWS, *An interactive environment for virtual manufacturing: the virtual workbench* (Computers in Industry Journal, v38, 1999, pp. 5-15).
17. V. RAGHAVAN, R. MOLINEROS, R. SHARMA, *Interactive evaluation of assembly sequences using augmented reality*, (IEEE Transactions on Robotics and Automation, v15, n3, 1993, pp. 435-449).
18. http://www.se.rit.edu/~jrv/research/ar/introduction.html., *Introduction to augmented reality* (2004).
19. B. SPEAR, *Virtual reality: patent review* (World Patent Information, v24, 2002, pp 103-109).
20. I. D. CARPENTER, J. M. RITCHIE, R. G. DEWAR, J. E. L. SIMMINS, *Virtual manufacturing* (Manufacturing Engineer Journal, Insitution of Manufacturing Engineers, 1997, pp. 113-116).
21. J. J. BERKLEY, *Haptic devices* (White Paper prepared by Mimic Technologies Inc., Seattle, WA, 2003, pp. 1-4).
22. J. S. BAO, Y. JIN, M. GU, J. Q. YAN, D. Z. MA, *Immersive virtual product development* (Journal of Materials Processing Technology, v129, 2002, pp 592-596).

23.C. CRUZ-NIERA, *Making virtual reality useful: a report on immersive applications at Iowa State University* (Future Generation Computer Systems, v14, 1998, pp 147-155).

24.G. GOMES DE SA, G. ZACHMANN, *Virtual reality as a tool for verification of assembly and maintenance processes* (Computers and Graphics Journal, v23, 1999, pp 389-403).

25.F. M. NG, J. M. RITCHIE, J. E. L. SIMMONS, R. G. DEWAR, *Designing cable harness assemblies in virtual environments* (Journal of Materials Processing Technology, v107, 2000, pp. 37-43).

26.F. M. NG, J. M. RITCHIE, J. E. L. SIMMONS, *The design and planning of cable harness assemblies* (Proceedings of the Institution of Mechanical Engineers, Part B, v214, 2000, pp. 881-890).

27.J. RITCHIE, J. SIMMONS, P. HOLT, G. RUSSELL, *Immersive virtual reality as an interactive tool for cable harness design* (Proceedings of PRASIC 2002 – Product Design, Univ. Transilvania, Brasov, Romania v3 (CD), 2002).

28.T. H. DANI, R. GADH, *Creation of concept shape design via a virtual reality interface* (Journal of Computer Aided design, v29, n8, 1997, pp. 555-563).

29.H. KAN, V. G. DUFFY, C. SU, *An Internet virtual reality collaborative environment for effective product design* (Computers in Industry Journal, v45, 2001, pp. 197-213).

30.J. M. RITCHIE, J. E. L. SIMMONS, I. D. CARPENTER, R. G. DEWAR, *Using virtual reality for knowledge elicitation in a mechanical assembly planning environment* (Proc. 12th Conference of the Irish Manufacturing Committee, Cork, 1995, pp. 1037-1044).

31.G. LANE, *VR at TWR* (Manufacturing Engineer, 2001, pp. 13-15).

32.M. SCHULTZ, T. REUDING, T. ERTL, *Analyzing engineering simulations in a virtual environment* (Journal of Computing Graphics and Applications, 1998, pp. 46-52).

33.M. J. RYKEN, J. M. VANCE, *Applying virtual reality techniques to the interactive stress analysis of a tractor lift arm* (Finite Elements in Analysis and Design, v35, 2000, pp. 141-151).

34.W. AKL, A. BAZ, *Design of quiet underwater shells in a virtual reality environment* (Department of Mechanical Engineering, University of Maryland, USA, http://www.enme.umd.edu/vrlab/Vrlab_Web/Publications.html, 2000).

35.R. GUPTA, D. WHITNEY, D. ZELTZER, *Prototyping and design for assembly analysis using multimodal virtual environments* (Computer-Aided design, v29, n8, 1998, pp. 585-597).

36.www.dartforms.com, July, 2003.

37.J. M. RITCHIE, J. E. L. SIMMONS, R. G. DEWAR, I. D. CARPENTER, *A methodology for eliciting product and process expert knowledge in immersive virtual environments* (Proceedings of PICMET '99, Portland State University, Portland, Oregon., 1999, paper 11-10).

38.R. HEGER, M. RICHTER, *Advanced Assembly Planning using virtual reality techniques* (Proceedings of the European Conference on Integration in Manufacturing, 1997, pp. 73-82).

39.J. M. RITCHIE, R. G. DEWAR, J. E. L. SIMMONS, *The generation and practical use of plans for manual assembly using immersive virtual reality* (Journal of Engineering Manufacture (Part B), Institution of Mechanical Engineers, v213, 1999, pp. 461-474).

40.P. D. GARDINER, J. M. RITCHIE, *Project planning in a virtual world: information management metamorphosis or technology too far?* (International Journal of Information Management, v 19, 1999, pp. 485-494).

41.S. JAYARAM, H. I. CONNACHER, K. W. LYONS, *Virtual assembly using virtual reality techniques* (Computer-Aided Design Journal, v29, n8, 1997, pp. 575-584).

42.G. BOWYER, G. BAYLISS, R. TAYLOR, P. WILLIS, *A virtual factory*, (International Journal of Shape Modelling, v2, 1996, pp. 215-226).

43.www.deneb.co.uk (2004).

44.C. F. CHANG, A. VARSHNEY, Q. J. GE, *Haptic and aural rendering of a virtual milling process* (Proceedings of ASME DETC'01, Pittsburg, USA, DETC2001/DFM-21170, 2001).

45. S. GILL, R. A. RUDDLE, *Using virtual humans to solve real ergonomic design problems* (International Conference on Simulation, IEE Conf. Publ. 457, 1998, pp 223-229).

46. S. CHUAN-JUN, L. FUNHUA, Y. LAN, *A new collision detection method for CSC-represented objects in virtual manufacturing* (Computers in Industry Journal, v4, 1999, pp 1-13).

47. V. G. DUFFY, G. SALVENDY, *Concurrent engineering and virtual reality for human resource planning* (Computers in Industry Journal, n42, 2000, pp. 109-125).

48. F. LIN, L. YE, V. G. DUFFY, *Developing virtual environments for industrial training* (Journal of Information Sciences, n140, 2002, pp. 153-170).

49. H. MORAVEC, *When will computer hardware match the human brain*, (Journal of Evolution and technology, v1, 1998, pp. 1-5).

50. J. M. RITCHIE, L. IOSIF, *The virtual artisan: a product development full circle theory* (Proceedings of PRASIC 2002 – Product Design, Univ. Transylvania, Brasov, Romania v3 (CD), 2002).

51. J. E. L. SIMMONS, J. M. RITCHIE, P. O'B. HOLT, G. T. RUSSELL, *Human in the Loop* (Proceedings of the 1st CIRP(UK) Seminar (University of Durham), v1, 2002, pp. 109-112).

A TRIZ APPROACH TO DESIGN FOR ENVIRONMENT

D. Serban[1], E. Man[2], N. Ionescu[3] and T. Roche[1]
[1]*Galway Mayo Institute of Technology, Galway, Ireland;* [2]*CIMRU, National University of Ireland, Galway, Ireland;* [3]*Politehnica University of Bucharest, Romania*

Abstract: The purpose of the research carried out was to identify ways in which tools and methodologies from Theory of Inventive Problem-Solving (TRIZ) might be used in Design for Environment (DFE) approaches. The aim was to develop an integrated methodology for environmentally superior product design at the conceptual stage of the product design phase. The paper commences with a review of the TRIZ methodology itself, followed by a discussion of the DFE strategies. A TRIZ approach to DFE is proposed as a new methodology and the paper concludes with a case study using the new TRIZ to DFE approach.

Key words: design process, theory of inventive problem-solving (TRIZ), design for environment (DFE).

1. BACKGROUND

In recent times, increasing attention has been given to the design process, both in industry and in research. A number of reasons can be identified for this trend. First, it is recognized that, even if the design process itself had only a minor contribution to the cost of product, a considerable portion of the cost to be made in later product life cycle phases is committed at the design stage. Furthermore, the earlier in the design process the decisions are made the higher the impact is on the final design. Secondly, manufacturers are faced with increasing demands from their customers to increase variety in their product types, reduce costs, increase quality, environmentally superior and compliant products (ESCPs) in reduced time-to-market. Compounding this, the complexity of products has increased and the typical life cycle has reduced, focusing specifically on the environmental aspects of product design.

D. Talabă and T. Roche (eds.), Product Engineering, 89–100.

With the emergence of new global policies (e.g. Integrated Product Policy), emerging legislation (e.g. WEEE and EEE) and environmental standards (e.g. ISO 14000) manufacturers are forced to move towards the development of ESCPs. According to WEEE and EOLV manufacturers are obliged to take responsibility for their waste management by implementing reuse, recycling and recovery for their products. The design of ESCPs (through DFE practices) is a strategy to support compliance with these environmental drivers (ISO 98).

Figure 1 shows the interaction between the design process and the life cycle of the product. Life cycle information is acquired through a set of life cycle design information loops, i.e. design for use, design for end of life. The design process transforms this information into product design characteristics, which are in turn embedded in the product.

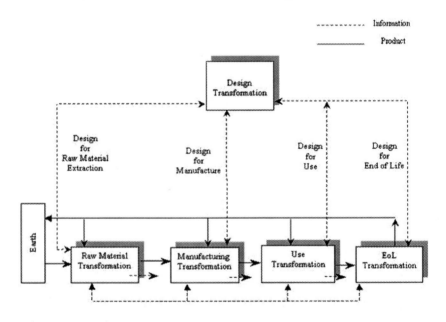

Figure 1. Life Cycle model [2]

Four generic and interrelated strategies for the development of ESCPs can be derived from the model as follows [2]:
a) Select low impact materials and processes over all life cycle phases.
b) Reduce life cycle resource consumption (Material and Energy).
c) Reduce life cycle waste stream (Material and Energy).
d) Resource sustainment by facilitating first life extension and post first life extension, i.e. reuse, remanufacture and recycling.

There is a need for a new design model to support the development of mew methodologies and tools to assist the designer in the creation of ESCPs.

2. DESIGN PROCESS METHODOLOGIES FOR DESIGN FOR ENVIRONMENT

Extensive research has been carried out in the Design Process field in the attempt to map DFE tools and methodologies into existing design process models trying to develop new approaches to DFE that are more appropriate for the design engineer.

There are many models and classifications of the tasks included in the design stage, but most of the experts agree that the design process must start with collecting information and defining requirements of the product and finishes with a complete and detailed description of the product, as shown in the model presented in Figure 1 [2].

The design process can be described by (a) the degree of embodiment and (b) the solution space. The degree of embodiment is described on the vertical axis, i.e. representing the transformation of information through four generic stages of design (requirement definition, functional definition, general design and detailed design) from qualitative to quantitative environmental information. In the early stages of design, the solution space is very large, however as the design evolves this solution space becomes narrower until there is one specific solution, as shown in figure 2. Decisions made in the earlier phases of the design process have the largest influence on the final design.

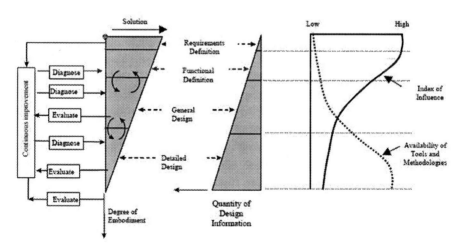

Figure 2. The degree of embodiment of design versus quantity of environmental information [2]

3.   THEORY OF INVENTIVE PROBLEM-SOLVING (TRIZ)

Theory of Inventive Problem-Solving (TRIZ) is a unique knowledge-based methodology, for accelerated development of design concepts. TRIZ is a creative problem-solving methodology especially tailored for scientific and engineering problems.

The TRIZ philosophy is based on the fact that the evolution of a design is not a random process. It correlates with the evolution of customer needs. Every field of engineering influences the evolution of other fields. Therefore, the process of the design evolution can and has to be studied.

The TRIZ's major discovery was revealing the origin of an inventive problem, which is defined as a *"contradiction"*. A contradiction arises when two mutually exclusive design requirements are put on the same object or system. When a designer faces a contradiction that cannot be solved by redesigning a technical system in a known way, this means that he faces the inventive problem, and its solution principle resides outside the domain the technical system belongs to. There are two ways to solve problems that contain contradictions: by finding a compromise between two conflicting parameters or by eliminating the contradiction. TRIZ is aimed at solving problems by eliminating the contradictions.

The main difference between the TRIZ methodology and all other innovative or creative methods is the reduction of ineffective solutions by using a purposeful and systematic procedure and by passing over the psychological inertia barrier. The method of TRIZ is to break the psychological barrier of abstracting the initial problem (see figure 3). This means, generalization of a specific problem to an analogous problem, comparison of this standard problem with analogous standard solutions well known in other scientific branches and industries and transferring back this analogous standard solution to a specific solution [3].

As the earlier stages are the most important in the design process and as the TRIZ methodology is focusing on the conceptual design stage, TRIZ might provide a systematic support for the following phases of conceptual design [4]:

- analysis of ill-defined design problems by describing functions between the system components and identification of core problems by formulating contradictions;
- generation of new solution concepts by using TRIZ problem-solving techniques: inventive principles, inventive standards and pointers to physical effects ;
- producing a technological forecast (prediction) of a particular design product using TRIZ technology evolution trends (see figure 3).

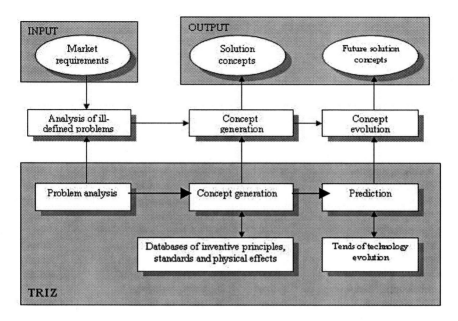

Figure 3. Conceptual design with TRIZ [4]

The most important tools and methods of the TRIZ methodology are the Contradiction Matrix, the 40 Inventive Principles, the 39 TRIZ Parameters, the ARIZ Algorithm, the 76 Standard Solutions, the Su-Field analysis, the laws and trends of technology evolution, the Separation Principles – to solve physical contradictions, the System of Operators, the Effects and Anticipatory Failure Determination (AFD) [5].

4. THE NEW METHODOLOGY FOR EARLY STAGES OF THE DESIGN PROCESS APPLYING TRIZ FOR DESIGN FOR ENVIRONMENT

Analysing and applying the TRIZ approach to problem-solving for Design for Environment, TRIZ principles for DFE were obtained, and the authors considered they could be implemented successfully in the early stages of the design process. There were two steps as follows: first, the TRIZ methodology was analysed in an attempt to identify how it can be implemented or adapted to fulfil the DFE strategies, especially focusing on the Inventive Principles of TRIZ.

Secondly, through a process of comparison between TRIZ principles and DFE strategies and by adoption, adaptation and elimination, the authors obtained the TRIZ principles for DFE, as presented in Table 1.

Table 1. The 40 Inventive Principles TRIZ for DFE [6]

1. Segmentation	Adopt	21. Skipping	Eliminate
2. Taking back	Adapt	22. "Blessing in disguise"	Adopt
3. Local quality	Adapt	23. Feed-back	Adopt
4. Asymmetry	Eliminate	24. "Intermediary"	Adopt
5. Merging	Adopt	25. Self-service	Adopt
6. Universality	Adopt	26. Copying	Adapt
7. "Nested doll"	Adopt	27. Cheap short-living objects	Adopt
8. Anti-weight	Adopt	28. Mechanical substitution	Adopt
9. Preliminary anti-action	Adopt	29. Pneumatics and hydraulics	Adopt
10. Preliminary action	Adopt	30. Flexible shells and thin films	Adapt
11. Beforehand cushioning	Eliminate	31. Porous materials	Adopt
12. Equipotentiality	Eliminate	32. Colour changes	Eliminate
13. "The other way round"	Adapt	33. Homogeneity	Adopt
14. Spheroidality	Adopt	34. Discharging and recovering	Adapt
15. Dynamicity	Adopt	35. Parameter changes	Adapt
16. Partial or excessive action	Eliminate	36. Phase transition	Adopt
17. Another dimension	Eliminate	37. Thermal expansion	Adapt
18. Mechanical vibration	Adopt	38. Strong oxidants	Adapt
19. Periodic action	Adopt	39. Inert atmosphere	Adopt
20. Continuity of useful action	Adopt	40. Composite materials	Adapt

A systematic approach for obtaining the TRIZ principles for DFE is presented in Figure 4:

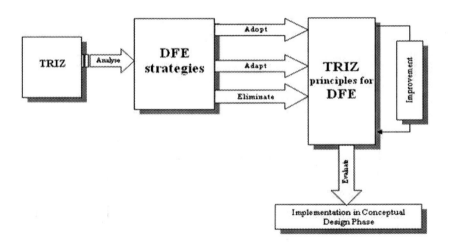

Figure 4. Systematic approach for obtaining the TRIZ principles for DFE [6]

The schematic representation of the proposed methodology for early stages of the design process applying TRIZ for DFE is presented in Figure 4, which shows the schematic way of using the TRIZ methodology for the DFE proposed for this study.

The algorithm of the new methodology (Figure 5) is presented in four steps as follows:

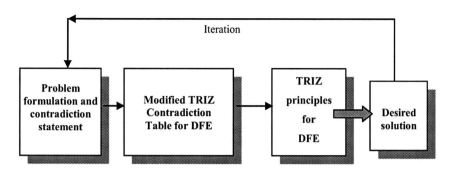

Figure 5. Schematic representation of using the proposed methodology [6]

Step 1. Problem formulation and establish contradictions: At this stage an analysis of the problem was made in order to identify the basic functions of the system, the sub-systems, super-systems, the environment, useful, unuseful and harmful systems. The technical contradiction (the conflict) ought to be solved when the analysed problem has been established.

Step 2. Establish specific DFE-TRIZ principles using the contradiction TRIZ matrix for DFE: Based on the technical contradiction established in step 1, product characteristics that should be improved are identified. Coupled product characteristics are also identified. The row of the Contradiction Matrix is entered with a product characteristic that it is desired to be improved, and this is intersected with the column of the coupled product characteristic that is producing an undesired result.

Step 3. Establishing generic solutions using TRIZ principles for DFE: The cell in the intersection gives the number of the inventive principles that are suggested as being able to resolve the contradiction.

Step 4. Establishing specific solutions: By obtaining the inventive principles needed, the designer applied them to the product for obtaining the desired solution. Data was then re-synthesised from the (new) model and the process begins again. This continuous improvement process continued by iteration until the best solution was found.

5. CASE STUDY

The automotive industry is among the most resource intensive of all major economic systems; for example, in USA it represents at least one-third of the consumption of iron, lead, platinum and synthetic and natural rubber.

Resource depletion for the automobile includes product component (all the replacement parts: tires, hoses, lights, belts, filters, batteries, etc), process component (fuel, fluids and associated packaging, highway infrastructure), distribution components (packaging associated with replacements parts).

But the most serious environmental issue facing the automobile is surely its enormous consumption of non-renewable energy. The most significant impacts on vehicle maintenance are too often not environmentally friendly.

From all the subassemblies composing a car, the authors decided to apply the proposed methodology on the car rear view mirror. The authors focused on the improvement of the case and the back-can subassemblies of the car-mirror assembly, using the proposed methodology of TRIZ for DFE.

The improvement was applied to the virtual prototype design model of the car-mirror assembly, as seen in the Figures 6, 7, 8 and 9. The advantage of using virtual prototyping included three dimensional visualisation, as well as the opportunity to extract information from the model for the purposes of evaluation and the use of information to manipulate the model prior to manufacture.

First the functions of the car-mirror assembly should be considered and start seeking out the relevant contradictions. Among the functions of a car-mirror assembly there are:

a) assuring rear and lateral visibility when driving a car and

b) the designing and manufacturing process of the mirror, in the context of DFE requirements, which should consider the ease of disassembly, the product waste minimization and the use of hazardous or undesirable materials.

Figure 6. Car-mirror assembly (front view)   *Figure 7*. Car-mirror assembly (back view)

Figure 8. Case subassembly   *Figure 9*. Back-can subassembly

The search for and elimination of physical contradictions is a fundamental principle of the TRIZ methodology. The search for design trade-offs, in other words to search for physical contradictions, is an often-potent means of defining the "right" problem to be solved.

In the case of the car-mirror assembly, the fundamental design trade-off may be seen to be one of compromise between a requirement for a functioning rear view mirror, in order to achieve the best visibility for driver and an environmentally superior mirror in order to fulfil the environmental requirements, from the manufacturer's point of view.

Seeking out the best compromise between the two extremes was not solving the "right" problem. From a TRIZ point of view, a physical contradiction results when a physical attribute should be increased to improve one function of the system, and decreased to improve another. Traditionally, trade-offs are used to handle contradictions. But TRIZ always seeks a solution without compromise. The right problem is more likely to be how to achieve a functioning mirror and an environmentally superior one at the same time.

Regarding just the back-can and case subassemblies, it has been identified that they are stuck together using adhesive, which makes them hard to separate for recycling, and a solution must be found to join them without using an adhesive. It has also been identified that they are made from plastic or different polymers. Degradation of plastics is more critical than that of the metals, since it is often difficult to ensure that different types of polymers were not mixed together. Only thermoplastic polymers can be recycled. And often, recycled material is used for a less critical application than its original use. Thermosets, which are degraded by high temperature, cannot be recycled. Composite materials consisting of mixtures of glass and polymer represent a problem in recycling.

In the case of redesigning the car mirror, the problems that should be solved must contain the improvement of the parameters to solve the technical contradiction, therefore two of the TRIZ instruments were selected: the Contradiction Matrix and the 40 inventive principles.

Thinking from the final product point of view, expressed in the terms of the Contradiction Matrix, the authors thought that the parameters to be improved for the car-mirror assembly are:

a) "Stability of the object composition" (parameter 13 from the Contradiction Matrix),
b) "Quantity of substance/matter" (parameter 26),
c) "Ease of manufacture" (parameter 32),
d) "Ease of operation" (parameter 33).

And the parameters that are getting worse when trying to improve the ones above are:

- "Shape" (parameter 12),
- "Strength" (parameter 14),
- "Device complexity" (parameter 36).

The principles extracted from the Contradiction Matrix are presented in

Table 2. Based on those principles that are actually generic solutions, 28 specific solutions of conceptual design were established among them new ideas for designing for environmental compliance .

All these specific solutions to improve the car mirror assembly can be applied to the virtual prototype design models, without any large cost, just the designer's knowledge experience and ability to make the changes, when the product is still in conceptual phase.

Table 2. Generic Solutions and Specific Solutions for car-mirror assembly

Principle No.	Freq. of appearance	TRIZ Principles	Specific Solutions
1	4	Segmentation	1. Provide the mirror easy to disassemble and recycle 2. Avoid embedding and non-dismantling assemblies 3. Avoid adhesives, welding and soldering constructions
10	3	Preliminary action	4. Possibility of mirror adjusting from inner car
26	3	Copying	5. Using virtual prototypes to design the assembly
27	3	Cheap short-living objects	#. No ideas;
28	3	Mechanics substitution	6. Instead of painted labels use inscriptions 7. Replace manual adjustment system with electrical/electromagnetic system
35	3	Parameter changes	8. Internal heating of the mirror against steaming to permit the hot air to enter through the acclimatization system of the car
13	2	The other way round	#. No ideas;
14	2	Spheroidality - curvature	9. Make aerodynamic shape of the back can; 10. Use spherical hinge for position's adjustment; 11. Use partially spherical shape glass in order to obtain a bigger image;
15	2	Dynamics	12. Make a position system easy to adjust from the inside of the car; 13. Make folding system to protect the mirror during the parking; 14. Make a system that "memories" the initial position after folding and brings it back to the "memorized" position;
22	2	"Blessing in disguise" or "turn lemons into lemonade"	#. No ideas;

continued

Principle No.	Freq. of appearance	TRIZ Principles	Specific Solutions
34	2	Discarding and recovering	15. The possibility of rapidly ejection of the glass for replacing without disassembling the entire mirror from the car; 16. Dismounting assembly between the housing and the back body using elastic elements;
2	1	Taking out	17. Elimination of the screws for assembly and use snap fits and spring clips for easy disassembling; 18. Minimise the quantity of material used and the number of components; 19. Minimise the number of materials used to manufacture the product; 20. Elimination of the non-recyclable or non biodegradable materials; 21. Choose compatible materials that don't need separation before recycling (ex. Use on single type of plastic); 22. Elimination of the adhesives, paints, inks and labels that can harm the environment during recycling and increase recycling cost 23. Use integrated labels (material inscriptions made by molding injection) instead of painted labels 24. Avoid thermoset materials and use thermoplast materials that are recyclable
3	1	Local quality	25. Against steaming, the fixing system shape should allowed the access of the hot air through the acclimatization system inside the mirror during the cold seasons
9	1	Preliminary anti-action	26. Use pre-strained spring to self adjust the mirror in case of accidental disturbing
18	1	Mechanical vibration	#. No ideas;
29	1	Pneumatics and hydraulics	27. Replace manual operation system with hydraulic or electrohydraulic operating system 28. Use hydraulic mini-damper to self adjust the mirror in case of accidental disturbing
40	1	Composite materials	#. No ideas;

6.   CONCLUSIONS

TRIZ is a logical, knowledge-based methodology for early stages of the design process. Guided by TRIZ, users not only overcome the psychological barrier but they also have the opportunity to analyse the best direction for improvement of products. The integration of TRIZ methodology into Design for Environment help the designers to maximise the utilisation of the resources of a system to meet the objectives of new product development with less cost and without any unwanted effects.

The application of traditional TRIZ accelerates the search for breakthrough solutions and gives users the ability to reach greater levels of product performance. The results of the research carried out were realised in the development of a new methodology for applying TRIZ to DFE, by adapting one of the most useful tools of TRIZ - the Inventive Principles, for Design for Environment. The results of the proposed methodology of TRIZ for DFE were also demonstrated by the case study.

7.   FURTHER WORK

The proposed methodology does not represent a ready-to-use methodology by the designers. The aim of the research was to present a proposal and to prove the viability of the idea. The proposal for the methodology has been done in a simple manner, graphically, to facilitate the understanding of the TRIZ methodology for Design for Environment. But high-level performance methodologies applying TRIZ for Design for Environment can be developed using Computer-aided TRIZ and CAD/CAM software. As further work, the aspect of integrating the TRIZ for DFE methodology with the DFE Workbench may be looked at.

REFERENCES

1. ***, *Life Cycle Assessment– Principles and Guidelines* (ISO FDIS 14040, ISO TC 207/SC5/WG1, 1998).
2. T. ROCHE, *Design for Environment Workbench* (PhD Thesis, CIMRU, NUIGalway, 1999).
3. G. MAZUR, *Theory of Inventive Problem Solving (TRIZ)* (2001, http://www-personal.engin.umich.edu/-gmazur/triz/).
4. V. SOUCHKOV, *TRIZ: A Systematic Approach to Conceptual Design* (Ideal Design Solution, The Netherlands, 1998).
5. E. DOMB, *40 Inventive Principles with Examples* (TRIZ- Journal, April, 1996).
6. D. SERBAN, *A Systematic Approach to Design for Environment* (BTech. IV, Galway-Mayo Institute of Technology, Ireland, 2003).
7. C. FUSSLER, *Driving Eco-Innovation* (Pitman Publishing, 1996).

ECO-DESIGN APPROACH FOR THE TRIPODE TYPE COUPLING

M. Lates and A. Jula
Transilvania University of Brasov, Romania

Abstract: The paper provides some aspects regarding the principles of eco-design for the
              machine elements in particular for the tripod type couplings. First, aspects
              regarding the eco-design principles are presented for the case of machine
              elements in general and as an application the case of a tripod coupling used in
              transversal transmission of passenger cars.

Key words: eco-design, tripod coupling.

1.   INTRODUCTION

During the last years some eco-design principles have been tested and implemented successfully in Europe, USA, Australia and Japan. However, in many parts of the world, particularly in newly industrializing countries in Asia, Latin America and Africa, the experiences with eco-design are scarce. The products oriented on environmental policies, market demands, activities of competitors, quality demands and pressure from the environmental organizations, can motivate companies to start eco-design [2, 8]. The technical literature show that many fields in industry need urgently to use the eco-design principles [2, 8]. Generally, the improvement of design can take place at various levels, e.g. optimizations of the existing product, re-design, searching for a new way of fulfilling the same function.

In this paper, an example for eco-design is presented, as far as it can be influenced by innovation and optimization, as a starting point for the case of a tripod coupling.

D. Talabă and T. Roche (eds.), Product Engineering, 101–110.

1.1   General eco-design concepts

The eco-design principles are applied in the product design processes using some external data, information, principles and legislation [2, 8]:
a) eco-design manuals;
b) environmental design rules;
c) national and European ecology and environmental laws;
d) sustainable development principles;
e) best practice examples and case studies.

A general design algorithm is presented in figure 1. The place of the eco-design is thus presented. In this general design algorithm, three elements are involved which are at the basis of eco-design process:
–   material cycle – material consumption; recyclable material;
–   energy use – energy consumption and output energy = efficiency;
–   toxic emissions = environmental health.

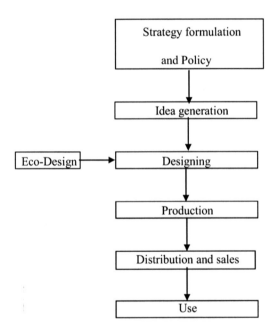

Figure 1. General design algorithm

1.2   Eco-design application

The eco-design application, presented in this paper, is characteristic for a tripod coupling. The tripod coupling is widely used in the power transversal transmission of the driving and steering wheels with independent suspension

of cars [1, 3, 6, 7]. At present two tendencies are acting:
a) Optimization of classic tripod coupling;
b) Design of some new similar couplings with superior features.

The present example belongs to the second tendency and is referring to an angular-axial tripod coupling with cylinder/cylinder contacts. This coupling has some new interesting features [1, 3, 4, 5, 6, 7] i.e.:
1. remove the sliding friction;
2. simplify the construction and the technology.

2. APPLICATION

2.1 Theoretical aspects

The tripod coupling with cylinder/cylinder contacts is presented in figure 2. Between the input shaft 1 and the output shaft 2 three pairs of cylinders are in mechanical contact: three of them are mounted on the input element and the other three on the output element.

Figure 2. Tripod coupling with cylinder/cylinder contacts

This solution reduces the production costs (because of the constructive simplicity) and has higher efficiency (it is removing the sliding friction which is replaced by rolling friction). To improve the loading capacity the solution with another type of elements being in contact can be identified – for example, hyperbolic type elements [1, 2, 4, 7, 8].

The mechanism associated to the tripod coupling is not a perfect synchronous mechanism (for a constant velocity at the input shaft, the output shaft will have a variable velocity). Therefore, it is necessary to make a dynamic simulation for the tripod mechanism to identify the variation of the output shaft's angular velocity. Figure 3 shows the block scheme of the tripod mechanism, where F represents the mechanism's mobility and L the number of the external links. The following parameters are given:
– the angular position φ_1 (or angular velocity) for the input shaft;

- the resistant torque at the output shaft, T_3;
- the angle between the axes of the input and output shaft, α.

Finally, the angular position φ_3 (or angular velocity) will be obtained for the output shaft. Also, it is possible to obtain the variation of the resistant torque T_2 applied to the element which gives the motion α.

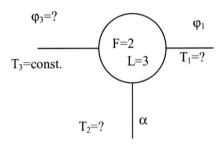

Figure 3. Block scheme for the tripod mechanism

To simulate the dynamic behavior of the tripod mechanism, the geometrical model of the mechanism is made using ADAMS 10.0 software, based on the multibody system's theory. First, the bodies are described, then the restrictions (the joints) between them or between them and the ground. In next step the external loads are applied by introducing the actuators and the brake models. The model of the tripod mechanism, defined in ADAMS 10.0 software, it is presented in Figure 4.

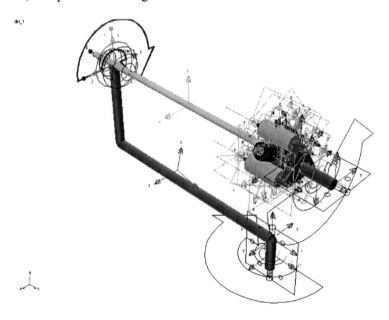

Figure 4. ADAMS 10.0 model for the tripod mechanism

As input data the following parameters are considered:

1. the angular velocity ω_1 for the input shaft, considered constant (Figure 5) to identify, in the easiest way, the variation of the angular velocity at the output shaft, ω_3.

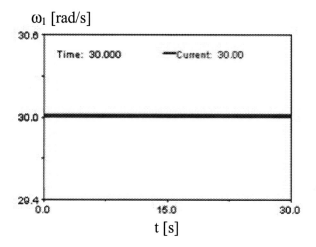

Figure 5. Angular velocity at the input shaft

2. the resistant torque T_3, it is considered constant, to decrease the dynamic influence of that in the system (Figure 6).

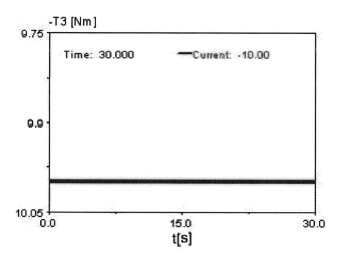

Figure 6. Resistant torque

3. A sinusoidal variation for the angle α is considered between the shaft's axes, with a maximum value of $30°$, at the middle of the simulation

period, to identify the variation of the output shaft's angular velocity (Figure 7).

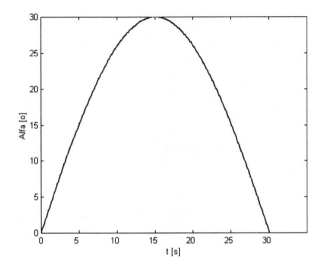

Figure 7. Angle between the shaft's axes

Prior to the simulation, the model check is performed with the following results:
a) 9 moving parts (not including ground),
b) 8 revolute joints,
c) 1 spherical joint,
d) 3 in plane primitive joints,
e) 2 motions,
f) 2 degrees of freedom.

As an hypothesis, the proposed dynamic model is considering as only the steady state functioning of the system and does not consider the whole dynamic behavior for the real machine with the transient regime; the aim of the simulation is just to identify the variation of the angular velocity at the output shaft of the tripod coupling, by considering the tripod mechanism implemented on a test rig. The next step (if the angular velocity of the output shaft will have small variation) is a practical problem: to identify the tripod mechanism efficiency, i.e. to identify if this tripod mechanism is eco friendly, according to the energy consumption.

Figure 8 shows the variation of the angular velocity at the output shaft's comparing on the variation of the angular velocity at the input shaft. It can be seen that the maximum difference between the values for the two velocities is obtained for the maximum value for the angle α and it is small (smaller than 0.5°/s), so the tripod mechanism can be considered a synchronous one (constant angular velocity at the output shaft when the angular velocity at the input shaft is constant); therefore it can be assumed that this tripod mechanism can be used in mechanical transmissions and also in car suspensions.

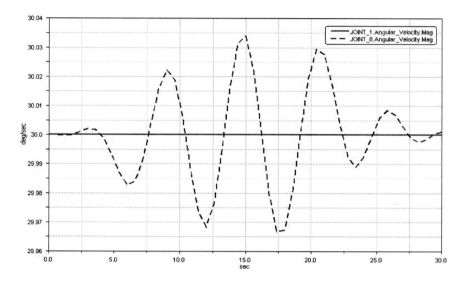

Figure 8. Variation of the angular velocities at the input and the output shaft

2.2   Practical aspects

Figure 9 and figure 10 shows practical examples for cylinder/cylinder contacts and hyperbolic/hyperbolic contacts, respectively.

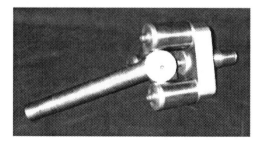

Figure 9. Cylinder/cylinder contacts

To make the experimental study, the experimental test rig with open power circuit was used as presented in figure 11. The test rig is installed at the National Institute for Automotive Comp., Braşov, Romania [7].

Figure 10. Hyperbolic/hyperbolic contacts

The experimental set up is driven by an electrical motor 1, the command being made by a command table 2. The motion is transmitted to the testing element 3 by the coupling 6 and the cardano type transmission 7. The test rig is loaded by the hydraulic brake 4, controlled through the command block 5. The load is transmitted to the tripod mechanism by the coupling 6 and by the elastic coupling 8. The test rig is equipped with incremental transducers 9. These transducers are measuring the torque, the angular velocity and the angular position at the input and the output of the system. The transducers are fed by the electric power sources 10 and 11.

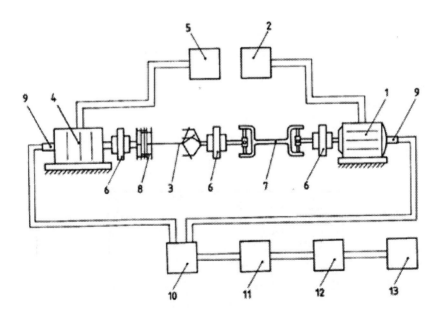

Figure 11. Scheme of the experimental stand

By using experimental methods the diagram for the variation of the instantaneous efficiency is obtained for the two types of couplings (Figure 12) with $\alpha=30°$.

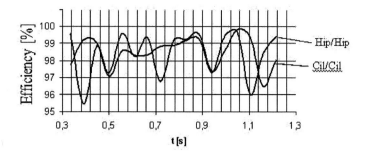

Figure 12. Tripod coupling efficiency

It can be observed that these types of couplings have higher efficiency, therefore less energy consumption is involved.

3. CONCLUSIONS

The example presented in the paper provides in a simple way, the design based on some of the eco-design principles (energy consumption reduction = efficiency improving) to develop of machine elements.

The dynamic modeling of the tripod coupling represents a modern way to obtain the dynamic behavior of the machine composed by an actuator, a brake and the tripod mechanism; using that type of modeling one can say that the analyzed mechanism is of homokinetic type (synchronous mechanism).

The experimental test is used to find out the efficiency of the tripod mechanism; A competitive open power flow testing rig is used and the results are certifying that the tripod mechanism has a higher efficiency.

As a conclusion, the tripod coupling can be used in mechanical transmissions because of the small variation of the angular velocity at the output shaft; also, this coupling, because of the higher efficiency, is eco-friendly, according to the energy consumption criteria.

REFERENCES

1. D. DIACONESCU, a.o., *Tripodekupplung mit Zylinder-Zylinder Podegelenken* (Bulletin of the Transilvania University of Brasov, Series A, vol. XXXII, 1990, p.21-28).
2. J. C. DIEHL, a.o., *Eco-design in Central America, ecodesign methodology: product improvement tool* (PIT) (The Journal of Sustainable Product Design, vol.1, 2001, p.197–205).
3. FL. DUDITA, a.o., *Cuplaje mobile podomorfe* (Trisedess Press publishing house, Sf. Gheorghe, Romania, 2001).

4. E. J. HAUG, *Computer aided kinematics and dynamics of mechanical systems* (Vol. I-II. Allyn and Bacon LTD, Massachusetts, USA, 1989).
5. H. HEISLER, *Advanced vehicle technology* (London, The bath press, 1997).
6. A. JULA, a.o., *Multibody modeling of the tripod coupling* (The Proceedings of Research and Development in Mechanical Industry – RaDMI 2002, vol.2, Vrnjacka Banja, Yugoslavia, 1 – 4 september, 2002, p.649-654).
7. M. LATES, *Cuplaje cu contacte mobile* (PhD thesis, Transilvania University of Brasov, 2003).
8. A. TUKKER, a.o., *Eco-design: the state of implementation in Europe* (The Journal of Sustainable Product Design, vol.1, 2001, p.147–161).

NATURAL SHAPES – A SOURCE OF INSPIRATION FOR ECO-DESIGN

V. Podborschi and M. Vaculenco
Technical University of Moldova, Republic of Moldova

Abstract: The study of nature forms, of cycles of existence of natural products as a source
              of inspiration and the familiarization for specialists, designers, constructors,
              technicians in the designing of a new generation of industrial products is
              outlined. In nature one can find even more constructive structures as a source
              of permanent creative inspiration in conceiving shapes and material products.
              By using the laws of nature evolution and harmonizing the function with the
              shape, modern design approaches suggest that humanity will succeed to
              conduct the ecological production of material goods.

Key words: design, ecology, bionics, nature, construction, production.

1.   INTRODUCTION

For centuries the nature has worked out and updated itself, creating forms and mechanisms of surviving, the analogies of which we may find within today's technical means: logo, planes, optical equipment, radiolocation equipment and navigation tools.

The material world surrounding us is made up of objects that have shapes and aesthetic peculiarities. This is due to the fact that any form is the result of one of the processes as described below:

a) Uncontrolled processes, in which the shape depends only on the conditions of the environment (e.g.: formation of mountains, rocks, river gravel),

b) Processes that depend on the laws of physics and chemistry of nature and of their formation environment (e.g.: ice crystals),

c) Processes guided genetically and by the conditions of the environment (e.g.: living organisms),

d) Processes guided by human demands, insects, and animals and by the

111

D. Talabă and T. Roche (eds.), Product Engineering, 111–120.
© 2004 *Springer. Printed in the Netherlands.*

conditions of the environment (e.g.: the shape of the industrial products, beaver dams, bird nests, Figure 1).

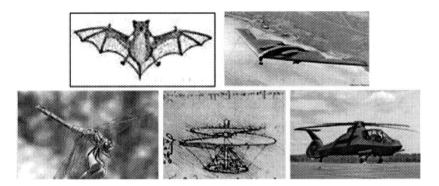

Figure 1. Inspiration from natural forms

Looking carefully, one can notice that all processes that contribute to the formation of the shapes of objects surrounding us are connected to a general factor – the environment where they take place. So what is the connection between the shapes of nature and the shapes of human created products? where does the border between nature environment and environment created for satisfying human demands by engineer, designers and architects lie? Economic problems, the tendency towards the utmost utility of products with minimal material losses, the necessity of organizing and harmonizing the material and vital environment with the biosphere, the development of advanced technologies and of technical potential have made us to pay close attention to the processes and phenomena that happen in nature (Figure 2).

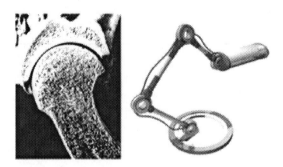

Figure 2. Inspiration from natural construction

Nature has been, is and will be an endless source of creative inspiration for humanity. By studying and analyzing nature's formal and constructive processes, humanity has always solved and continues to solve many of its vital problems.

Between the years 50's – 60's, a new science emerged, the basis of which was the research for modeling of different viable systems. The emergence of this science that has been called "bionics" (from a Greek word meaning element of life") is the result of the active development of biophysics, biochemistry, cybernetics and cosmic biology.

2. DEFINITIONS IN BIONICS

Bionics is the science of studying the basic principles of nature and the application of these principles and processes for finding solutions for the problems that humanity encounters [1]. Major Jack Steele, of the US Air force, used the term "Bionics" in 1960 to describe what was then an emerging research into interface between natural and synthetic systems. He defined bionics as "the analysis of the ways in which living systems actually work and having discovered nature's tricks, embodying them in hardware" [5]. The Concise Columbia encyclopedias define bionics as follows:

Bionics study of living systems with the intention of applying their principles to the design of engineering systems.

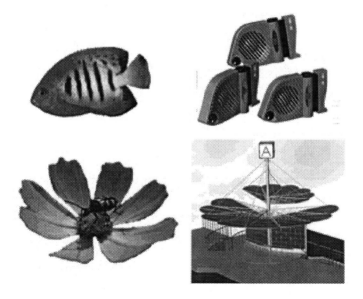

Figure 3. Design for engineering and architectural structure alike to natural forms

Bionics can be classified in five main categories as follows:
– <u>Total mimicry</u> - an object material chemical structure that is indistinguishable from the natural product e.g. early attempts to construct flying machines.

- <u>Partial mimicry</u> - a modified version of the natural product, e.g. artificial wood.
- <u>Non-biological analogy</u> – functional mimicry, e.g. modern planes and use of airfoils.
- <u>Abstraction</u> – the use of an isolated mechanism, e.g. fiber reinforcement of composites.
- <u>Inspiration</u> – trigger for creativity, e.g. design for architectural and engineering constructions alike to plants, animals and insects (Figure 3).

Bio-design is probably the oldest methodology of designing with real examples from all over the history of humanity. Probably the greatest beneficiary of this design methodology is the area of transportation and architectural design. Democritus (460 – 370 BC) wrote: "The spider taught us to weave, the swallow – to build houses".

3.   STAGES OF BIONICS DEVELOPMENT

The process of using the shape creating laws of nature has always evolved through change and modification. There can be underlined three chronological stages of this process as a predecessor of the modern stage. The first stage is the oldest and is characterized by the spontaneous use of constructive and special-functional means of birds, insects, and animals for building primitive houses. It is difficult to speak of the esthetic value of these usages. Evident enough is just their functionality. Often, the artificial shape of constructions together with their function has copied the natural shape too, so there was no great difference between, for instance, a South-American Indian house and a termite hill (Figure 4).

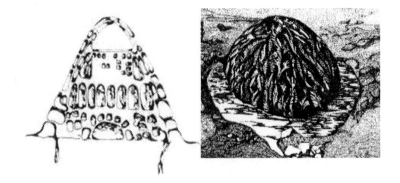

Figure 4. Analogy of South-American Indian house and a termite hill

The second stage lies between the first intentions of humanity to conceive esthetically the shapes of nature for material goods and the middle of the 19th century. Though this is a long period of time and comprises a variety of stages and styles, it is still based on one principle - the principle of nature mimicry. Decorative shapes of nature have been actively used for embellishing buildings, tools, arms, all along this period. Nevertheless, when studying this period one can notice the interpretation of some constructive and tectonic principles of nature. For example, the tectonics of columns mimed the tectonics of tree trunks, the constructive logics of styles of Greek temples actually repeats the stems of plants or the backbone of animals, and the decorative and constructive ribs from gothic architecture – the ribs of a green leaf.

The intentions of applying the constructive methods of nature in the tectonics of material goods have been ineffective because of the insufficiency of technical possibilities. It was easier to imitate the shapes of nature in rock and clay with artistic purpose than to create a constructive system similar to the natural one (Figure 5).

Figure 5. Bionic structure in architectural forms

The third stage (the end of the 19th century – the beginning of the 20th century) evolved under the auspices of the style "Modern", in which the natural principles have more or less appeared in the constructive functional-structural decorative concepts as a complex of solutions of the products shapes.

Natural constructive principles of biology have influenced industry in construction (e.g. the invention of ferro-concrete, the intensive use of metal constructions and ceramics).

The spatially developed buildings characteristic for the style "Modern" look very similar to certain structures of the nature. The traditional decorative elements of nature were present not only in the shapes of products, but also served the constructive functional structures for it.

Teams of specialists of different professions (biologists, engineers, architects, designers, IT-specialists) are always in search of methods of harmonization between the shape and the function of industrial products that are natural in the shapes and structures of nature.

4.   UNITY OF FORM AND FUNCTION

The change of seasons, days and nights, the periodicity of plant and animal evolution, their disappearance and revival had conjured up the notions of rhythm, symmetry – asymmetry, proportions, tectonics which became basic means of damping shapes within designers' creation (Figure 6).

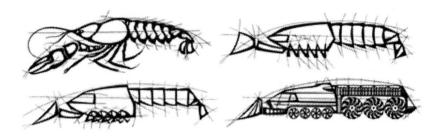

Figure 6. Principles of the bionic forming in machine designing

It is well known that there is no shape without function in nature, as well as function without shape. Harmony between function and shape in the material world is one of the most important tasks of the today's designers. The shape of the product requires tectonization, i.e. matching the constructive functional structure and the used materials. The advanced technologies and the revolution that takes place in the study of materials allow us to create such shapes of industrial products that are in harmony with the environment where they are used or placed.

It is in the nature where we find a large range of constructive tectonic systems. We shall outline just some of them, the ones that are connected to the shape of the products.

5.   BIONICAL CONSTRUCTION SYSTEMS

5.1   Constructive systems of column type

In nature one could find many plants with a great height and a small surface of support which is still resistant to different actions of the environment. The stem of rye ear, for instance, has a relation between the diameter of the straw and its height of 1:500. The weight of the ear outruns the weight of the stem by 1.5 times; the cane has a height of approximately 3m and a stem diameter of 15mm (Figure 7).

The durability and stability of these natural constructions can be explained through a range of peculiarities: the reciprocal arrangement of solid and soft textures within the stem, their capacity to react to compression as well as to stretching. The stem of cereals has the shape of an ankle of bearings, and its knots represent articulations with elastic dampers.

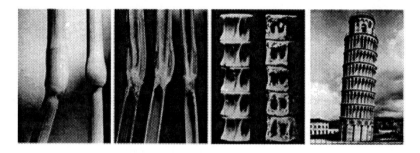

Figure 7. Constructive systems of column type

A strong wind just bends the feeble stem of cereals, whereas a tree is withdrawn together with its roots or broken. On the basis of studying these principles sky-scrapers are being built (Figure 8).

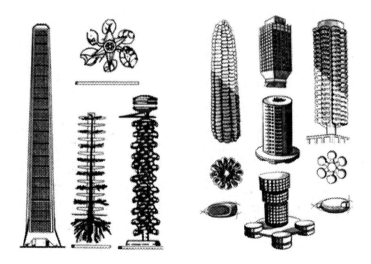

Figure 8. Bionic structure in forming of architectural constructions

5.2   Structures of tegument type

In nature's workshop one may often find constructions in the shape of cupola (egg and nut shell, testaceous animals, leaves and petals of plants). These structures spatially bent, with thin walls, due to the shapes of flowing, linear character, have the capacity of homogeneous distribution of forces over the whole section.

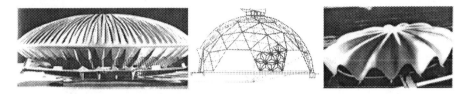

Figure 9. Tegument structure in forming of architectural constrictions

A unique construction, ideal from the point of view of durability, is the egg shell. The peculiarity of this structure consists not only of its geometrical shape. Even if the thickness of the egg shell is of only 0.3 mm, it has 7 layers, each of them having its own function, and the elastic coat that covers the egg on the inside transforms this shell in a construction with preventive tension.

This type of structures is perhaps one of the most wide-spread one in building huge spaces, with great distances between the main stays, (e.g.: exhibition pavilions, cinema theatres, sport grounds) and requires little building materials, they are light, the depth of walls being of just several millimeters (Figure 9).

5.3   Constructions with elastic cable-stayed

The spider thread is a constructive miracle of nature. They are much more resistant than the steel wires of the same diameter, with an elasticity that allows them to stretch 1.25 times (Figure 10).

These light, elegant and resistant constructions have attracted engineers' attention on them. Engineers used them to create the conception of elastic cables. The spider threads served as prototype for constructive structures of suspended bridges, which are a creation of engineer art through their diversity.

Besides the spider the same constructive models can be seen within other natural models as well, such as palmipeds, fins, and bat wings, where the constructive spread ribs are tied between them with a membranes surface.

In construction with elastic cables the basic carrying element is the "steel web" – cables or systems of steel cables on which membranes of different materials can be placed. Such constructions are very effective for covering spaces with large distances between the support points (e.g. the membrane thickness of the roof of the Olympic stadium is of just 5mm and the surface without intermediary pillars is of 30.000 m^2, Figure 10).

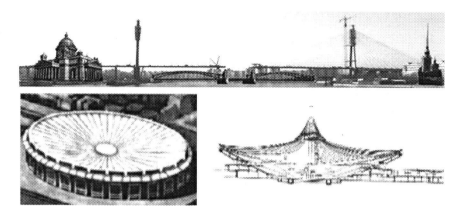

Figure 10. Constructions with elastic cable-stayed

Figure 11. Students projects of Industrial Design Department
from Technical University of Moldova

The combination of these types of constructive-functional structures in the design and creation of product shapes may provide huge savings in material resources. In nature one could find even more constructive structures that supply us a source of permanent creative inspiration in conceiving shapes of material products.

By using the laws of nature evolution, of harmonizing the function with the shape, humanity will succeed to conduct the ecological production of material goods. Not only designing the product itself is necessary, but also the designing of its cycle of existence – from its production to its disposal.

Nature offers us such great lessons and we must brilliantly assimilate them – design for the environment.

REFERENCES

1. V. PAPANEK, *Design for real world, Design pentru lumea reală* (Edit. Tehnică Bucureşti 1997).
2. J. S. LEBEDEW, *Arhitectur und Bionic* (Moscow, Berlin 1983).
3. LE CORBUSIER, *Sur les quatres, routes* (Paris, 1965).
4. J. S. LEBEDEW, *Arhitectural bionics* (Moscow 1990).
5. F. LODATO, *Bionics: Nature as a tool for product development* (Cambridge USA 2001).
6. E. TJALVE, *A short Course in Industrial Design* (London-Boston Moscow 1984).
7. V. PODBORSCHI, *The bionic design for eco-design* (Scientific Conference "Modern technologies. Quality. Reconstruction – T.C.M.R. 2002", 23rd – 25th May 2002, Iasi, Romania).
8. I. VOLKOTRUB, *The bases of artistic design* (Kiev, 1982).
9. ***, Students projects of Industrial Design Department from Technical University of Moldova

Part 2

COMPUTER AIDED DESIGN

1.   INVITED LECTURES

2.   CONTRIBUTIONS

ON SOME CRUCIAL ISSUES OF COMPUTER SUPPORT OF CONCEPTUAL DESIGN
What to consider in order to be successful

I. Horváth
Delft University of Technology, The Netherlands

Abstract: The attention in the development of design support systems is gradually shifting to conceptual design because of its importance in product realization. However, the fundamental mechanisms of conceptual design have only been partially explored. There are many open issues from cognitive, epistemological, methodological and computational points of view. This gave the motivation for this paper, which summarizes the essential characteristics of conceptual design and investigates what to consider in the development of computer aided conceptual design systems in order to be successful. It also gives an account on the industrial use of the computer-based methods and tools developed by academia. Its general conclusion is that much has been done, but even more has to be done towards genuine solutions. As a first result of the authors, nucleus-based concept modeling is discussed in the paper.

Key words: computer aided conceptual design, cognitive aspects, methodological aspects, computational aspects, relation-based modeling, and nucleus technology.

1. INTRODUCTION

Conceptual design of products has been regarded as the most influential part of each product realization process. It plays a particularly essential role in the development of new products. For this reason the attention of researchers and developers is gradually shifting to conceptual design, which offers new opportunities in computer support of design [1]. As far as the current situation is concerned, there are at least as many not identified challenges as known ones. Actually, the current situation is somewhat paradoxical since the need for an effective computer support has already

D. Talabă and T. Roche (eds.), Product Engineering, 123–142.
© 2004 Springer. Printed in the Netherlands.

emerged, but conceptual design as a phenomena and a problem-solving asset is just partially explored and understood [2]. Given that the design problems, the platform of problem solving, the processed information, and the representation of products largely differ from what is typical in detailed design, completely different approaches are needed.

These facts gave the motivation for my studies whose results are reported in this paper. The focus of investigation was put on what should be considered in the development of computer aided conceptual design systems in order to be successful. Section 2 summarizes the essential characteristics of conceptual design. It elaborates on the most important notions and analyses the current situation with a view to the development of computer-based methods and tools by the academia and their utilization in the industry. Section 3 analysis the differences of the views of the industry and the academia on computer support of conceptual design. Sections 4, 5, 6 and 7 investigate the cognitive, epistemological, methodological and computational issues. Section 8 proposes a new approach to computer support of conceptual design. The key element of the proposal is relation oriented modeling based on the nucleus concept. The paper shows that many notorious problems can be eliminated or reduced based on this approach.

2.   WHAT IS THIS THING CALLED CONCEPTUAL DESIGN?

There is no unambiguous definition of what conceptual design is. It has different goals and appears in different forms in the various sub-disciplines, e.g., in mechanical design, architectural design, industrial design, and interior design. It can also be seen from multiple aspects. There are however some common elements in all observable forms of conceptual design [3].

Conceptual design is understood as the front end of product development processes. It is typically preceded by a market-product-technology (MPT) research, which gives the motivation for a company to define a new product, or to decide on redesigning of a marketed product. The actual product conceptualization builds on the results of product definition, and is guided by the specification of goals and requirements [4]. The objective is to generate and test a set of alternative solutions, that is, product or system concepts in order to facilitate finding the best one. The concepts are typically represented by some sort of schemas that serve as the basis of the embodiment of a product, or of a system. Usually various levels of abstraction, incompleteness and uncertainty characterize them. Conceptual design always deals with the intended product attributes rather than with product characteristics. The results of conceptual design are directly used in the embodiment, detailing and first physical realization (prototyping) of the product ideas.

From a methodological point of view, conceptual design is a creative problem solving process, enabled by human knowledge, intuition, creativity and reasoning [5]. It can also be seen as a cognitive process, in which ideation, externalization, synthesis and manipulation of mental entities, called design concepts, takes place in symbiosis in a short-term evolutionary process [6]. From the aspects of information science and technology, conceptual design is an iterative search process in which designers gather, generate, represent, transform, manipulate, and communicate information and knowledge related to various domains of design concepts. The whole process is characterized by an inherent abstraction - a property which computer aided conceptual design has to cope with. It can be observed not only in the information and the schemes that designers use to represent their first ideas about products, but also in their thinking [13]. Finally, from the viewpoint of computer technology and computation, conceptual design is the source of requirements for, and the target field of application of a set of dedicated computer based tools and techniques to assist or automate concept generation, modeling, evaluation and documentation.

Computer support of conceptual design is not straightforward due to its non-mechanistic (intuitive, heuristic, reflective and intellectual) nature. Human problem solving plays an important role, but it also introduces a lot of incidental actions and informal knowledge. The purposeful creativity, heuristic thinking and past experiences of the designers mean a large challenge for computer support since it is difficult to explain them by formal theories and to cast them into formalized procedures. It became obvious that an algorithmic formalization of conceptual design towards a complete automation does not have a strong rational basis, and does not even make sense from a practical point of view. There are however several sub-problems in conceptual design in which human problem solving (thinking and creativity) can be, and have been efficiently supported by computational techniques [7].

3.   DIFFERENT VIEWS ON COMPUTER SUPPORT OF CONCEPTUAL DESIGN

In the last decade, university researchers proposed many methods and tools for computer-aided conceptual design, but the interest of companies to use these methods and tools in their daily routine has been fairly law. An overview of the virtual models generated by conceptual design tools is shown in Figure 1. The academia and the industry have their own views on

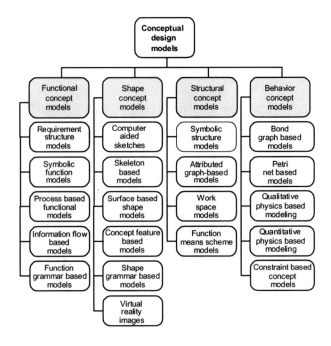

*Figure 1.*Taxonomy of virtual conceptual models

computer support of conceptual design. For long time, computer mediated product conceptualization has been a mystery for the industry. The comprehension of the emergence, exteriorization and manipulation of ideas and concepts towards a computer support were not considered important by average companies. They are profit orientated, therefore, pragmatic in thinking and doing. The overwhelming majority of the product developer companies usually delegate the responsibility over conceptualization to some distinguished designers. They usually prefer using the intuitive methodologies and apply less sophisticated, sometimes primitive, but convenient techniques such as sketching and physical modeling to solve conceptualization problems ...

   In the industry, by and large, product conceptualization and conceptual design are simply reduced to, but supported seemingly successfully by, personal ingenuity, inventive group meetings, activation of experiences, and using the analogies from existing products. Obviously, the industry feels that computer support of conceptual design is a necessity, but they always have more urgent tasks to accomplish. Due to the methodological and technological immatureness, they have serious reservations about the achievable advantages and the usefulness of the computer aided conceptual design tools developed by the academia. In view of the premature methodologies and disintegrated technologies they think twice their investments in this field. Most of the chief technology officers would sooner pay for an excellent human 'ideator' than for the most sophisticated knowledge-based conceptual design tool. Of course, there are exceptions and success-stories as always, but the global picture of using dedicated conceptual design tools is thought provoking. An average company seldom uses the design methods and the computer aided conceptual design tools of the academia on a daily basis. It happens even more rarely that a company requests the development and/or implementation of a particular conceptual design tool according to their own ideas. Conversely to the market of computer aided design systems, the market of the computer aided conceptual design tools is dormant.

We can easily identify many reasons why the methodological and technological developments of the university researchers or developers do not receive wide acceptance in the industry. One of them is that practically all techniques developed for computer support of conceptual design embed a given level of abstraction. It means that the designers should learn this abstract thinking in order to be able to work with the abstract representations (Figure 2). This is however inconsistent with the concrete thinking that is typically needed in detail design and in the application of CAD/E systems. Notwithstanding, the designers are supposed to provide the missing information and to make efforts whenever the results of conceptualization need to be transferred to the downstream design processes (i.e. to embodiment or detailing). The problem of automatic conversion of product concepts into fully detailed artifact models seems to be unsolvable.

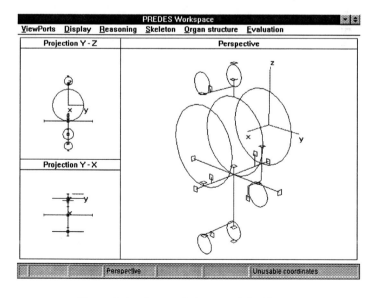

Figure 2. An abstract skeleton model with ports

The second probable reason of the low success rate of academic proposals is the disparagement or ignorance of the importance of the human role and factors in conceptual design. In the starting phase of product development, in particular in conceptual design, designers should clarify not only the generic but also the specific technical, economic and environmental requirements for a product and define its (initial) functions, structure, shape, materials, interfaces, behavior and appearance. What makes conceptual design different from embodiment and detail design is the creative leap whose roots are in human knowledge, intuition, heuristics, analytic reasoning, expressing, creativity, synthesis and reflections. There is no any comprehensive theory or methodology for the development of conceptual solutions for wide-ranging, complex and under-constrained conceptualization problems.

Nevertheless, there have been many computational approaches proposed, which try to apply algorithmic solution generation for well-defined sub-problems in bounded and properly constrained solution spaces [8]. In the field of artificial intelligence (AI) research many efforts have been devoted to the theoretical understanding, mathematical underpinning, and practical implementation of various inference and creative mechanisms that enable human beings to be successful in conceptual design [9]. But the vagueness or lack of information, the under-determined design constraints, rapid evolution of concepts, multiple aspects of synthesis, and the possibility of alternative solutions also mean challenges for the concerned computational tools. In addition, it is difficult, if not impossible, to integrate them with each other.

The third likely reason of the limited industrial proliferation of computer support of conceptual design is the wide variety of non-convertible presentation, analysis and simulation tools. While in computer aided design the overwhelming majority of the application models, for instance, finite element models, process planning models, assembly models, use models, and maintenance models are based on a kernel geometric model, in computer aided conceptual design it is not the case for two reasons: a) the initial geometric model in this stage does not have any priority relative to the other initial models (function models, structural models, process flow models, or image models), b) the aspect models are evolving in close interaction with each other. Since the information content, the representation means, and the processing actions related to the aspect models are different, the aspect models cannot be incorporated in one comprehensive model.

Though conceptualization of mechanical systems proved to be one of the most fertile application areas, it was also found that all pieces of knowledge needed for conceptual design cannot be included in any single model at the same level of abstraction. Consequently only aspect models can be generated, each of which has its advantages and shortcomings. For semantic and complexity reasons, thinking of meta-models is also useless. Hence, each non-transitive model is a procedural 'blind-alley' and conceptual design based on these models is a sack race from a practical point of view.

In order to resolve a significant part of the problems mentioned above, we have two tasks: a) to achieve a better understanding of what conceptual design is and what its proper computer support means, and b) to study and develop new modeling concepts that take into consideration the above analyzed characteristics of computer aided conceptual design, provide opportunities for integration of, or elimination of the need for aspect models, comply human thinking, as well as the methodological, technological and procedural requirements. The intent of the next three Sections is to provide contribution to the first goal through an investigation of the cognitive, epistemological, methodological and computational issues. The eights Section will present our ideas and results related to the second group of goals.

4. COGNITIVE ISSUES

As we already touched upon this, the conceptual part of product design dominates human intuition, heuristics, experiences, believes, soft theories, vague models, verbalism, graphics, and common sense reasoning, but also perception, preferences, and experiences. These informal elements of conceptual design processes make computer support difficult. From a research point of view, these altogether form an unmanageable complexity. From this complexity, we are going to investigate those concerns here, which are typically referred to as the cognitive issues. They are grouped around human design intellect, mental processes, obtaining and processing knowledge and information, perception and memory, concept formation and experiences, and goal driven problem solving.

It has to be mentioned that many researchers believe that conceptual design does not actually start with cognition, but with perception. It concerns the apprehension of the world through the action of various sensory systems and can be either sensual, or intellectual, or both. Sensual perception relies on the sensors, whereas intellectual perception is based on the mind (memory). Perception provides us with an immediate automatic grouping of the visual and mental information, without acknowledgement, reasoning or judgment. Perception happens in a tiny fragment of time, while an image is retrieved, a stimulus is exerted, and a chunk of information is created.

Through the contribution to visual grouping and compliance grouping respectively, image perception and intellectual perception are important for the more complex mechanisms of cognition. Cognitive processes start as reactions to perceptions. One of the most characteristic cognitive processes is conceptualization. In epistemology the term 'conception' indicates a) beginning of a process of existence, and b) deriving or forming an idea of something. Research explored that conceptual design is not only supported by, but also depends on human cognitive capabilities such as a) conjectures, b) hypothesizing, c) ideation, d) generalization, e) abstraction, f) creativity, and g) analysis. It is not yet known how the ideas happen, but imagination definitely triggers conceiving design concepts.

Based on the aforementioned fundamental notions, the various cognitive theories as well as psychological and artificial intelligence research suggest that conceptualization is a blend of:

a) application of intuitions and heuristics in a semi-rational problem solving in a target area,
b) externalization of human mental images in the form of observable representations,
c) creative composition driven by human perceptions, conjectures, experiences and reasoning,
d) triggered mental activities towards generating, learning and using design concepts as some chunks of knowledge, and
e) creation of semantic and contextual associations between intuitive and learnt design concepts [10].

The term concept originates in Latin word *conceptus*. In epistemology, 'conceptus' (i.e., a concept) means several things, for instance, a) a general notion related to cognitive knowledge, b) a mental impression or image of human mind, and c) an abstract or generalized idea of a class of particles. To give a proper circumscription of what a design concept is, we take these interpretations as a basis. Thus, a design concept

– is a mental image or abstract reflection of reality in the human mind,
– represents an individual logical unit of cognitive reasoning, which evolves and may go through metamorphosis,
– establishes links between the logical space of reasoning of humans and the metric-temporal space of artifact manifestation,
– expresses dependencies among requirements, functions, principles, structures, forms and attributes, and
– is a notional element of communication amongst designers and amongst applications.

In addition to these, the notion of 'concept' in knowledge processing also denotes a new invention to help create a commodity.

Based on the above semantic analysis, we can conclude about what to take into account for a proper computer support of conceptualization, and in capturing, describing and processing design concepts. If conceptualization is a selection and composition of design concepts, then the modeling entities, dedicated to conceptual design, have to be able to represent design concepts. The modeling entities should have the potential to create concrete and formal relationship between the logical space of reasoning of humans and the metric-temporal space of artifact manifestation. In this respect we must not forget that design concepts, generated systematically or intuitively by designers, can be very vaguely bordered and complex. They can be decomposed, but not beyond any limit.

The last issue we are going to address in this Section concerns the cognitive scheme of conceptualization (Figure 3). This scheme explains the sub-processes and how they happen, or are supposed to happen, in conceptual design [11]. The cognitive scheme covers the entire process of converting ideas to models and indicates the time normally elapsed by the activities of conceptualization. As it is shown in Figure 3, the average speed in the internal (ideation) loop of activities comprising ideation, presentation and reasoning is 10^{-1} to 10^{0} s. It is 10^{1} to 10^{0} s in the external (modeling) loop involving presentation, reasoning and model building. Knowing the time requests of the current product modeling systems, we can claim that these are in conflict with the cognitive performance values that are natural for human beings. This fact also has influence on the acceptance of tools.

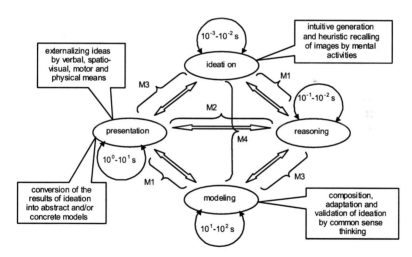

Figure 3. Cognitive scheme of conceptualization

The techniques that lag behind these values hold back design thinking, externalization and creativity. Actually, this is the reason why, for instance, hand sketching became a standard presentation technique for shape conceptualization. Not only the actions are in harmony with the cognitive mechanisms of conceptualization, but also the time needed for a stroke-based presentation complies with the duration of the mental actions of ideation and reasoning. The message from these is that computer aided conceptual design tools will be natural for the designers, and most probably also the most effective, if the speed of modeling approximates the speed of human thinking and ideation.

5.   EPISTEMOLOGICAL ISSUES

In philosophy of science, the epistemological issues are: a) the sources of knowledge, b) the way of obtaining knowledge, and c) the structuring of knowledge. Computer support of conceptual design has to do with at least two specific epistemological issues: a) obtainability of the knowledge related to, so called, design mappings, and b) decision making based on imperfect (uncertain and incomplete) knowledge.

Design mappings are actions in the process of transforming a set of design parameters of a specific semantics to a set of design parameters of a different semantics. Design parameters of particular semantics are, for instance, specifications, requirements, functions, principles, shapes and behavior. The epistemological problem comes from the fact that there is no generic theoretical explanation on how the semantically different design parameter sets are interlinked. Conventionally, the mappings involved in product conceptualization are completed based on human domain knowledge and cognitive reasoning. In specific cases, when certain correspondence rules are already known, the general design mappings reduce to association problems (e.g., selection solution principles for functions from an existing catalogue).

The following major design mappings are still missing a deep scientific understanding and explanation:
a) converting requirements to a system of functions or potential operations,
b) mapping design functions to physical principles and controllable processes,
c) deriving structures from first principles and physical phenomena,
d) transforming functions or potential operations to shapes,
e) converting on purposeful operation system to another ones,
f) integrating syntagmatic and paradigmatic shape definition, and
g) interactions of concepts of different semantic contents.

Another epistemological issue is the nature of information that is to be processed in conceptual design. This information is extremely diverse: it concerns nature, sciences, technology, humans, society, artifacts, processes, methods, tools, attributes, and values. Design information appears in various representation forms in the course of product conceptualization, ranging from verbal, textual, numeric, mathematical, symbolic, graphical, visiospatial, virtual and physical data, through information recorded on audio and video, to structured symbolic and procedural models. In addition to the variety of the kinds and representations, the amount of information also means an extra challenge for computer support of conceptual design.

6.   METHODOLOGICAL ISSUES

Conceptual design sets off with a specification, which circumscribes the idea of a requested product, the technical and non-technical requirements, the conditions of realization and the constraints/opportunities. To start with, the requirements are converted to ideas about functions, then the functions are covered by first principles, the first principles are used to generate structural arrangements, the structural elements are characterized for its material and initial shape, the morphological and functional integrity of the concept product is achieved, and finally the proper behavioral processes are established by the designers. This apparent sequential nature of design mappings gave the basis to describe the conceptual design process as a waterfall (Figure 4).

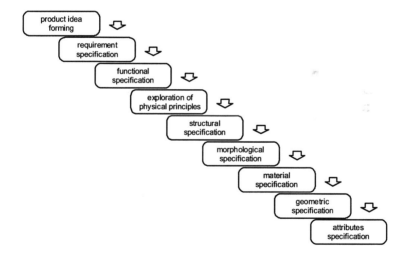

Figure 4. Cognitive scheme of conceptualization

The origins of the waterfall model are in the German school of methodological design thinking [12]. The goal was to introduce a systematic framework to guide designers, especially the inexperienced ones, to find optimal solutions for the optimization problems. This process model arranges the activities in an order that reflects a logical way of acquiring and processing knowledge about concepts and concept structures. The sequence of activities that must be fulfilled corresponds to that is vaguely indicated by the cognitive model of product conceptualization.

The waterfall model does not offer solution for the afore-mentioned epistemological problems. It leaves the tasks of semantic mapping on the designer, and controls the process of conceptualization only. An important element of this systematized conceptual design is the reasoning about the functions, function structures, physical principles and function carriers. This makes its application in the field of mechanical design more appropriate, than in the field of industrial design, where form giving and appearance design play a distinguished role. In the mechanical design field, application of this systematized conceptual design methodology is supported by various design-enablers such as function libraries, principle catalogues, and morphological charts [15]. We look for conceptual solutions in standardized catalogues at all levels of the waterfall model. The functions are mapped onto function carriers through possible physical principles and structural arrangements.

The waterfall model is a useful contribution to the understanding of the main aspects of a conceptual process design. The related systematic methodology increases the level of knowledge intensiveness and facilitates the inclusion of computer-based conceptualization support tools [16]. As a critique of this methodology we have to mention that the arrangement of the activities in a pure sequential order is against the concurrence that is typical in the practice. The real-life conceptual design processes always show procedural recurrences, that is, the best fitting concepts and compositions are found in multiple iterations. Therefore, it makes no sense to force the kernel activities in a strict procedural framework.

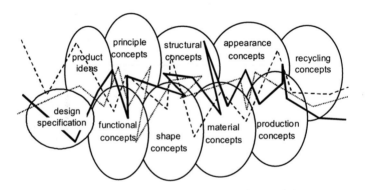

Figure 5. The pathfinder model of conceptual design

For the above reason the so-called pathfinder model has been proposed, which depicts the process of conceptual design as a search zigzagging over the various subspaces of design concepts and a series of recurrent compositions (Figure 5). The pathfinder model explains conceptual design as a designer and problem dependent concept aggregation and composition process, i.e., as a set of purposefully arranged actions towards the intended goals, both explicit and implicit. The former ones originate in the artifactual and behavioral requirements, and latter ones are set by the general FTQC (function, time, quality, and cost) requirements of product development.

7.   COMPUTATIONAL ISSUES

Conceptual design is often described as the process of synthesizing design concepts, which are simultaneously mental images and knowledge chunks for the designers. Obviously, a large number of computational issues can be identified in conjunction with computer aided conceptual design. In principle, there are two main approaches to the application of computational means: a) developing soft computational means that assist the designers to solve conceptual design problems in their designerly way [14], and b) developing robust mathematical theories, formal languages, methods and techniques to solve conceptualization problems in an autonomous way. The former is referred to as interactive support, and the latter as algorithmic support. Both approaches have their own strengths and limitations. There are four main computational issues: a) elicitation, representation and processing of human knowledge assets, b) development and validation of algorithms for conceptual design, c) building and representation of conceptual models, and d) interaction between designers and computer aided conceptualization systems.

Development of computational tools is challenged by the interconnection with human creativity. The best computational approach to conceptual design would be the one based on a high fidelity modeling of the human mind. Therefore current research pays a lot of attention to the understanding of human thinking, creation, decision making and appreciation processes, and tries to realize them in, or support them by, design tools. Being interwoven with elements of human individuality, it is very difficult to transfer ideation and creative synthesis completely to digital computers. The major issue is how to replicate the physical, mental and sensory capabilities that enable and authorize humans to conceptualize. Many theories and computational tools have been proposed on the basis of formal (mathematical) logic, analogies, cases, rules, constraints, and models, but the results are typically inferior to those of average designers. In addition to the problems of generating concepts and creative building of models, problems related to context dependent decision-making and complexity management have to be also addressed.

Assuming that engineering design is a computable function, design automation research deals with the reproduction of design knowledge and problem solving capabilities in and by artificial systems. It has to cope with uncertainty, incompleteness, modality, complexity, and subjectivity. Together with scientific and technological knowledge, common sense reasoning is used as a strong means of validation and optimization of design concepts and conceptual solutions in traditional conceptual design. However it is extremely challenging to transfer this asset to computer tools or systems. Current research is dealing with both qualitative and situational reasoning, and creates quantitative models for early behavioral simulation. A branch of research is focusing on tools for expressing aesthetic reflections and use experiences in conceptual design. Development of formal tools for computer-based validation of design concepts is still in its infancy.

Research in computational concept generation intends to implement synthesis processes and mechanisms to more than one design concept by evolutionary computational techniques. Typically AI principles and techniques such as qualitative reasoning, neural networks, genetic algorithms, rule based computational tools, computer learning strategies, and agent-based problem solving are used for procedural inference and concept synthesis in conceptual design. Many results have not reached beyond initial benchmark tests. They either did not have the capacity to address complex industrial problems, or created a closed world in which concept generation appears as a well-formed problem. Computational approaches are weak to consider social, scientific, technological, financial and intellectual conditions. Many industrial designers consider design automation an educated academic exercise without any strong theoretical support and practical expectation. They are convinced that interactive approaches have more potential to consider intrinsic practical rules of product conceptualization. They would prefer interactive and knowledge-intensive approaches, which exploit the best of designers and tools.

8.   A NEW APPROACH TO COMPUTER AIDED CONCEPTUAL DESIGN

Our research team has been working on a new approach to computer aided conceptual design, which tries to consider the above-discussed issues in an interactive and knowledge-intensive conceptual modeling methodology [17]. It places the design concepts into the focus, and operates with the associations (relations) between design concepts (DC). Our methodology introduces a new modeling entity, called nucleus, based on which it is possible to capture, model and manipulate design concepts. The new modeling entity is supposed to facilitate building knowledge-intensive conceptual and application models.

It has been hypothesized that any design concept can be abstracted as a composition of interacting particles and multiple physical relations appearing in various situations. If the particles, relations and situations are missing, the abstraction becomes meaningless. Actually, this is the reason of calling the triplet $N = \{\Pi, \phi, S\}$, the nucleus of a design concept. Likewise atomic nucleus, it serves as a central core and forms the basis of the realization of (the functions of) a design concept. The particles incorporated in a nucleus are metric entities, which are supposed to be finite and self-contained. Depending on the expected form of manifestation, they manifest either as a boundary particle, or a volume particle in a nucleus. Boundary particles are characterized for their reference point(s), contact surface, surface normal vector and volume. The contact surface of a boundary particle ($B\Pi$) is represented by a half space indicating the material domain of the particle.

Thus, a nucleus containing boundary particles can be formally described as $N = (B\Pi_i, B\Pi_j, \phi, S)$, where $B\Pi_i$ and $B\Pi_j$ are the boundary particles. Volume particles ($V\Pi$) are characterized for their reference point(s) and volume. A nucleus expressing relationships between one boundary particle and one volume particle can be formally represented as described as $N = (B\Pi_i, V\Pi_j, \phi, S)$, where $B\Pi_i$ is as above and $V\Pi_j$ is the volume particle. A nucleus can represent the relationships between one-one particle of two artifacts, formally, $N = (B\Pi_{p,i}, B\Pi_{q,j}, \phi, S)$, where $B\Pi_{p,i}$ is a boundary particle of artifact p and $B\Pi_{q,j}$ is a boundary particle of artifact q. $B\Pi_{p,i}$ is called a native boundary particle of artifact p, and a complement boundary particle of artifact q.

By defining artifacts as sets of finite and self-contained particles we establish a discrete modeling [18]. The discrete model makes it possible for us to specify relations between two boundary particles, a boundary particle and a volume particle, two volume particles of the same or different artifacts, keeping in mind the constraints origination in the physical reality. Relations are special sort of 'objects' that connect particles but they are ontologically independent and functionally distinct from them. A nucleus can describe finite number of relations. Relations can be assigned if and only if at least one particle has been defined. If one particle is defined, the reflexive physical relations can only be assigned. For physical modeling and simulation, arbitrary number of relations can be specified between pairs of particles. Particles acting as 'environment' must have at least one reflexive relation to result in a non-limitless system. Incorporating physical relations lends itself to a physically based conceptual modeling.

Relations are existential, manifestation and behavioral associations, dependencies and interactions between humans, artifacts and environments. Relations can be different from the viewpoint of semantics. We identified two general categories of relations: unary and binary. Unary relations are a) existence, b) reference, and c) substance relations. Binary relations are a) connectedness, b) positioning, c) morphological, d) kinematical, e) deformation, f) kinetic, g) physical effect, and h) field effect relationship. Connectedness as a relation expresses that a particle belongs to an artifact, even if it is only logically defined. Different relations can be specified for the different types of the particles (i.e., whether $B\Pi$ or $V\Pi$).

Relations can be internal (i.e., within one artifact) and external (i.e., between two artifacts). If Π_i stands in relation to Π_j, but neither its identity nor its nature depend upon Π_j, the relation ϕ is external. If the opposite is true, then ϕ is internal. Specification of the functional relations includes definition of the parameters, the mathematical formulas over the parameters, the constraints, and the value domains. Mathematical formulation of relations makes it possible for us to simulate the behavior of a nucleus, an artifact, or a system of artifacts. Each physical relation implies a set of elementary processes that contribute to a computable behavior. Actually, the time-dependent changes described by the relations between particles are used to simulate the behavior, B, of a nucleus in a given situation:

$$B(N) = \Gamma \{S_k (\Pi_i \phi_{ij} \Pi_j)\}, \tag{1}$$

where Π_i, $\Pi_j \in \Pi$, ϕ_{ij}, S_k are as earlier, and Γ is a behavior generator function, which takes into consideration the interaction of various nuclei and the influences on each other's behavior. The introduction of Γ to handle the interactions of nuclei is necessary, since the observable behavior of a modeled design concept is the aggregation of the elementary behaviors of the nuclei. Since all nuclei might interact in a composition, this aggregation should be represented as a Descartean product rather than as a Boolean union of the observable elementary operations:

$$B(DC) = B(N_i) \times B(N_j), \text{ or } B(DC) = \Pi (B(N_i), B(N_j)), \tag{2}$$

where Π denotes a mathematical product. Our goal is to simulate the conceived behavior of artifacts and to be able to control the simulation of the behavior.

A situation is a given arrangement around a nucleus. The word 'setting' has been introduced to describe a structure of situations, in which the operation and interactions of nuclei take place. In a simulation process, the situations are changed and the changes of the nuclei are computed based on the specified relations and constraints. Computationally, the arrangement of situations is governed by so called scenarios [19]. A scenario, Σ, prescribes a sequence of situations, in which the interactions of the nuclei incorporated in a design concept happen. That is,

$$\Sigma = \cup \, (S_k). \tag{3}$$

With these, the behavior of a DC on the level of nuclei is $B(DC) = \Gamma \, (\Sigma \, \{ \, N_i \, \})$, or, on the level of relations is:

$$B(DC) = \Gamma \, (\, \cup \, (S_k \, (\Pi_i \, \phi_{ij} \, \Pi_j))). \tag{4}$$

In the development of the database scheme of nucleus-based modeling we relied on the fact that the relations can be arranged in a hierarchical structure according to their content (meaning). The developed database scheme is shown in Figure 6. It favors to the application of object oriented programming of the nucleus-based conceptual design system. On the lowest level of the database scheme, nuclei are defined as couplings of two particles, boundary or volume. Figure 6 shows the data fields directly stored in the data structure of a nucleus. The access to the coupled particles is implemented through pointers to the data fields. The geometric data are specified through the substance relation to the existing nuclei of the concerned artifact (mechanical part). It means that the reference vectors of the particles, the surface normal vectors, the mass and material attributes can be specified using reflexive relations. The couplings between distinct artifacts are described by binary relations, which are stored in a common list of relations.

Apart from itself, a nucleus can represent a particle cloud (i.e., a natural surface), a mechanical component, an assembly of components, and a complete system. To handle these five levels of artifacts in the database, nuclei are stored in five layers of triangular matrices. In theory, $n \cdot n - n/2$ non-reflexive and n reflexive nuclei can be defined. Hence, each matrix contains $n \cdot n + n/2$ elements, where n is the number of particles being in relation in a given level. The reflexive relations are stored in the main diagonal of the matrix. Relations between the nuclei of different levels are defined by "has a" connectedness relations.

The nucleus-based modeling methodology has been tested in various applications. The first experiences show that it can manage with incompleteness without losing the ability to give feedback about the product appearance and its behavior. It is significant difference from the existing conceptual modeler tools, which tend to treat the defined model as complete. Nucleus-based modeling can easily deal with multi-level representations without the need to fully specify every level. It can represent complex components (e.g., supplied subassemblies, such as an electromotor) as a 'black box' without the need to model its internals. It can handle vague shapes, since the geometry of the particles can be vaguely defined.

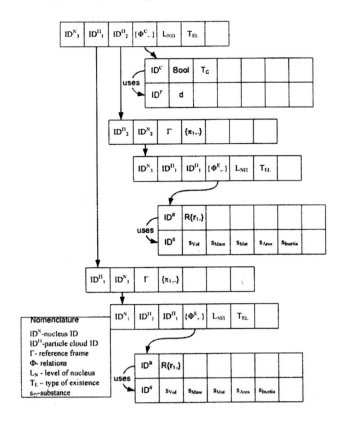

Figure 6. Database scheme of nucleus-based modeling

By specifying a minimal and a maximal surface overlaying the vague particles, the designer can develop an interval representation of a family of shapes.

The basic laws of mechanics or physics can be incorporated in the nuclei as default relations. This makes it possible for designers to run simulations without having to specify all the relations between the modeling entities individually and without having to derive the equations in question. This is unlike, for instance, constraint-based modeling, where the user has to define all the relations in the model. Nucleus-based modeling allows conformity between the visual feedback from the modeling and simulation system, and the physical appearance of the intended artifact. The behavior of an artifact can be simulated in a particular setting by a set of scenarios. A set of scenarios can be specified as a branching tree with alternative logical arrangement of Σ, and parameter values of sets of initial conditions (IC), boundary conditions (BC), and procedural conditions (PC). Each branch of the tree represents an alternative scenario, which the designer might want to realize or avoid. The significance of a scenario based behavioral simulation is that it enables us to model any logically and procedurally possible series of happenings in various settings.

9. CONCLUSIONS

Conventional CAD tools have not been proposed and do not have all necessary capacities to support designers from the beginning of conceptual design. However, due to its influential nature, conceptual design cannot miss an effective computer support. Our understanding has been that due to its highly intellectual nature, an exhaustive algorithmic formalization of conceptual design cannot be the goal. As for now, computer support has to be implemented by considering the interactions between the designers and the knowledge intensive design system. We proposed the theoretical framework and methodology of nucleus-based modeling as a possible approach.

Nucleus-based conceptual models have the potential to expand the reach of computer support by offering low-threshold modeling facilities for simulations that tolerate incompleteness and vagueness, and do not require the user to develop a deep understanding of the underlying algorithms. The models are versatile in that they are aimed at supporting multi-physics, multiple-resource applications and logic. At the same time geometries can be obtained as by-products of the relational modeling. The nucleus facilitates structural and behavioral representation of artifacts of various complexity levels. Our further work will include the development of a highly interactive (hand motion and verbal controlled) user interface.

REFERENCES

1. W. HSU, B. LIU, *Conceptual Design: Issues and Challenges* (Computer-Aided Design, Vol. 32, No. 14, 2000, pp. 849–850).
2. M. J. FRENCH, *Conceptual Design for Engineers* (Design Council, London, 1985).
3. I. HORVÁTH, *Conceptual Design: Inside and Outside* (Proceedings of the 2nd International Seminar and Workshop on Engineering Design in Integrated Product Development - EDIProD 2000, Rohatynski, R. ed., UZG, Zielona Gora, Vol. 1, 2000, pp. 63-72).
4. G. PAHL, W. BEITZ, *Engineering Design – A systematic Approach* (2nd Ed., Springer, Berlin, 1996).
5. N. F. M ROOZENBURG, *On the Pattern of Reasoning in Innovative Design* (Design Studies, Vol. 14, No. 1, 1993, pp. 4-18).

6. J. S. GERO, *Computational Models of Creative Design Processes, Artificial Intelligence and Creativity* (ed. by Dartnell, T., 1994, pp. 269-281).

7. D. SERRANO, D. GOSSARD, *Tools and Techniques for Conceptual Design* (Artificial Intelligence in Engineering Design, Vol. 1, 1992, pp. 71-116).

8. S. POTTER, *Artificial Intelligence and Conceptual Design Synthesis* (Ph.D. Thesis, University of Bath, Bath, UK, 2000).

9. E. CHARNIAK, D. McDERMOTT, *Introduction to Artificial Intelligence* (Addison-Wesley, Reading, MA, 1985).

10. I. HORVÁTH, GY. KUCZOGI, J. S. M. VERGEEST, *Development and Application of Design Concept Ontologies for Contextual Conceptualization* (Proceedings of 1998 ASME Design Engineering Technical Conferences DETC'98, September 13-16, 1998, Atlanta, Georgia, CD-ROM: DETC98/CIE-5701, ASME, New York).

11. I. HORVÁTH, *Investigation of Hand Motion Language in Shape Conceptualization* (Journal of Computing and Information Science in Engineering, Vol. 4, No. 1, 2004, pp. 37-42).

12. K. H. ROTH, *Design with Design Catalogues* (Springer Verlag, Berlin, 1982, pp. 1-475, in German).

13. V. ADZHIEV, M. BEYNON, A. CARTWRIGHT, Y. P. YUNG, *A New Computer-Based Tool for Conceptual Design* (Proceedings of 1994 Lancaster International Workshop on Engineering Design - *CACD'94*, ed. by Sharpe, J., Oh, V., Lancaster, Engineering Design Centre, 1994, pp. 171-188).

14. K. D. FORBUS, *Qualitative Reasoning* (CRC Computer Science and Engineering Handbook, Tucker, A. ed., CRC Press, Boca Raton, FL, 1997, pp. 715-733).

15. F. ZWICKY, *Discovery, Invention, Research-Through the Morphological Approach* (Macmillan Publisher, New York, 1969).

16. R. H. STURGES, K. O'SHAUGHNESSY, R. REED, *A Systematic Approach to Conceptual Design* (Concurrent Engineering: Research and Applications, Vol. 1, 1993, pp. 93-105).

17. I. HORVÁTH, W. F. VAN DER VEGTE, *Nucleus-Based Product Conceptualization: Principles and Formalization* (Proceedings of ICED '03, Stockholm, August 19-21, 2003, pp. 1-10).

18. Z. RUSÁK, *Vague Discrete Interval Modeling for Product Conceptualization in Collaborative Virtual Design Environments* (Ph.D. thesis, Millpress, Rotterdam, 2003).

19. W. F. VAN DER VEGTE, I. HORVÁTH, *Consideration and Modeling of Use Processes in Computer-Aided Conceptual Design* (Transactions of the SDPS - Journal of Integrated Design & Process Science, Vol. 6, No. 2, 2002, pp. 25-59).

EXPERT SYSTEM FOR THE TOTAL DESIGN OF MECHANICAL SYSTEMS WITH GEARS

G. Mogan and E. V. Butilă
Transilvania University of Brasov, Romania

Abstract: The CAD of mechanical systems involve activities specific to a given expertise domain and operate with two information categories: knowledge and data. Expert rule based knowledge processing generally works with qualitative information and involves searching for suitable solutions principles and their combination into concept variants. Data processing is based on computational models and it is supposed to be inter-related with reasoning in the knowledge process. An Intelligent Integrated System is proposed in this paper to design gearboxes as independent products. For the proposed expert – CAD/CAE/CAM system for gearbox design, the CATIA package is used to allow the integration of knowledge processing activities with solid modeling, performance analysis and manufacturing aspects.

Key words: expert systems, computer aided design, gearboxes, total design.

1.   INTRODUCTION

"Expert Systems" is a branch of artificial intelligence that makes extensive use of specialized knowledge to solve real problems which normally would require a specialized human expert. At a basic level an expert system is an intelligent computer program that uses knowledge and reasoning procedures. To build an expert system, first one needs to extract the relevant knowledge of problem domain (knowledge acquisition) in a way that facilitates the introduction in computers as a Knowledge Base. This is a difficult task and it is the job of the knowledge engineer. Expert systems have been used to solve a wide range of problems in fields as medicine, mathematics, engineering, geology, business, law, education etc. Types of problems can involve activities of prediction, diagnosis, design, planning,

143

D. Talabă and T. Roche (eds.), Product Engineering, 143–162.
© 2004 *Springer. Printed in the Netherlands.*

monitoring, interpretation etc. In the field of mechanical engineering design, expert systems are produced for a wide range of purposes e.g. design for specific products, design for assembly, diagnostics, design for disassembly and design for environment.

The traditional generations of CAD/CAE programs focus primarily on the aspects associated with representation of form and secondly on the simulation of the functionality performance capabilities. This is in contrast with tendencies included in modern design software which have possibilities to take decisions on whether the solution is acceptable and, if not, how to make it adequate.

On the other hand the traditional CAD/CAE programs usually are not containing in clear way the conceptual design and embodiment design phases which usually are parts of the design process. The integration of expert systems in CAD/CAE systems generate new capabilities in the direction of reasoning taking into account specific knowledge that consist in integration of product design specification, conceptual design, detail design, and manufacture. In order to generate the Knowledge Base for design it is necessary to describe the domain knowledge using the following representations: features as a set of quantifiable attributes [3, 4]; relations as possible correlation between features; rules heuristic learned from experience or derived from generalizations of analytical results, knowledge of analysis methods, knowledge of design problem types, knowledge of design methodologies and knowledge of decision methods.

The traditional CAD technologies do not efficiently perform functional design because no functional reasoning starting from the available qualitative information is systematically implemented. The conceptual design, usually characterized by information that is often imprecise, inadequate and unreliable, is an important stage and has a crucial influence on the structure and efficiency of the final product. Developing methodologies that integrate various stages in total design process, including conceptual design, detail design and manufacture with the aim to reduce production cost and time to market by blending the Artificial Intelligence techniques and CAD/CAE/CAM packages lead to the Intelligent Integrated System (IIS) [1].

This paper proposes an Intelligent Integrated System that links expert system stages to a CAD/CAE/CAM environment to facilitate total design of mechanical systems with gears. In order to take into account the quantitative and the qualitative information about the design of gearboxes, specific algorithms have been generated that allow computerized representation of these data, knowledge and specific programs. The following software is used in this approach: Expert System shell modules (PROLOG flex), Data Base modules (Excel) and a CAD/CAE/CAM Package (CATIA).

2. EXPERT SYSTEMS IN MECHANICAL ENGINEERING

The use of expert systems technology makes possible to incorporate reasoning design knowledge within the CAD/CAE/CAM environments resulting in an Intelligent Integrated System that enables the integration of various stages in a total design process. This could be obtained by further development of CAD/CAE software including a Knowledge Base and Inference Engine modules that are meant to act as design assistants. They contain a vast amount of theoretical and empirical information including access to any detail about materials properties, available parts and relevant design standards. Expert systems modules integrated or linked with CAD/CAE/CAM systems bring the following benefits:

a) a better design and improved overall design performance, being able to consider more features and a larger database,
b) shorter design time and consequently the design costs would be reduced,
c) make knowledge available to a wider audience,
d) can improve safety of human workers,
e) synthesizes the knowledge of many human experts.

In Figure 1 the structure of a complex Expert - CAD/CAE/CAM System is illustrated, which incorporates two main elements: *Knowledge Processing* and *Data and Algorithms Processing*

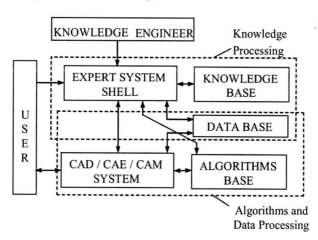

Figure 1. Expert - CAD/CAE/CAM System structure

Knowledge Processing represents the knowledge problem with solutions according to the terminology and experience in the focused domain and fixes the conditions of execution of the data processing by categorizing solutions as being good or not good. Knowledge processing is based on the *knowledge* *model* which groups expertise on explored domain and allows the description and evaluation of the problem in terms of the expert.

The *knowledge modeling* involves two stages: pre-processing and processing. First, the knowledge pre-processing includes the following tasks:

- Generate a topology of *features* as a "semantic group characterized by a set of parameters, used to describe a non-decomposable object useful for reasoning about one or several activities of design" [9]. Features are defined by sets of *attributes* defining various measures of physical characteristics such as load, speed, size, weight, color and form.
- Generate a topology of possible *relations* between features. The definition of a relation includes the possible type of related features, and the relation consequence for related features. A relation is defined by qualitative functions of attributes from the related features.

Secondly, knowledge *representation* involves two steps:

- Generate a set of rules used to create the knowledge problem model that is a translation of the problem data in terms of the features and relations. The rules are in the general form IF <condition> THEN <consequence>, where the condition is in terms of problem data and as a consequence define the number and type of features and relations and/or values for feature attributes.
- Generate a set of *evaluation criteria* rules to quantify, what a good knowledge is, what is a good problem model and what is a good solution for the examined domain. A criterion is defined by a set of attributes from the problem model features.

Thus, the knowledge processing is a transformation of the problem input knowledge, using the rule set, in a knowledge model, in terms of features and relations, which is itself processed and evaluated by criteria functions to obtain the *Knowledge Base* as a computer representation.

A *knowledge engineer* has the job of extracting the relevant knowledge from human expert or existing algorithms and methodologies in literature in a way that can be used by a computer, and also building the knowledge base.

Data and Algorithms Processing is the automatic evaluation of the input data, which contain implicit alternatives and the systematic exploration of them to obtain the desired output data (it is an application of a given algorithm to the input data). The computational model uses the algorithms and work satisfying quantitative constraints and it is defined by:

a) a set of problem variables with different possible values for these variable that define the state of a problem,

b) a set of problem constraints as a relation between the sets of values that have to be satisfied.

The aim of data processing is to find the optimized state of the problem and generate the *Data Base* as sets of input data that are used by the Expert

System Shell to reach a conclusion and by the CAD/CAE/CAM software to reach an optimized solution.

Because the design activities involve finding a number of variables describing the problem it is necessary to use a lot of computation modules. The dimensioning and checking algorithms implemented in specific subroutines are stored in an *Algorithms Base*. The reasoning process uses adequate algorithms that usually imply considering knowledge from the Knowledge Base.

In order to efficiently develop an expert system an independent structure was created named Expert System Shell that usually separates specific domain knowledge from more general aspects about knowledge representation and reasoning. The Expert System Shell programs usually contain the following modules: User Interface, Knowledge Base Editor, Inference Engine and Explanation Subsystem. The *Inference Engine* is used to reason with both the knowledge, typically being in the form of a set of "IF <condition> THAN <consequence>" rules, and specific data base, provided by the user. Partial conclusions, based on this data, are temporally stored in Working Memory, in order to solve the particular problem. The main role of the Inference Engine is to search for the most appropriate item of knowledge and to apply it at any given moment. Using Expert System Shells as PROLOG flex or CLIPS to develop an expert system, generally significantly reduces the cost and time.

3.   DESCRIPTION OF PRODUCTS' MECHANICAL SYSTEMS

Technical systems as independent products, usually include a mechanical system and use qualitative and quantitative flows conversion of input energy, materials and/or information (Figure 2) into similar output entities [6]. Information is more specifically expressed as signals that describe either flows of material or energy. The mechanical systems as distinct units of a product or as independent products may be described as independent structures usually as a black box type diagram like in Figure 3.

Products are defined by their function. A designer typically starts with an objective which is a functional description of the product. A *functional model* is a description of a product in terms of functions and flows that are required to achieve its overall purpose [10, 11]. A functional model represents a concise group of knowledge that in addition contains product and component specific data, such as component to function map, performance specifications and/or customer needs.

The *global (overall) product* function is defined as input/output relationship having the purpose to perform the *overall task* [6]. An overall function may be divided in *subfunctions* corresponding to subtasks. The combination of the subfunctions into overall function produces the *structure function*. The subfunctions may be *main functions* that contribute to the overall function directly, and *auxiliary functions* that contribute to the overall function indirectly. Functions are usually defined by statements consisting of a verb and a noun (e. g. for a simple gear: the main function of the gear is to *transform* the *revolution motion*; the subfunctions that can be derived are: *transmit torque* for each gear, *support* for each shaft, *connect* for each of the shafts to maintain relative distance).

Figure 2. Technical products black box scheme

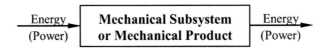

Figure 3. Mechanical products black box scheme

The operation of the overall functions is expressed by *flows* as changes in material, energy or information inside the product and consequently *auxiliary flows* are defined as part of the *main flows*.

The conceptual design phase involves the establishment of functions structures, the search for suitable solutions principles and their combination into conceptual variants. The resulting structures are presented in a graphic form using symbols called *functional diagrames,* as illustrated in Figure 4 where the functional scheme of a gearbox with helical gear is presented.

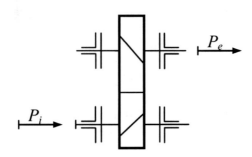

Figure 4. One stage gearbox functional diagram

A mechanical system may consist of any number of components connected in a variety of configurations. The individual components of mechanical systems are configured interactively in order to achieve the desired performance characteristics during operation cycle, considering aspects such as power transmissibility, loading, speed rating, dynamics response and life [2].

The structural decomposition of mechanical systems, at the detail design stage consists in general in identification of the assemblies, subassemblies and individual elements (parts). *Assemblies* are independent functional entities that comprise a number of subassemblies and elements. *Subassemblies* are independent entities (not always with a functional role) that contain a number of other subassemblies and elements grouped and taking in account the assembly and disassembly technologies. The *elements* (parts) are individual entities of subassemblies or assemblies (mechanical systems) connected with other elements in various configurations.

An important characteristic of mechanical systems decomposition, which distinguishes it from product components, is the representation of dependencies between assemblies, subassemblies and elements (parts). The subassemblies and elements are in physical contact through a large variety of direct mechanical contact type and sometimes are joined by intermediate elements (keyed joints, pins, gripsprings). For instance, in Figure 5 a gearbox with one stage is presented. Because this product has only one functional unit, it may be considered as a distinct assembly and a possible first level decomposition for this product is presented in figure 6. Thus, the subassemblies S1 (input shaft), S2 (output shaft), S3 (principal housing), S4 (secondary housing), S5 (input casing cover), S6 (output casing cover) and elements 1 and 2 (casing cover) are identified.

The structure of a complex mechanical system that contains a number of assemblies, subassemblies and elements can be visualized on an *assembly scheme*. First, a high level decomposition is possible for an assembly containing large subassemblies and elements. In the next steps it is possible to decompose the subassemblies identified in the *subassembly scheme*. The assembly and subassembly schemes are configured as a system template that describes the order of connectivity between the elements.

In Figure 7 the assembly scheme of the gearbox from Figure 5 is presented. The divisible subassemblies are symbolized with boxes, the elements with circles and their connections with straight lines. According to their connections and function performed, they are evaluated and classified as unitary, binary and multiple elements. The unitary elements have only a single connection and usually have an auxiliary function that does not participate on the flow. The binary elements possess two connections with other mechanical elements and transmit a principal or auxiliary flow.

Multiple elements are main elements that possess more than two connections and involve full data propagation from all connected elements. Also, a connection between two elements may be single, double or multiple and are visualized with one, two or three cross lines.

On the other hand, the components (subassemblies and elements) of mechanical systems may be *specific components* that are used only for this product, and *specialized components*, which have a general overall function.

The decomposition has to be continued at inferior levels and it is stopped when the resulting subassemblies are specialized and are non-divisible for the respective product case. Thus, in the lowest level assembly scheme all components symbolized using circles are entities that are designed separately.

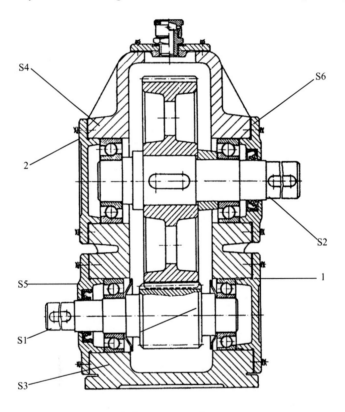

Figure 5. One stage gearbox structure

In Figure 8 the decomposition of the subassembly S2 is presented and the following elements are obtained: wheel (3), shaft (4), ring (5), and specialized subassemblies ball bearings (S7, S8). In the subassembly scheme, associated with this decomposition, presented in Figure 9, the input and internal links between these components are shown.

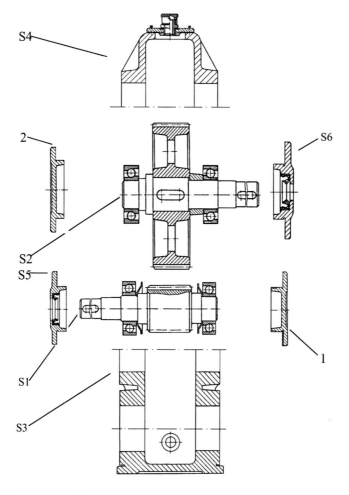

Figure 6. Components (subassemblies and parts) of one stage gearbox structure from Figure 5

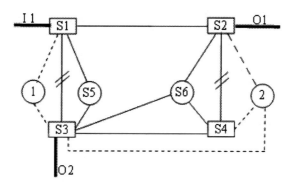

Figure 7. Assembly scheme of gearbox from Figure 5

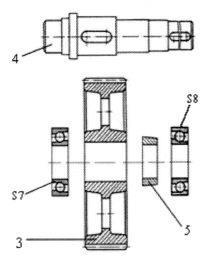

Figure 8. Components (subassemblies and parts) of subassembly S1 (input shaft)

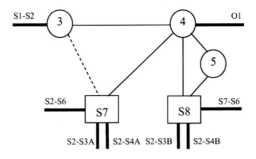

Figure 9. Scheme of subassembly S1 (input shaft)

Table 1 describes the components of first level of decomposition presented in Figure 5 and Figure 6 and with scheme from Figure 7. The external connections (input and output) and internal connections between components, are described in Table 2.

Table 1. Components of first level decomposition of gearbox assembly

No	Cod	Name of component	Type of component	Main and auxiliary function
1	S1	Input shaft	Specific	Transmit motion and forces
2	S2	Output shaft	Specific	Transmit motion and forces
3	S3	Principal housing	Specific	Transmit forces
4	S4	Secondary housing	Specific	Transmit forces
5	S5	Input casing cover	Specific	Transmit forces and mobile tightening
6	S6	Output casing cover	Specific	Transmit forces and mobile tightening
7	1,2	Casing cover	Specific	Transmit forces and fix tightening

Table 2. Connections of first level decomposition of gearbox assembly

No	Cod	Type of joint	Physical parameters
1	I1	Contact on cylindrical surface contact and key joint	Input rotational speed Input torque
2	S1-S2	Line contact and gearing	Gearing speed. Gearing force
3	S1-1	Contact on plane surface	Axial surface
4	S1-S3A	Contact on cylindrical surface contact	A Radial force
4	S1-S3B	Contact on cylindrical surface contact	B Radial force
5	S1-S5	Contact on plane surface	Axial force
6	S2-S4A	Contact on semi-cylindrical surface	C Radial force
7	S2-S4B	Contact on semi-cylindrical surface	D Radial force
8	S2-S6	Contact on plane surface	Axial force
9	S2-2	Contact on plane surface	Axial force
10	S3-1	Contact on plane surface and screwed joints	Axial force
11	S3-S5	Contact on plane surface and screwed joints	Axial force
12	S3-S6	Contact on plane surface and screwed joints	A Axial force
13	S3-S4	Contact on plane surface and bolted joints	Resultant force
14	S3-2	Contact on plane surface and screwed joints	A Axial force
15	S4-S6	Contact on plane surface	B Axial force
16	S4-2	Contact on plane surface	B Axial force
16	O1	Contact on cylindrical surface contact and key joint	Output rotational speed Output torque
17	O2	Contact on plane surface and screwed joints	Support resultant force

Also, in Tables 3 and 4 the components and associated connections of S2 subassembly are presented.

Table 3. Components of S2 subassembly

No	Cod	Name of component	Type of component	Main and auxiliary function
1	S7	Output ball bearing	Specialized	Sustain mobile elements and transmit forces
	S8	Output ball bearing	Specialized	Sustain mobile elements and transmit forces
2	3	Gear	Specific	Transmit motion and forces
3	4	Output shaft	Specific	Transmit motion and forces
4	5	Bushed	Specific	Transmit force

Table 4. Connections of S2 subassembly

No	Cod	Type of joint	Physical parameters
1	3-4	Contact on cylindrical surface and key joint	Intermediate speed Intermediate torque
2	S7-4	Contact on semi-cylindrical surface	A Radial force
3	S8-4	Contact on plane surface	B Radial force
4	4-5	Contact on plane surface	Axial force
5	S8-5	Contact on plane surface	Axial force

Using the assembly and subassembly schemes it is possible to decompose complex engineering design problems into a manageable number of sub-problems that can be solved independently [4]. These schemes can be implemented using the Product Function Definition module of the CATIA package (Figure 15).

4.   DESIGN OF GEARBOXES' MECHANICAL SYSTEMS

4.1   Formulation of Product Definition Specification

The Product Definition Specification (PDS) consists in a set of product attributes (inputs, outputs) and constraints associated with the product and optimization goals [7]. The PDS attributes can be included in three groups of information:

–   Requirements e.g. transmission power, input revolute motion, overall speed ratio, orientation of inputs/outputs shafts and distance between centers (Figure 10).
–   Evaluation criteria: size, manufacture cost, ease of manufacture, life, etc.
–   Environment: temperature, humidity, etc.

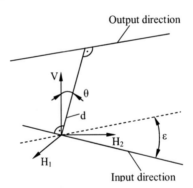

Figure 10. Layout of input/output

The constraints refer to the possible variation of attribute values and are represented as a set of relations among the attributes. Three types of constraints can be defined: relational constraints, causal constraints and spatial constraints. Relational constraints are direct function attributes (e.g. the speed increases by a factor 5.5).

The PDS is an evolutionary and dynamic document that contains all aspects of product such as functional requirements. The input module allows adding and modifying PDS items. PDS can be implemented using devoted packages like CATIA Knowledge Advisor module and Knowledge Expert.

4.2   Conceptual design

For the mechanical systems with gears four basic types of function have been derived. They are related to the concepts of :

POWER (electrical, mechanical, hydraulic, pneumatic) that involve action like transmit, dissipate, modify;

MOTION (rotary, linear, oscillatory) associate with verbs: create, convert, modify, transmit;

FORCES (torque, linear forces, elastic forces) that can be: generated, modified, transmitted, adjusted;

AUXILIARY (enclosure) that use the following action: cover, view, protect, support, attach, connect, guide, limit, aerate, ventilate and measure.

In order to explicitly describe the conceptual approach one has to identify:

a) The arrangement (position) of axes (parallel/rectangular orientation of input/output shafts (Figure 10):

IF $\varepsilon = 0$ and $d = 0$ THAN the structure is with identical input/output direction,

IF $\varepsilon = 0$ and $d > 0$ THAN the structure is with parallel direction

IF $\varepsilon = 90$ and $d > 0$ THAN the structure is with spatial rectangular direction,

IF $\varepsilon = 90$ and $d = 0$ THAN the structure is with planar rectangular direction.

b) The number of stages and types of gears (spur, helical, double helical, bevel, hypoid, Figure 11):

IF the structure is with parallel direction and overall speed ratio < 5 THEN the structure is with one stage and the structure of gears is spur or helical,

IF the structure is with identical direction and overall speed ratio < 5 THEN the structure is with two stages and the structure of gears is spur or helical,

IF the structure is with planar rectangular direction and overall speed ratio < 4 THEN the structure is with one stage and the structure of gears is bevel,

IF the structure is with identical direction and overall speed ratio < 5 THEN the structure is with two stages and the structure of gears is spur or helical,

IF the structure is with parallel direction and overall speed ratio > 5 THEN the structure is with two stages and the structure of gears are spur and helical.

c) Functional scheme:

IF $\theta = 0$ THEN the structure has a vertical axis with input above of input

horizontal plane,
IF the structure is with two stages and the gears are spur and helical
THEN the first stage is with helical gears.

After the mechanical system identification it is necessary to populate it
with components (assemblies, subassemblies and elements) and then to check
its overall performance capability.

4.3   Detailed design

The intelligent functional reasoning strategy in the design of mechanical
systems is based on matching the physical behavior with a desired
function/behavior corresponding to the functional requirements of a
component. Selections of the assemblies and subassemblies reveal the
possible couplings and matching with respect to their performance
capabilities and possess a synergy effect that achieves the desired overall
performance parameters. Thus, the designer has to find the components in
order to ensure that they are compatibles, having in view their performance
capabilities and connectivity parameters (speeds, loadings, geometric
attributes).

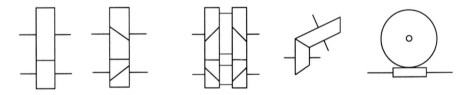

Figure 11. Gears types

Components (assemblies, subassemblies or elements), are selected to
perform the specific subfunctions. Each of these subfunctions can become a
root function for a given subsystem.

The detailed design stage includes calculation of design parameters,
selection of material and fabrication method (process), determination of
shapes, dimensions, and tolerances of product and its components, and
drawings of the product and its components

Starting from the functional diagram and from the function structure, the
embodiment diagrams are established that contain the first level of
subassemblies and elements only defined as symbols. Thus, for the gearbox
with functional diagram from Figure 4 and with the structure from Figure 5
the embodiment diagram presented in Figure 12 is derived. This diagram is
structured taken into account decomposition into subassembly and elements,
presented in Figure 6, and included in assembly diagram from Figure 7.

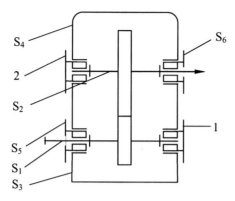

Figure 12. Embodiment diagram of one stage gearbox

The operation of a function structure is expressed by *flows structure* as changes of power components (motion and forces) inside of the product. In Figure 13 the flow structure associated to the embodiment diagram (Figure 12) is presented mentioning *main flows*:

a) input rotational speed →S1 →gearing speed →S2 →output rotational speed,
b) input torque →S1 →gearing force →S2 → output torque, and auxiliary flows:
c) S1 → A, B Radial force → S3 → Support force,
d) S1 → Axial force → S5 → Axial force → S3 → Support force etc.

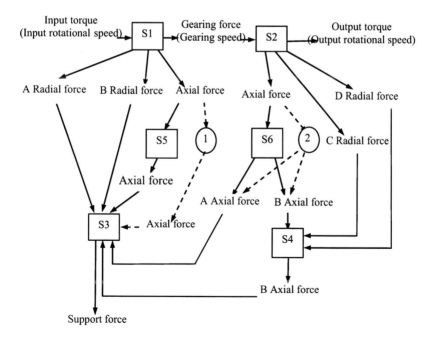

Figure 13. Flow structure of gearbox first level decomposition

The main and auxiliary flows of the associated subfunctions of subassemblies can be showed separately. For instance, in Figure 14 the flows inside of the subassembly S2 are presented:

- Gearing speed → 3 → Intermediate rotational speed → 4 → Output rotational speed;
- Radial force → 4 → C Radial force → S7 → C Radial force → D Radial force → S8 → D Radial force;
- Axial force → 4 → Axial force → 5 → Axial force OR Axial force → 4 → Axial force → S7 → Axial force.

The assembly and subassembly diagram and flows diagram using CATIA Product Functional Definition module can be considered as a start for a specialized Intelligent Integrated CAD system. In Figure 15 the functional scheme is presented using the CATIA module. The user identifies the functional objects involved in the product (seen as a system) and the interactions between them. Furthermore, the user can define several associations and several links to the process of product design items of any kind: part, sketch, wireframe element (in a sketch, rough design, preliminary design, detailed design, etc.). Consequently, more flexibility and more accuracy are given to the user, who can fully define functionally and physically a product in the CAD environment.

Finally, for each indivisible subassembly, a component can be defined and three parameter groups: geometrical, physical and financial. For instance, for a rolling bearing these groups can contain:

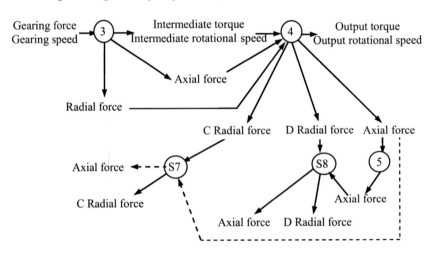

Figure 14. Flow structure of subassembly S1 (input shaft)

a) geometrical parameters: inner diameter, out diameter, width;
b) physical parameters: radial force, axial force, inner speed, outer speed;
c) financial parameters: price, guaranty, service.

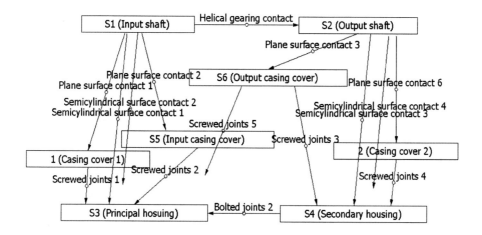

Figure 15. CATIA assembly diagram

Also, for each joint that is associated with a link between subassemblies and elements the similar parameter groups can be defined and evaluated. For instance, for a helical gearing joint the characteristics parameters are:

a) geometrical: tooth width, module, pressure angle, helix angle, number of teeth in the pinion, number of teeth of the wheel, addendum coefficient for teeth length modification, addendum modification coefficients, rack tip radius;

b) physical: speed ratio, normal force, radial force, axial force, velocity;

c) financial: manufacture cost, lubricant cost.

The associated data from components and joints can be modeled using Knowledge Advisor module and Knowledge Expert module available in CATIA. The parameters of a component or a joint have known values (input parameters) or unknown values (output parameters). Usually, the number of unknown parameter is greater than the number of known physical equations. Thus, some of the parameters have to be selected based on quantitative and/or qualitative information.

In order to determine the unknown parameter values for each type of component (subassembly, or elements) and for each type of joint a calculus algorithms or selection algorithms are associated (in the case of specialized components). The calculus tasks that are performed in detail design phase include: gear strength analysis, bearing selection, shaft design, design optimization, parametric design of parts, part and assembly modeling, finite element analysis. Also, the calculus algorithms are implemented using Knowledge Advisor module available in CATIA.

Furthermore, in the case of a contradictory constraint it is possible to use optimization algorithms. For instance, using the Product Engineering optimization module available in CATIA in the case of gearbox design it is possible to solve a model that has as objective function the minimization of the weight structure of the helical gear.

The basic algorithms contain a program module that is included in the knowledge process using production rules as shown below.

IF goal is *shaft design* THEN (execute *Preliminary Design* module AND execute *Forces* module AND execute *Dimensioning* module AND execute *Check* module AND execute *Solid modeling* module AND execute *Finite Element Analysis* module AND execute *Redesign* module)

IF goal is *gear design* THEN (execute *Preliminary Design* module AND execute *Optimization* module AND execute *Check module* AND execute *Output data* module)

IF goal is *bearing selection* THEN (execute *Selection* module AND execute *Calculus* module AND execute *Assembly* module)

Many modules used have traditional algorithms that are based on data but they can be based on knowledge as well. For instance, the module *Selection* use reasoning with qualitative knowledge [5]. In order to build up and share the knowledge stored in rule bases and to use it to ensure design compliance with established standards the Knowledge Expert module from CATIA was used.

The general algorithm to process design is used to perform the following algorithms: Gears design, Wheel design 1, Wheel design 2, Shaft design 1, Shaft design 2, Bearing design 1, Bearing design 2, Housing design 1, Housing design 2, Plane surface contact joint design, Semi-cylindrical surface contact design, Screwed joint design, Bolted joint design, Assembly design. In Figure 16 the solid model associated with the subassembly input shaft including finite element analysis results are presented.

Figure 16. Structure of S1 subassembly including finite element analysis results

5.   CONCLUSIONS

The Intelligent Integrated System approaches has been developed to integrate various stages in the total design process (conceptual design, embodiment design, detail design and manufacture) in order to reduce time to market and to decrease product costs.

The automatic design of mechanical systems, including design of process algorithms, knowledge processing, and data processing was integrated in a commercial CAD/CAE/CAM package.

Structural decomposition of mechanical systems as an independent formal method allows computer implementation of an intelligent system that integrates multiple cooperative knowledge and data. Thus, during the design process it is possible to ask and receive information and details about the adopted solutions.

The automation of the embodiment process enables the designer to consider more configurations, components types within a shorter time and allows the development of better solutions which might not have been obtained if the traditional procedures were used. It also provides optimization algorithms which use objective functions and constraints to increase product performance.

A further benefit of this approach is that the design model can be permanently modified taken into account changed supplier or different component specification.

On the other hand the intelligent design system can be used in education: teaching students how to design is a difficult task; enabling them to clearly distinguish and work with elemental functions, conceptual design, embodiment design and detailed design becomes much easier.

ACKNOWLEDGEMENTS

This work was supported by the European Commission under the Growth programme scheme, through the Project ADEPT, contract no. G1MA-CT-2002-04038/ADEPT, for which the authors express hereby their full gratitude.

REFERENCES

1. S. DAIZHONG, *Design Automation with the Aids of Multiple Artificial Intelligence Techniques*. (Concurrent Engineering: Research and Applications, vol 7, No. 1, 1999, pp. 23-29).

2. B. J. HICKS, S. J. CULLEY, *An integrated modeling of environment for the embodiment of mechanical systems* (Computer-Aided Design, ELSEVIER .vol 34 (2002) pp. 435-451).

3. C. F. KIRSHMAN, G. M. FADEL, *Classifying Functions for Mechanical Design* (Journal of Mechanical Design (1998), vol. 120, pp 475-482).

4. A. KUSIAK, N. LARSON, *Decomposition and Representation Methods in Mechanical Design.* (Transactions of the ASME. Special 50th Anniversary Design Issue (1995), vol. 117, pp 17-23).

5. G. MOGAN, E. V. BUTILA, *Expert Systems in Design of Mechanical Systems.* (Part II. Design of Rolling Bearings Subsystems. Ovidius University Annals of Mechanic Engineering – Tehnonav 2004, vol.6, 2004).

6. G. PAHL, W. BEITZ, *Engineering Design. A Systematic Approach.* (Springer-Verlag, 1988).

7. U. ROY, N. PRAMANIK, R. SUDARSAN, R. D. SRIRAM, K. W. LYONS, *Function to form mapping: model representation and applications in design synthesis.* (Computer Aided Design 33, 2001, pp 699-719).

8. Y. SHIMOMURA, M. YOSIOKA, H. TAKEDA, Y. UMEDA, T. TOMIYAMA, *Representation of Design Object Based on the Functional Evolution Process Model.* (Journal Of Mechanical Design, 1998, vol. 120, pp 221-229).

9. F. SPRUMONT, P. XIROUCHAKIS, *Towards a Knowledge-Based Model for the Computer Aided Design Process* (Concurrent Engineering: Research and Applications vol. 10, 2, 2002, pp 129-141).

10. B. R. STONE, K. L. WOOD, *Development of a Functional Basis for Design.* (Journal of Mechanical Design, vol. 122, 2000, pp 359-369).

11. W. Y. ZHANG, S. B. TOR, G. A. BRITTON, Y. M. DENG EFDEX, *Knowledge-Based Expert System for Functional Design of Engineering Systems.* (Engineering with Computers, 2001, 17, pp 339-353).

VIRTUAL DESIGN PROTOTYPING UTILIZING HAPTIC IMMERSION

U. Künzler and M. Iseli
Berne University of Applied Sciences, Berne, Switzerland

Abstract: This paper presents a description of haptic technology and its applicability to free-form design prototyping. In particular we report on the results of an international EU/IMS research project to develo3p an interactive Immersive Design Application (IDA) using Virtual Reality and Haptics immersion technologies. The application uses a virtual sculpting metaphor to interactively define arbitrary formed shapes as they are used in the domain of industrial design prototyping. The paper concludes with the presentation of the results of an IDA prototype usability evaluation through a team of design experts.

Key words: virtual sculpting, haptics, free-form design prototyping, virtual reality, human computer interaction.

1. BACKGROUND

The research results presented in this paper have been elaborated within the context of an EU/IMS research project for the development of a Configurable Virtual Reality System for Multi-purpose Industrial Manufacturing Applications (IRMA). The four year project involved international partners from Europe, the Newly Associated States of Eastern Europe (NAS), Japan and Switzerland and has been successfully completed in February 2004. The objective of the project was to develop configurable low-cost, PC-based generic modules, which will integrate and enhance existing technology with modeling and simulation tools to create a suite of industrial Virtual Reality software solutions for primary use in the Small to Medium Manufacturing Enterprise [4].

The Swiss contribution to the IRMA project finds itself at the very beginning of the manufacturing process chain. In collaboration with an industrial design partner (Iseli Design und Partner AG) the Berne University

D. Talabă and T. Roche (eds.), Product Engineering, 163–174.
© *2004 Springer. Printed in the Netherlands.*

of Applied Sciences has developed an interactive Immersive Design Application (IDA) for free-form industrial design prototyping using Virtual Reality and Haptics immersion technologies.

2.   VIRTUAL DESIGN PROTOTYPING

The rational to develop a virtual design prototyping application is due to the fact, that the typical design prototyping process currently does not incorporate modern software systems. We therefore started our research with an analysis of the conventional design prototyping and the industrial design requirements needed for an effective application of software tools for design prototyping.

2.1   Conventional Design Prototyping

For initial, conceptual work, most industrial designers still prefer to use only paper and pencil to sketch a number of drawings ("Scribbles") for the initial design of a new consumer electronic or home appliance device. Although these methods provide a very direct and immediate response when designing a new shape, the resulting Scribbles are only 2D and therefore are very limited in their 3D expressiveness and offer no 3D viewing and manipulation options. As a result, the designer additionally often needs to hand-build a foam model for design evaluation with customers and prospective users of the envisioned product. Rapid design prototyping thereby becomes very time consuming and expensive. Furthermore it is almost impossible to accurately digitize these design sketches or the foam-based model into a standard CAD/CAM system that will be used to engineer and manufacture the product.

A recent study of a Japanese research team conducted a quantitative analysis of different tools used for the initial design process. Their study confirms that designers using paper and pencil for design prototyping are about ten times as productive as designers using a conventional CAD system, regarding the number of design ideas evaluated within a given time [5]. This shows that designers are quite reluctant to move away from the paper and pencil approach, because using a CAD system for design prototyping is less productive and it rather hinders initial conceptual design creativity.

2.2   Industrial Design Requirements

For a software system to effectively support the initial design prototyping phase, it must be highly interactive and needs to provide powerful functions and tools for 3D shape visualization and manipulation, but nevertheless should be similar in usage and feedback to conventional tools a designer is familiar with. Additionally such a free-form design application must fit seamlessly into the product manufacturing process by simple integration into existing CAD/CAM environments. Maintenance and setup efforts needed for such a design system must be comparable to a standard PC system and the total purchase price must be within an SME affordable range.

Besides these general system requirements, the functional software requirements have been elaborated through a comprehensive Use Case Analysis, where we conducted a detailed case study for the actual design of a wireless phone. The results of this analysis and a description of the individual use cases with their categorization can be found in an earlier paper [3].

2.3   The need to utilize Virtual Reality and Haptics

As our requirements analysis proved the need to provide a software design environment for realistic 3D shape visualization, it was obvious to use Virtual Reality (VR) techniques for state of the art 3D graphics display. As VR technology has reached a very mature level, it is very well suited for real industrial applications which require a high level of robustness.

In order to improve the designer's 3D response and shape perception our requirements analysis identified the additional need to utilize haptic force feedback. This allows for intuitive and more accurate interaction control while the designer is creating, manipulating and refining 3D free-form design shapes. Furthermore it permits us to provide additional helper functions such as the easy drawing of planar objects in virtual 3D space or a precise 3D point or line snapping mechanism. Through recent advances in Haptics technology and the wider availability of haptic devices, prices are getting in a range which makes this technology ready for mainstream industrial applications.

To effectively use Virtual Reality and Haptics for free-form design tasks, these technologies must be seamlessly integrated into one multimodal Virtual Environment (VE) specifically optimized for Virtual Design Prototyping. To achieve a high level of user acceptance, such a system must be very intuitive to use even for occasional users having only little training and VE experience.

After careful evaluation, we decided to use SensAble's PHANTOM Desktop product (www.sensable.com), which has a stylus-shaped tool for haptic interaction and therefore closely resembles the handling of a pencil a designer is used to work with. This device provides three degrees of freedom

(DOF) haptic force feedback and at the same time serves as a mechanical six DOF tracker which delivers very accurate 3D position (~0.023 mm) and orientation input, needed for precise shape drawing and manipulation.

3.   IMMERSIVE DESIGN APPLICATION (IDA)

Based on our requirements analysis we elaborated the software specification for the development of an interactive Immersive Design Application (IDA). For seamless VR and Haptics integration the software architecture needed to provide a uniform visual and haptic scene graph rendering structure. The focus of our research was to develop a suite of virtual tools for the interactive creation and manipulation of free-form design shapes, which allow for intuitive and efficient design prototyping within a virtual design workplace.

3.1   Immersive Virtual Design Workplace

To increase a designer's 3D shape perception and to ease navigation and tools operation within a three-dimensional work environment our goal was to develop a Virtual Environment, where the user would be visually and haptically completely immersed into an interactive virtual design workplace.

Special attention was given to software usability aspects, which led us to the conclusion to use a haptic enhanced Virtual Sculpting interaction metaphor for free-form design prototyping. Through the availability of a broad range of virtual design tools with haptic interaction behavior, the designer is able to interactively create, manipulate and refine 3D shapes while having enhanced tool operation control through haptic feedback.

In order to realize a complete visual and haptic immersion work environment, the use of a conventional 2D graphical user interface with its 2D interaction concepts would not have been appropriate for 3D virtual environment interaction. We therefore needed to develop a novel 3D Immersive User Interface (IUI) widget set. Only such an interface allowed us to fully immerse the user into the envisioned virtual design workplace. The following chapters give a description of the developed IUI widgets and their interaction concepts.

3.2   Immersive User Interface

The basic virtual design workplace interaction widget is made of a *Z-Depth Adaptive Menu System* for convenient 3D accessibility. Depending on the current stylus z-position, the menu bar is faded in just slightly behind the stylus. Thereby the user's 3D stylus movement is minimized and menu interaction becomes more efficient (Figure 1).

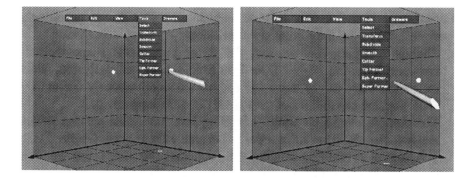

Figure 1. Z-Depth Adaptive Menu System

The menu buttons themselves not only have a visual representation but also provide the haptic feedback of a touchable button. To activate a specific menu function the user needs to use the stylus to physically press the button with the chosen function. As menu buttons are visually transparent and real 3D haptic objects, they can be operated from the front or the back, depending which is more convenient in regard to the user's current stylus position. The menu bar appearance or disappearance is triggered through pressing an invisible, but touchable button, which extends over the entire ceiling of the virtual design workplace. The menu bar can also be operated through the cursor keys and most menu functions are also accessible through keyboard shortcuts.

The virtual design workplace boundaries are visualized through a so called *Cube Grid* with magnetic haptic walls extending along the -x, -y, -z planes of the workplace (Figure 1). The grid's position and orientation can be freely changed as does the workplace and the design primitives contained therein. Additionally the haptic planes may be used for planary drawing or alignment tasks.

To start a new design, the user can choose from a library of predefined *Design Primitive (DP)* shapes (sphere, box, pyramid, cylinder, etc.), which are selectable through the "Drawers" menu. Alternatively the user can also import an existing CAD design, either in STL or VRML data format, which gets converted into the IDA internal triangle surface mesh representation.

3.3   General Visual and Haptic Helper Functions

To ease 3D virtual environment user interaction we developed a variety of different visual and haptic helper functions to support the user's general

IDA operation tasks:

a) Every shape is surrounded by a yellow Proximity Bounding Box (Figure 2). It appears only for one design primitive at the time. Thereby the proximity bounding box indicates which particular design primitive is closest, regarding the current stylus position.

b) Through grasping (i.e. pressing the stylus button) the shape with the proximity box, the Haptic Transform Manipulator gets activated, which means the shape can be translated and rotated simultaneously due to the 6 DOF tracking capabilities of the PHANTOM Desktop device.

c) The corners of the proximity bounding box are visual spheres with a magnetic haptic behavior in order to become easy to grasp (Figure 2). By grasping one of the bounding box spheres, the shape can be non-uniformly scaled until the desired distortion is achieved.

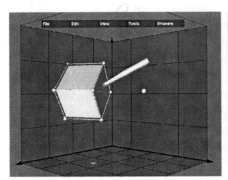 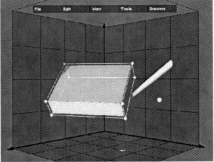

Figure 2. Haptic Transform Manipulator for non-uniform object scaling

d) The surface of each DP has a touchable checkbox button behavior. By simple pressing of the shape surface, the DP can be intuitively switched from transform mode to select mode, which is used for subsequent drawing or refinement operations. Multiple shapes can be selected at the same time, which is indicated through a green Select Bounding Box.

e) The transparency of each object (i.e. DP shapes and IUI widgets) in the virtual design workplace is carefully adjusted to minimize potential mutual object obstruction which is often apparent in a 3D virtual environment.

f) For improved visual 3D orientation and navigation the PHANTOM Desktop stylus tip is projected onto the x-, y-, z-planes of the cube grid, where three position markers are displayed and continuously updated in "real-time" as the user moves the stylus around.

3.4 Virtual Sculpting Tools

To implement the Virtual Sculpting metaphor, we developed an assortment of Virtual Sculpting Tools the user can choose from for a certain task to be applied to the currently selected shape(s). Each virtual tool has its specific visual representation and haptic behavior, which is defined by the purpose and mechanism of interaction of the tool. The following enumeration describes some of the IDA virtual tools and shape manipulation functions:

- The *Tip-Former Virtual Tool* (Figure 3) is used for point like deformation of a design shape. By grasping a selected shape with the tip-former, the surface mesh can be pushed or pulled whereas a continuously recalculated haptic feedback is applied to the tool. The force feedback is proportional to the deformation displacement and therefore gives the user a sense of the deformation extent. The surface model used for this operation is based on a spring-mass mesh model, which permits a very fast and continuous "real-time" update of the visual shape geometry.

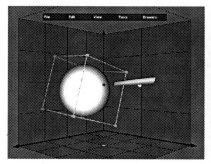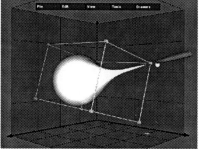

Figure 3. Tip-Former Virtual Tool with continuous update of visual geometry

- The Mesh Smoothing Function is used for uniform smoothing refinement of the shape surface. The function is typically used to remove unwanted sharp edges or spikes after a deform manipulation. The smooth algorithm repeatedly moves the mesh vertices to the center of gravity of their neighbors. By applying this refinement function multiple times, the user is able to control the degree of mesh surface smoothing as needed. The mesh topology and subdivision level is not altered during this operation.
- To get accurate shape deformation or manipulation results, the density of the mesh (i.e. the number of triangles) is a crucial factor for those operations. Through the implementation of a *Uniform Mesh Subdivision Function* the number of mesh triangles can be gradually increased by replacing every triangle through four newly inserted triangles such that the overall shape of the mesh remains unchanged.

- The *Adaptive Mesh Subdivision Function* (Figure 4) is used for selective mesh refinement and works very similar to the uniform mesh subdivision algorithm, except that the subdivision is only applied to certain triangles. Based on the length of each triangle perimeter only those exceeding a predefined threshold are subdivided. This adaptively increases the mesh density on surface regions where the deformation has been excessive. Careful attention must be given to the boundaries of the subdivision regions as the mesh topology needs to remain connected for the partially inserted triangles. Using an adaptive subdivision algorithm keeps the overall number of mesh triangles low, which improves user interaction responsiveness and visual and haptic rendering performance.

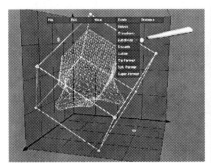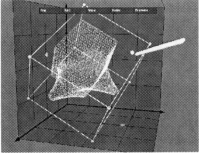

Figure 4. Adaptive Mesh Subdivision Function

- The *Mesh Cutter Virtual Tool* (Figure 5) is used for planar mesh cutting operations. The cutter tool displays a large helper plane so that the user can visually adjust the position and orientation of the shape part to be cut away. The cut plane itself is not limited through the size of the visual helper plane, but rather has an infinite extension, which is shown through continuous calculation and display of the actual cut line intersecting a shape as the user moves the virtual cutter tool around.

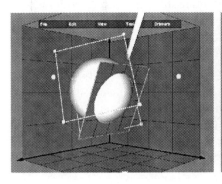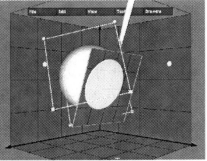

Figure 5. Mesh Cutter Virtual Tool with continuous update of cut line

- *Boolean Mesh Operations* (Figure 6) are used for boolean aggregation of design primitives. The IDA provided boolean functions are *Union*, *Difference* and *Intersection* of shapes. These functions only work if exactly two design primitives are selected and intersect each other. In the case of a boolean difference operation, the two shapes need to be selected in the appropriate order.

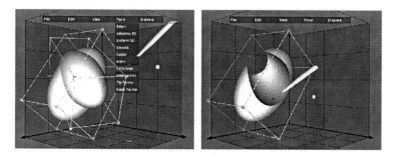

Figure 6. Boolean Difference Operation of two design primitive objects

- The *Super-Former Virtual Tool* is used for shape deformation and is based on a superellipsoid equation (1) [2].

$$\left(\left(\frac{x}{a_1} \right)^{\frac{2}{\varepsilon_2}} + \left(\frac{y}{a_2} \right)^{\frac{2}{\varepsilon_2}} \right)^{\frac{\varepsilon_2}{\varepsilon_1}} + \left(\frac{z}{a_3} \right)^{\frac{2}{\varepsilon_1}} = 1 . \tag{1}$$

Through user controllable variation of the ε_1, ε_2 equation parameters, the geometry of the super-former tool can be continuously changed to create a multitude of tool shapes.

An additional user controllable parameter allows the smooth scaling of the tools size. For convenience, a set of predefined parameter settings is stored to easily recall a specific shape through keyboard function keys (Figure 7).

Figure 7. Some Shapes of the Super-Former Virtual Tool

The shape of the super-former tool is used for the subsequent geometry dependent deformation of a design primitive's mesh surface (Figure 8). Based on a superellipsoid equation function, we are able to use a tool geometry independent collision detection algorithm, which is used for the surface mesh vertex displacement operation of the super-former tool. A haptic restriction is used to prevent the user from penetrating the design primitive thereby restricting the tool interaction to the mesh surface.

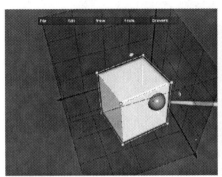 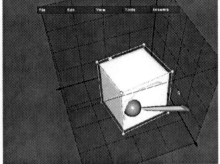

Figure 8. Spherical Super-Former Tool shape deformation

3.5   Implementation Technologies

The Immersive Design Application (IDA) is designed and built as an object oriented C++ application, which allows for easy extensibility and maintenance. The IDA system uses the Reachin API (www.reachin.se) software library as its basic virtual reality framework. This library provides an object oriented C++ interface for unified and synchronized graphic and haptic scene graph rendering. By use of the OpenGL graphics interface, high performance graphics hardware can be fully utilized for rendering and stereo display. The Reachin API is highly extensible by enhancing existing shape or surface nodes through customized geometry or special haptic surface effects. As the library internal scene graph representation is very similar to the VRML scene graph model, import of VRML data is readily supported.

The IDA internal data model for design shapes is represented through a triangle manifold mesh data structure. For the implementation of this data structure we used the OpenMesh library developed from the Computer Graphics Group at the RWTH Aachen [1]. This object oriented library is implemented in C++ and can easily be customized for triangle meshes. It is built upon a highly efficient half-edge data structure, which provides very fast access to mesh items (i.e. vertices, edges and faces) through the use of special iterator and circulator functions.

4.   DESIGN EXPERT USER EVALUATION

The IDA expert evaluation has been conducted by a team of designers from Iseli Design und Partner AG. The evaluation methodology used, was based on a practical, qualitative software usability study. The system evaluation lasted one week and was carried out by five subjects, two women and three men, with different expertise in using IT technology and CAD modeling systems. A typical IDA design workplace system was setup, which used a PHANTOM Desktop device for haptic feedback and input tracking and a 20" CRT monitor with infrared controlled CrystalEyes shutter glasses from StereoGraphics for VR stereo display. The designers were intentionally given no training or instructional manuals for the system evaluation experiment, which was conducted in three steps:

a) Free experimentation with the application (warm up).
b) Solving of specific free-form design modeling problems.
c) Assessment of the above steps by predefined criteria.

The assessment criteria used for the evaluation were hardware-ergonomics, software-ergonomics in general and the Immersive User Interface (IUI) and haptic interaction mechanisms in particular.

The results raised some concerns about the hardware ergonomics, especially for the long term use of the haptic device and some visual congruence problems of the used VR stereo display system.

Concerning the area of software ergonomics, all subjects were able to use the virtual tools immediately and without a manual. The selection of design primitive objects and their manipulation has been evaluated to be very user-friendly. Navigation in three-dimensional space is very good, even without shutter glasses, through the stylus-tip projection. Additionally, several suggestions for usability improvements of IUI widgets and tools have been made, which will be considered for a future release of the IDA system.

The expert evaluation concludes that the IDA system can be handled very easily and intuitively. The interactive Virtual Design Workplace and the provided virtual sculpting tools allow a designer to explore, evaluate and change new 3D free-form shape variants within one integrated environment, similar to the conventional design prototyping process ("VR Scribbles").

Through the ease of use and the multimodal visual and haptic feedback, this application for the first time not only supports but also inspires the creativity process of an industrial designer. Optimized through the suggested improvements, the IDA system will be a tool which makes the process of industrial design prototyping significantly easier and more effective. The time and costs needed for rapid design prototyping and the consecutive CAD refinement will be considerably reduced.

5.   FUTURE WORK

Besides the expert evaluation suggested usability improvements for existing widgets and virtual tools, several enhancements are planned for a future release of the IDA system. Some of these are the extension of the IUI widget set and the refinement of the existing Virtual Sculpting Tools through a surfaces region selection function. This will allow applying available operations only for a selected surface region of a design primitive. Additional virtual tools such as a mesh mirror shall also be implemented. The runtime performance of boolean operations needs to be improved through the use of a BSP-Tree data structure. And finally the mesh quality for shape intersection operations shall be enhanced through the elimination of degenerated triangles.

ACKNOWLEDGEMENTS

The work presented in this paper is supported by Intelligent Manufacturing Systems (IMS), Project 97007: Project IRMA - "Virtual Reality for Manufacturing Applications" and sponsored within the EU-FP5 Program under contract G1RD-CT-2000-00236 and from the Swiss Federal Office for Education and Science (OFES) under contract BBW 99.0663. The authors wish to acknowledge the European Commission, IMS and the OFES for their support.

REFERENCES

1. M. BOTSCH, S. STEINBERG, S. BISCHOFF, L. KOBBELT, *OpenMesh - a generic and efficient polygon mesh data structure* (First OpenSG Symposium, Darmstadt, 29th of January 2002).
2. A. JAKLIC, A. LEONARDIS, F. SOLINA, *Segmentation and Recovery of Superquadrics* (Computational Imaging and Vision 20, Kluwer, Dordrecht, 2000).
3. U. KÜNZLER, R. WETZEL, M. ISELI, *Interactive Immersive Design Application: Analysis of Requirements* (INTERACT'03, M. Rauterberg et al. (Eds.), IOS Press, 2003).
4. J. P. SOUNDERS, *Project IRMA: The Technology* (Proceedings of the First International IMS Project Forum, Ascona, Switzerland, IMS International, 2001, pp. 179-185, for additional information see project website - www.project-irma.com).
5. S. TANO, et. al., *Godzilla: Seamless 2D/3D Sketch Environment for Reflective and Creative Design* (INTERACT'03, M. Rauterberg et al. (Eds.), IOS Press, 2003).

A NEW GEOMETRY REPRESENTATION FOR MULTIPURPOSE MODELING IN CONCEPTUAL DESIGN

Z. Rusák and I. Horváth
Delft University of Technology, The Netherlands

Abstract: To be able to model products in conceptual design, typically aspects such as functional, structural, shape, behavioral, sustainability and service are considered. It is a problem for research and development how to support these aspects of conceptualization with tools and methods in an integrated manner. Tools usually focus on individual aspects. The author proposes a new modeling approach called vague interval discrete modeling (VDIM) that integrates shape, structural, and behavioral modeling. The integration is achieved by the introduction of a multipurpose modeling entity called particle. VDIM offers three means for the representation of a cluster of shapes, for instances of shapes, and for physically-based manipulation of shapes. Interval modeling allows representing uncertainty of shapes, which is a characteristic property in shape conceptualization. In addition, particle systems can be applied to model the mechanical behavior of the product. This constructive modeling approach makes it possible to describe procedural models of incomplete geometries and to capture the structural relations between components. The paper reports on the representational and computational issues related to VDIM.

Key words: conceptual design, computer aided conceptual design, multipurpose modeling, vague discrete interval modeling.

1. INTRODUCTION

Conceptual design precedes detail design with the aim to generate product concepts for further processing [5]. According to the classical interpretation, conceptual design of industrial products integrates functional, shape, structural and behavioral modeling. Current research in conceptual design has extended these aspects with sustainability and services. In the

D. Talabă and T. Roche (eds.), Product Engineering, 175–186.
© *2004 Springer. Printed in the Netherlands.*

process of concept generation, these aspects are either simultaneously or sequentially considered by the designer. Contrary to this, conceptual design tools typically provide on the individual support for these aspects or at most try to integrate functional, structural, and behavioral modeling [3]. Nevertheless, integration achieved by these tools follows a sequential design approach, where simultaneous handling of these conceptual design aspects is neglected. Aspect models are generated in a sequential order and are converted to each other to provide a platform for an integrated system. However, the conversion assumes well defined and compatible representations of the various models but it is rather time consuming.

To achieve better integration of design tools supporting various aspects of design, multipurpose models are introduced. For instance, multipurpose geometric models in detailed design have already been addressed by researchers, who introduced the concept of application features and multiview feature management into CAD [1]. To support conceptual design of industrial products, the author proposes a new approach, called vague discrete interval modeling (VDIM) that combines shape, structural, and behavioral modeling but can also be used as a basis of design for sustainability. The paper focuses on the representational issues of VDIM. First, a set of requirements was collected from the literature to identify the needs of computer aided conceptual design. Based on these requirements, the functionality of the VDIM has been determined and it is presented in section 3. In section 4 a unique geometric representation is introduced, which enables modeling of shape, structure and behavior of a product concept with the modeling entity, called particle. Application of VDIM in conceptual design is demonstrated through a comprehensive design scenario.

2.   REQUIREMENTS FOR NEW MULTIPURPOSE MODELING FOR CACD

A study of the modeling means offered by the current modeling systems as well as the requirements of industrial designers for new modeling approaches supporting shape conceptualization has been carried out. The results, which were based partly on publications and partly on experimenting with commercial and research systems, were summarized in [8]. As far as the opinions of other researchers are concerned, they indicated the need for:
a) handling impreciseness in modeling the geometry of the shape [2],
b) tolerating incompleteness in structural modeling of products [7],
c) intuitive model building tools to create and modify shape models that allows the use of natural input means [10],
d) managing alternatives of a shape in terms of size and geometric variation [4].

e) applying physical principles to manipulation of shapes [6].

Based on these requirements, the idea of vague discrete interval modeling (VDIM) was developed. VDIM is a modeling technique dedicated to the support of shape conceptualization. It is vague in the sense that it models (a) a cluster shapes with single representation allowing the integration of a nominal shape with its domain of variance, (b) describes the structural relations between shape components, which represent the shape either completely or incompletely, and (c) supports manipulation of an evolving shape by means of dedicated modeling methods and tools. It is discrete since the representation of the geometry is composed from discrete entities. Finally, VDIM is an interval modeling technique, since it describes the domain of variance of the shape by a finite interval of the geometry. VDIM consists of three design phases i.e. vague modeling, instantiation, and physically based manipulation. The vague modeling phase is dedicated to represent imprecise and incomplete shapes in that form as the idea appears in the designer mind. Instantiation aims at sampling alternative solutions from a vague model. Physically based manipulation is applied to refine and evaluate the selected shape alternatives.

3. FUNCTIONALITY OF VDIM

The lower level functionality is presented in Figure 1. Introducing vagueness in modeling helps fulfilling the first three requirements, the fourth one implies instantiation, and the last requirement can be met by physically based modeling. Vague models can be generated from nominal shapes.

Therefore, the first sub-function of vague modeling is importing shape models. Nominal shapes can be imported in continuous (NURBS) and discrete forms (point-sets, meshes). Discrete point sets can be generated, for instance, by 3D digital scanning, tracking the hand's motion, or geometric interpretation of verbal control. To represent the imprecision of shapes, particle clouds are used, which can be generated either from a single point-set by considering noise in the points, or from multiple point-sets by determining the variation interval between some corresponding points. Particle clouds can also be generated by applying sweep operators to another particle cloud.

Following the principle of constructive modeling, particle systems are generated by volumetric and position operators from a finite set of particle clouds. Structural relations of the particle clouds are described in the procedural model, based on which even incomplete models can be handled. The elements of the vague interval models can be selected in the modeling space by means of dedicated identification tools. Parts of the model with semantic meaning can be defined and identified as regions. Geometric transformations can be applied to move, rotate, and scale the selected particle clouds, particle systems, or regions.

Figure 1. Functionality of the VDIM system in relation with phases of design process

Instantiation of vague models supports the generation of variations of shapes based on shape formation rules. The shape formation rules are generated by the user as a composition of elementary modifications of a target region of a shape. Due to size limitations of the paper, shape formation rules are not discussed here (interested readers are referred to [9]). When the shape formation rules are applied in the process of shape instantiation, the rules have to be mapped to effect parameters. To this end, the system interprets logical operations between the rules, as well as constraints on the effect parameters. Particle clouds can be instantiated by single or by compound instantiation functions. In the case of a particle system, the instantiation operator can take into account positional and shape constraints.

Physically based manipulation of shape models makes it possible for the designer to deform shapes based on physical laws and effects. To compute movements and deformations, and to simulate the behavior of artifacts, physically based particle systems can be created either from discrete boundary models or from discrete solid models. The support conditions are specified by various mechanical supports and the physical effects can be specified as forces, moments and torques. Stretch, squash, twist and bending effects can be applied to deform the shape. Computation of the shape deformations is based on the principles of solid mechanics.

4. GEOMETRIC REPRESENTATION IN VDIM

The fundamental modeling entity of VDIM is particle, or more precisely, coupled pairs of particles. Figure 3 shows the fundamental entities of VDIM. Particles take care of providing positional and morphological information for the geometric representation. Coupling of particles makes it possible to introduce various physical relationships and constraints in order to provide the means for a physically based manipulation of shapes and for the investigation of their behavior as physical objects. A vague discrete interval model is vague since the reference points of the particles are uncertainly specified. In the practice it means that a finite distribution space called metric occurrence is defined for each particle and merged in the shape model represented as a particle system. Note that internal particles do have metric occurrences, but they are not used in the vague representation for the reason that internal particles only play role in physically based modeling. The distance between the neighboring particles is the characteristic discreteness of the vague discrete model. Discreteness of the model comes from the fact that the particles themselves never coincide.

The minimal and maximal overlaying surfaces are generated on the extremes of distribution specified by the boundary particles of the vague discrete model. The distribution interval is a subspace that is in between the minimal and maximal overlaying surfaces. Thus, it represents a cluster or a family of possible shapes rather than a single nominal shape. The images (actually, image points) of the particles on the two overlaying surfaces of the distribution interval are connected by so-called distribution trajectories. This is a sort of simplification of the representation that enables to have 1D metric occurrences. Having a 1D metric occurrence reduces the required information over a vague shape as well as it regulates the discreteness of each shape that is represented in the vague interval model. The distribution interval represents the morphology of the described cluster or family of shapes, and is used to derive instance shapes of the same morphological character. A vague discrete model can represent global shapes, local shapes, and any composition of them. The vaguely defined particles contained in a VDIM support multi-resolution manipulation of shapes. For a volumetric representation, the interior of the closures is filled in with particles. Specifying physical relationships between all neighboring particles leads itself to a physical representation of the object. In multipurpose modeling the particle concept avoids the need for entity conversion between phases of conceptual design.

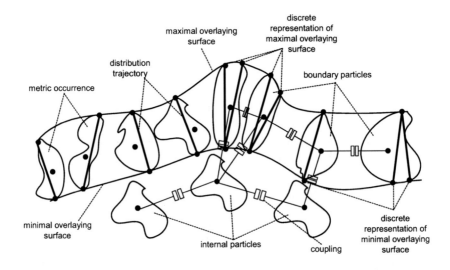

Figure 2. Fundamental entities of VDIM

The realization of the concept of VDIM as a practical method of shape conceptualization is rather simple. Take a point-set and attach some uncertainty to the location of the points to define particles. These particles describe a probable interval, where the shape of a concrete object may exist. Let a finite set of discrete particles form a particle cloud that defines either a region of an uncertain global shape, or an uncertain shape feature. Generate the particle clouds either by describing the uncertainty of the contained points, or by deriving distribution trajectories between the corresponding points of two (or more) non-coincident discrete point-sets. By composing some particle clouds, generate the representation of a cluster of objects of similar shapes, coined as particle system. Actually a system of particles is a vague discrete interval model. From a geometric point of view, the vague discrete representation can be complete or incomplete. To facilitate the generation of particle systems, volumetric operators have been introduced, which are able to calculate the Boolean union, difference, intersection, and Minkowski sum of vague shapes.

If a varied shape instance is required, the concept of rule based shape instantiation offers the necessary means for it. Assuming that the generic shape is defined in the form of an interval model, the modeling engine derives instance shapes by means of shape formation rules. These rules are defined based on the knowledge related to a particular field of interest (such as ergonomic comfort, aesthetic appeal, manufacturing, etc.) and converted to shape instantiation functions. An instance shape is actually a system of discrete particles of zero metric occurrences (or zero distribution trajectories). This approach indicates the consistency of the representation of vague interval shapes and nominal instance shapes. The instance shape is always inside the interval. It can be generated by mapping a particle system into an instance particle system either as a whole, or as a composition of instance regions. Based on one instantiation function, multiple shape instances can be generated. Instantiation specifies the position of the reference point of the particles along the distribution trajectories. It is in turn derived from shape formation rules relevant to the application at hand.

There are two fundamental issues here to be solved. First, the qualitative or quantitative shape formation rules have to be converted to shape instantiation functions. Second, the VDIM may represent a complex shape whose regions need to be instantiated individually by different functions and the instantiated regions have to be reunited. To efficiently convert qualitative shape formation rules to mathematical functions, the concept of effect function has been elaborated. Mathematically, the instantiation functions are composed from a set of effect functions that introduce alterations in a region of shapes through the coefficients of the employed bi-cubic bi-parametric polynomials. To enable region-oriented instantiation, three techniques have been considered, namely, simple, compound and constrained instantiation. Simple instantiation is applied when the whole particle system can be mapped to an instance by a single effect function. Compound instantiation applies multiple effect functions on the same particle cloud and tries to smoothly connect the instantiated regions by fuzzifying them in the transitions strips. Constrained instantiation applies multiple effect functions on multiple particle systems as well as constrained fuzzyfication to merge the instance shapes.

Because one of the application fields of VDIM is conceptual design, it is beneficial to integrate physically based modeling tools to VDIM. These tools facilitate physically based reshaping and the evaluation of product concepts by simulating their mechanical behavior for instance in use processes. The geometric representation of the object can be directly used in physically based modeling and can be extended with internal particles and the physical relations between the particles. In the physically based model, the particles have additional attributes, i.e., mass and velocity, but they are not supposed to behave as a rigid body. The material properties of an object are set by the coefficients of internal forces, i.e., damping constant, Hookean constant, and friction coefficient. In order to facilitate the application of specific physical effects on a particle system, physically based operators have been implemented.

5.   APPLICATION OF VDIM IN CONCEPTUAL DESIGN

As an application of VDIM, three popular designs of seats have been reproduced. The photos of Figure 3 illustrate the targeted seats. To produce the targeted shapes with VDIM, first a vague model was created as a composition of planar point-sets. The vague model had to be carefully chosen to be able to derive each targeted shapes from it. This required generating a large interval. The left top image of Figure 3 presents the vague model of a seat. Next, regions representing all features of the targeted shapes have been selected. To derive the instance shapes, shape formation rules of curving, offsetting, and tilting were applied to the regions. The derived instance seats are presented on Figure 3.

To demonstrate physically based modeling in VDIM, the seat presented in the right-bottom image of Figure 3 has been further developed. First this chair has been completed by adding legs and armrests to the model. The result is shown in Figure 4. Now, the task to create a more comfortable sitting support as well as an aesthetically pleasing product by redesigning an existing chair has been assigned to a designer. The targeted shape is presented on Figure 4. First, bending was applied to the leg of the chair. The designer used the bend operator, which requires him to define the effect region of the force and the needed fix support. Figure 5a shows the initial settings for the bend operator. The deformation process is shown in two stages in Figure 5b and Figure 5c.

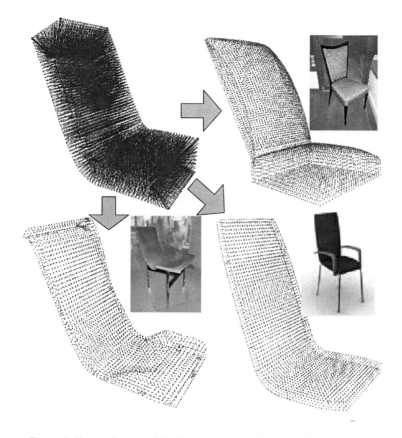

Figure 3. Vague shape model of a seat and some instance shape variances

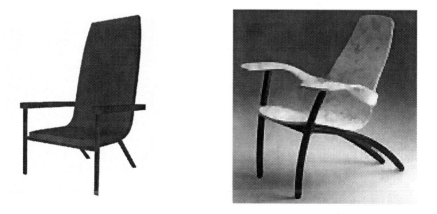

Figure 4. The initial model of a chair and the targeted chair

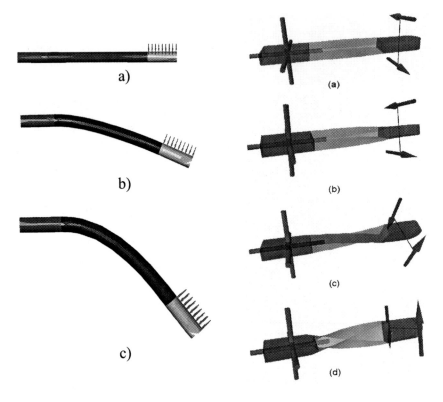

Figure 5. Bending a leg of the chair	*Figure 6.* Application of the twist operator on the armrest

Figure 7. The resulting chair

In the next step, the designer applies a torque to deform the shape that will be used for the armrest. In Figure 6a-d, the deformation process is illustrated in three stages. Finally, the unchanged parts of the chair and the parts that have been modified by the physically-based operators are reassembled with volumetric operators. Figure 7 shows the redesigned chair.

6.   CONCLUSIONS

This paper introduced a new geometric representation, called Vague Discrete Interval Modeling, with the aim to support the conceptual phase of the design process. A pilot system has been built that goes beyond the current methods and tools by involving a multi-purpose design method, which follows the advancement of design. Definition of a solution space for shapes by interval models, selection of solution alternatives by instantiation, manipulation of vague and instance shapes by volumetric and physically-based operators, and fast investigation of the mechanical behavior as an object have been defined as phases of conceptual design with VDIM. The inherent flexibility of VDIM makes it easy for designers to express their shape ideas and combine them in an initial model that can be modified and detailed, as needed. The real novelty of the presented technique is in its capability to (a) explicitly represent the distribution interval of a cluster of shapes, (b) fulfill the overwhelming majority of the modeling and presentation demands by discrete representation, (c) simulate the physical behavior by assigning physical properties, (d) cope with modality without the involvement of any geometric parameterization and. (e) allow convenient handling of geometric uncertainties in a design/modeling process. The capability of handling uncertainty and multiplicity gives freedom for the designer at the same time guides solution finding. The developed vague discrete modeling is sufficiently natural and interactive, and can efficiently support conceptual shape design. It can be adapted to various virtual environments, and the models can be converted to conventional CAD models straightforwardly.

REFERENCES

1. A. NOORT, *Multiple-view feature modeling with model adjustment* (Delft: DUT, Ph. D. Thesis, 2002).
2. X. GUAN, A. H. B. DUFFY, K. J. MACCALLUM, *Prototype system for supporting the incremental modeling of vague geometric configuration* (Artificial Intelligence for Engineering Design, Analysis and Manufacturing, 1997, 11(4), pp 287-310).
3. J. K. GUI, *A Function-Behaviour-Structure Machine Design Model and its Use in*

*Assembly Sequence Planning (*Journal of Engineering Design, 1990, 1(3), pp 239-259).
4. J. JIAO, M. M. TSENG, *Understanding product family for mass customization by developing commonality indices Journal of Engineering Design* (2000, 11(3), pp 225–243).
5. K. K. KUAN, *Conceptual designing concurrent engineering. Concurrent engineering: A Global Perspective*, (CE95 Conference, 1995, pp. 223-229).
6. KT. MCDONNELL, H. QIN, RA. WLODARCZYK, *Virtual clay: a real-time sculpting system with haptic toolkits* (Proceedings of the 2001 symposium on Interactive 3D graphics March 2001, pp. 179 – 190).
7. A. MUKERJEE, *Qualitative geometric design* (Proceedings of the First Symposium on Solid Modeling Foundations and CAD/CAM Applications, 1991, pp. 503 – 514).
8. Z. RUSÁK, *Vague Discrete Interval Modeling for Product Conceptualization in Collaborative Virtual Design Environments* (Ph.D. thesis, Millpress, Rotterdam, 2003).
9. Z. RUSÁK, I. HORVÁTH, *Deriving product variances by Rule based instantiation of Vague Discrete Interval Models* (Proceedings of TMCE 2004 Symposium, 2004).
10. D. WEIMER, SK. GANAPATHY, *A Synthetic Visual Environment with Hand Gesture and Voice Input* (Proceedings of the ACM Conference on Computer Human Interfaces, 1989, pp 235-240).

Part 3

SIMULATION TECHNOLOGIES

1. INVITED LECTURES

2. CONTRIBUTIONS

AN OVERVIEW OF SEVERAL FORMULATIONS FOR MULTIBODY DYNAMICS

P. E. Nikravesh

Department of Aerospace and Mechanical Engineering, University of Arizona, Tucson,USA

Abstract: In this paper several formulations for automatic generation of the equations of motion for rigid and rigid-deformable multibody systems are reviewed. The rigid-body formulations are the body-coordinate formulation based on Newton-Euler equations, a non-conventional point-coordinate formulation where rigid bodies are represented as a collection of particles, and the joint coordinate formulation that employs relative coordinates between bodies. A transformation process based on velocity relations is described for easy transformation of one formulation into another. Finally, a brief overview of the equations of motion for deformable bodies in a multibody environment is presented.

Key words: multibody dynamics; rigid bodies; deformable bodies; body coordinates; point coordinates; joint coordinates; equations of motion.

1.   INTRODUCTION

Derivation of equations of motion for computational multibody dynamics has been the topic of many research activities. The scope of these activities has been quite broad. Some techniques allow us to generate the equations of motion in terms of a large set of differential-algebraic equations. Other techniques yield the equations of motion as a minimal set of ordinary differential equations. Many other *in-between* approaches provide us with various alternatives. Each formulation has its own advantages and disadvantages depending on the application and our priorities.

Possibly, the simplest method to construct the equations of motion is in the form of a large set of differential-algebraic equations. The configuration of a rigid body is described by a set of translational and rotational

189

D. Talabă and T. Roche (eds.), Product Engineering, 189–226.

coordinates. Algebraic constraints are introduced to represent kinematic joints connecting bodies, and then the Lagrange multiplier technique is used to describe joint reaction forces. Although these formulations are easy to construct, one of their main drawbacks is their computational inefficiency. One such formulation is called the body-coordinate formulation and it is the first method discussed in this paper. This formulation has also been referred to as the absolute or Cartesian coordinate formulation [1].

The equation of motion from body-coordinates can be transformed to other forms in terms of other generalized coordinates. In this paper a systematic process based on a simple velocity transformation is reviewed. If the new set of generalized velocities contains a smaller number of velocities than the original set, then the process yields a smaller number of equations. In the process, some of the kinematic constraints and their attributes vanish from the transformed equations of motion. If the transformed set contains as many generalized velocities as the number of degrees-of-freedom, then the resultant equations will be in the form of ordinary second-order differential equations without any algebraic constraints. This transformation process is either partially or fully employed in two of the formulations that are discussed in this paper.

Another description of multibody systems that is discussed in this paper is the point-coordinate (or natural coordinate) formulation [2, 3]. This method takes advantage of a rudimentary idea of describing a body as a collection of points (and vectors [3]). The coordinates of these points are dependent on each other through constant-length constraints due to the assumption of rigidity of a body. Most kinematic joints do not introduce any algebraic constraints in the formulation if we allow bodies to share some of their points. However, some joints may require a few simple algebraic constraints. Similar to the body-coordinate formulation, this formulation also yields a large set of loosely coupled differential-algebraic equations of motion.

Unlike the first two formulations that are based on absolute coordinates of bodies or points, the third formulation that is discussed in this paper uses relative coordinates. This method, called the joint-coordinate formulation, is based on describing a set of joint coordinates (or velocities) as the generalized coordinates and, then, transforming the body-coordinate equations to a new set [4-6]. This process yields a much smaller set of differential-algebraic equations or even a minimal set of ordinary second-order differential equations.

The equations of motion for a deformable body in a multibody environment are also presented in this paper. It is assumed that the finite element technique is used to construct the nodal mass and stiffness matrices for the deformable body. Furthermore, it is assumed that the reader is familiar with the fundamentals of the finite element formulation. Issues associated with how to attach a reference frame to a moving deformable body are discussed. Finally, the equations of motion for a deformable body are appended to the equations of motion of other bodies in order to describe a multibody system.

In this paper matrix notation is used in order to keep our attention on concepts without any loss of details. Readers should find the notation to be effective in multibody formulation.

1.1 Nomenclature

Total Number of:
  n_b : bodies
  n_p : points or particles
  n_{nodes} : number of nodes
  n_c : constraints
  n_{dof} : degrees-of-freedom
  n_v : variables, coordinates
  n_e : equations of motion,
     velocities, accelerations

Vectors and Arrays:
  Lower case characters, boldface:
    Roman (contains *x-y-z*
      components; i.e., inertial)
    Italic (contains $\xi - \eta - \zeta$
      components; i.e., body-
      fixed)

Matrices:
  Upper case characters, boldface:
    Roman (described in *x-y-z*
      components)
    Italic (described in $\xi - \eta - \zeta$
      components)

Right Superscripts:
  Index of a point, a particle, or a
    node
  T : transpose
  b : boundary node(s)
  u : unconstrained node(s)

Right Subscripts:
  Index of a body or an arbitrary index
Over-Scores:
  ~ : transforms a 3-vector to a skew-symmetric matrix
  ^ : stacks vertically 3-vectors or 3x3 skew-symmetric matrices
  - : repeats a 3x 3 matrix to form a block-diagonal matrix
Stacked Vectors (Arrays) constructed from 3-Vectors:

$$
\mathbf{s} = \left\{ \begin{array}{c} s_1 \\ \vdots \\ s_{n_b} \end{array} \right\}, \mathbf{b} = \left\{ \begin{array}{c} b^1 \\ \vdots \\ b^n \end{array} \right\}, \hat{\mathbf{0}} \equiv \left\{ \begin{array}{c} 0 \\ \vdots \\ 0 \end{array} \right\}
$$

Null Vector/Matrix, Identity Matrix: Skew-Symmetric Matrix:

$$\mathbf{0} \equiv \begin{Bmatrix} 0 \\ 0 \\ 0 \end{Bmatrix}, \mathbf{0} \equiv \begin{bmatrix} 0 & 0 & 0 \\ 0 & 0 & 0 \\ 0 & 0 & 0 \end{bmatrix}, \mathbf{I} \equiv \begin{bmatrix} 1 & 0 & 0 \\ 0 & 1 & 0 \\ 0 & 0 & 1 \end{bmatrix}. \qquad \mathbf{a} \equiv \begin{Bmatrix} 1 \\ 2 \\ 3 \end{Bmatrix} \Rightarrow \tilde{\mathbf{a}} \equiv \begin{bmatrix} 0 & -3 & 2 \\ 3 & 0 & -1 \\ -2 & 1 & 0 \end{bmatrix}.$$

2.   BODY-COORDINATE FORMULATION

Body coordinates, also known as the absolute coordinates, provide the simplest formulation to construct the equation of motion for multibody systems for computational purpose [1]. This formulation is simple to learn and easy to implement in either special- or general-purpose computer programs, however, the computational efficiency is not the best. Several well known commercial multibody programs such as ADAMS and DADS employ this type of formulation. This formulation yields a *large* set of *loosely coupled* equations of motion.

2.1   Configuration of a rigid body

The fundamental assumption in this formulation is that a body is assigned a set of translational and rotational coordinates. Therefore, as shown in Figure 1, a typical rigid body i is positioned in an inertial (non-moving, global) x–y–z frame by vector $\vec{r}_i$. This vector locates the origin of a body-attached (moving, local) $\xi_i - \eta_i - \zeta_i$ frame that is at the mass center of the body. This reference frame may or may not coincide with the body principal axes.

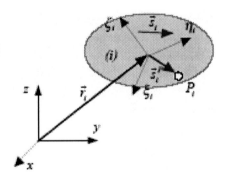

Figure 1. A rigid body positioned in an inertial frame

The x–y–z components of vector $\bar{r}_i$ represent the translational coordinates of the body and they are denoted by the algebraic vector $\mathbf{r}_i = \{x\ y\ z\}_i^T$. Angular orientation of $\xi_i - \eta_i - \zeta_i$ frame with respect to x–y–z frame can be described by a set of Euler angles or Euler parameters [1]. These rotational coordinates provide a 3×3 rotational transformation matrix, denoted as $\mathbf{A}_i$ where $\mathbf{A}_i^T = \mathbf{A}_i^{-1}$. This matrix transforms components of a vector from $\xi_i - \eta_i - \zeta_i$ to x–y–z.

The translational and rotational velocity vectors for the body are defined respectively as $\dot{\mathbf{r}}_i = \{\dot{x}\ \dot{y}\ \dot{z}\}_i^T$ and $\boldsymbol{\omega}_i = \{\omega_{(x)}\ \omega_{(y)}\ \omega_{(z)}\}_i^T$ or $\boldsymbol{\omega}_i = \{\omega_{(\xi)}\ \omega_{(\eta)}\ \omega_{(\zeta)}\}_i^T$, where $\boldsymbol{\omega}_i = \mathbf{A}_i \boldsymbol{\omega}_i$. Similarly, the acceleration vectors are defined as $\ddot{\mathbf{r}}_i = \{\ddot{x}\ \ddot{y}\ \ddot{z}\}_i^T$ and $\dot{\boldsymbol{\omega}}_i = \{\dot{\omega}_{(x)}\ \dot{\omega}_{(y)}\ \dot{\omega}_{(z)}\}_i^T$ or $\dot{\boldsymbol{\omega}}_i = \{\dot{\omega}_{(\xi)}\ \dot{\omega}_{(\eta)}\ \dot{\omega}_{(\zeta)}\}_i^T$ In order to simplify the discussion, we adopt the x–y–z components of angular velocity and acceleration vectors in our formulations.

A vector or a point can be defined attached to a rigid body. As shown in Figure 1, vector $\bar{s}_i$ is defined on a body representing a particular axis. However, vector $\bar{s}_i^P$ locates point P from the origin of the body. Local components of these vectors are constants and are respectively denoted as $s_i = \{s_{(\xi)}\ s_{(\eta)}\ s_{(\zeta)}\}_i^T$ and $s_i^P = \{\xi^P\ \eta^P\ \zeta^P\}_i^T$. Obviously, the following transformations can be performed in order to compute the global components of these vectors:

$$\mathbf{s}_i = \mathbf{A}_i\, s_i \ \text{(or } \mathbf{s}_i^P = \mathbf{A}_i\, s_i^P). \tag{1}$$

The velocity and acceleration of a body-fixed vector, either $\mathbf{s}_i$ or $\mathbf{s}_i^P$, are computed as:

$$\begin{aligned} \dot{\mathbf{s}}_i &= \tilde{\boldsymbol{\omega}}_i \mathbf{s}_i = -\tilde{\mathbf{s}}_i \boldsymbol{\omega}_i \\ \ddot{\mathbf{s}}_i &= -\tilde{\mathbf{s}}_i \dot{\boldsymbol{\omega}}_i + \tilde{\boldsymbol{\omega}}_i \tilde{\boldsymbol{\omega}}_i \mathbf{s}_i \end{aligned} \tag{2}$$

Position, velocity, and acceleration of a point P attached to body i are expressed as:

$$\begin{aligned} \mathbf{r}_i^P &= \mathbf{r}_i + \mathbf{s}_i^P = \mathbf{r}_i + \mathbf{A}_i\, s_i^P \\ \dot{\mathbf{r}}_i^P &= \dot{\mathbf{r}}_i + \dot{\mathbf{s}}_i^P = \dot{\mathbf{r}}_i - \tilde{\mathbf{s}}_i^P \boldsymbol{\omega}_i \\ \ddot{\mathbf{r}}_i^P &= \ddot{\mathbf{r}}_i + \ddot{\mathbf{s}}_i^P = \ddot{\mathbf{r}}_i - \tilde{\mathbf{s}}_i^P \dot{\boldsymbol{\omega}}_i + \tilde{\boldsymbol{\omega}}_i \tilde{\boldsymbol{\omega}}_i \mathbf{s}_i^P \end{aligned} \tag{3}$$

2.2   Kinematic joints

In a typical multibody system, some of the bodies may be connected to each other by kinematic joints. A kinematic joint presents conditions on the motion of the two connected bodies. These conditions can be described by algebraic equations that are written between coordinates of points and components of vectors defined on different bodies. In this paper we denote a constraint by the character Φ. This character, for reference purposes, may carry a right-superscript denoting the type and the number of algebraic equations it represents.

Different conditions can be defined between vectors attached to a body; e.g., vectors $\vec{s}_i$ or $\vec{s}_j$ in Figure 2, and a vector that connects a point on one

body to a point on another body, such as vector $\vec{d}$. Components of vector $\vec{d}$ can be computed as:

$$\mathbf{d} = \mathbf{r}_j^P - \mathbf{r}_i^P . \tag{4}$$

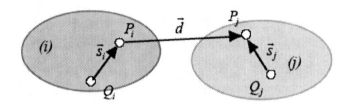

Figure 2. A vector connecting two bodies

For vectors $\vec{s}_i$ and $\vec{s}_j$ to remain perpendicular, we must enforce the condition (normal-type 1 containing 1 algebraic equation):

$$\Phi^{(n1,1)} \equiv \mathbf{s}_i^T \, \mathbf{s}_j = 0 . \tag{5}$$

Similarly for vectors $\vec{s}_i$ and $\vec{d}$ to remain perpendicular, we write (normal type 2 containing 1 algebraic equation):

$$\Phi^{(n2,1)} \equiv \mathbf{s}_i^T \, \mathbf{d} = 0 . \tag{6}$$

For two vectors such as $\vec{s}_i$ and $\vec{s}_j$ to remain parallel we can write (parallel-type 1 containing 2 independent algebraic equations):

$$\Phi^{(p1,2)} \equiv \tilde{s}_i \, s_j = 0 .$$

This vector-product equation results into three algebraic equations where only two are independent. This requires us to select the *best* two out of three equations. This selection issue can be avoided if we define two vectors on body j perpendicular to $\vec{s}_j$, and then enforcing $\vec{s}_i$ to be perpendicular to these two vectors.

The most common kinematic joint between two bodies is a spherical (ball) joint, as shown schematically in Figure 3. This joint requires point P_i on body i, and point P_j on body j to remain coincident. This condition, containing three algebraic equations, is expressed as:

$$\Phi^{(s,3)} \equiv r_i^P - r_j^P = 0 . \tag{7}$$

By combining the constraints in Eqs. (5)-(7), we can represent other types of kinematic joints. As an example, consider a revolute (pin) joint shown schematically in Figure 4. Point P_i on body i and point P_j on body j are defined along the joint axis, where they must remain together. Vector $\vec{s}_i$ is defined along the joint axis on body i and vectors $\vec{a}_j$ and $\vec{b}_j$ are defined on body j perpendicular to the joint axis. The following constraints can be written:

$$\Phi^{(r,5)} \equiv \begin{cases} \Phi^{(s,3)} \equiv r_i^P - r_j^P = 0 \\ \Phi^{(n1,1)} \equiv s_i^T a_j = 0 \\ \Phi^{(n1,1)} \equiv s_i^T b_j = 0 \end{cases} .$$

Similarly, other types of kinematic constraints, such as cylindrical, prismatic, universal can be constructed.

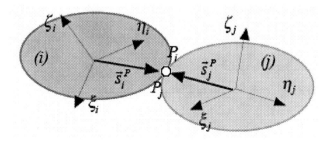

Figure 3. A spherical joint

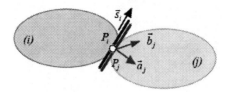

Figure 4. A revolute joint

Another useful constraint is the spherical-spherical joint shown in Figure 5. The rigid link between the two bodies is not represented as a body—it is represented as the following constraint in order to lower the number of coordinates in a model:

$$\Phi^{(s-s,1)} \equiv \frac{1}{2}(\mathbf{d}^T \mathbf{d} - l^2) = 0. \tag{8}$$

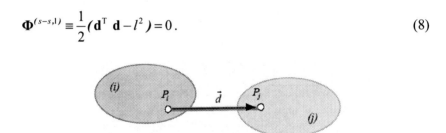

Figure 5. A spherical-spherical joint

The coefficient 1/2 is recommended in order to eliminate the 2 coefficients in the time derivatives. This constraint can be combined with Eqs. (5)–(7) to model more complex joints.

The first time derivative of a position constraint yields the corresponding velocity constraint. For our four fundamental constraints, the velocity constraints are:

$$\dot{\Phi}^{(n1,1)} \equiv -\mathbf{s}_j^T \tilde{\mathbf{s}}_i \boldsymbol{\omega}_i - \mathbf{s}_i^T \tilde{\mathbf{s}}_j \boldsymbol{\omega}_j = 0$$

$$\dot{\Phi}^{(n2,1)} \equiv -\mathbf{d}^T \tilde{\mathbf{s}}_i \boldsymbol{\omega}_i + \mathbf{s}_i^T (\dot{\mathbf{r}}_j - \tilde{\mathbf{s}}_j^P \boldsymbol{\omega}_j - \dot{\mathbf{r}}_i + \tilde{\mathbf{s}}_i^P \boldsymbol{\omega}_i) = 0$$

$$\dot{\Phi}^{(s,3)} \equiv \dot{\mathbf{r}}_i - \tilde{\mathbf{s}}_i^P \boldsymbol{\omega}_i - \dot{\mathbf{r}}_j + \tilde{\mathbf{s}}_j^P \boldsymbol{\omega}_j = 0$$

$$\dot{\Phi}^{(s-s,1)} \equiv \mathbf{d}^T (\dot{\mathbf{r}}_j - \tilde{\mathbf{s}}_j^P \boldsymbol{\omega}_j - \dot{\mathbf{r}}_i + \tilde{\mathbf{s}}_i^P \boldsymbol{\omega}_i) = 0$$

We consider an array of velocities for the two bodies containing the translational and rotational velocities for bodies *i* and *j*, in that order. The coefficient matrix for the velocity array will be referred to as the *Jacobian* matrix. Each velocity equation, whether it contains one or several algebraic equations, can be described in a general form as:

$$\mathbf{Dv} = \mathbf{D}_i \mathbf{v}_i + \mathbf{D}_j \mathbf{v}_j = 0 , \tag{9}$$

where $\mathbf{v}_i$ and $\mathbf{v}_j$ each contain six velocity components, and $\mathbf{D}_i$ and $\mathbf{D}_j$ are the corresponding sub-Jacobians. Table 1 shows the entries for the sub-Jacobian matrices for our four fundamental constraints.

The time derivative of a velocity constraint yields the corresponding acceleration constraint. In the acceleration constraint the coefficient matrix of the acceleration array is the same *Jacobian* matrix as in the velocity constraints. The acceleration constraint for a joint can be expressed in the following general form:

$$\mathbf{D}\dot{\mathbf{v}} + \dot{\mathbf{D}}\mathbf{v} = 0 . \tag{10}$$

Acceleration constraints contain quadratic velocity terms that are moved to the right-hand-side of the equations. Table 2 shows the right-hand-side quadratic terms for our four fundamental constraints.

Table 1. Sub-Jacobians for four fundamental constraints

Constraint	Dimension	D_i		D_j	
		$\dot{\mathbf{r}}_i$	$\boldsymbol{\omega}_i$	$\dot{\mathbf{r}}_j$	$\boldsymbol{\omega}_j$
n1	1×12	$\mathbf{0}$	$-\mathbf{s}_j^{\mathrm{T}}\tilde{\mathbf{s}}_i$	$\mathbf{0}$	$-\mathbf{s}_i^{\mathrm{T}}\tilde{\mathbf{s}}_j$
n2	1×12	$-\mathbf{s}_i^{\mathrm{T}}$	$\mathbf{s}_i^{\mathrm{T}}\left(\tilde{\mathbf{d}}+\tilde{\mathbf{s}}_i^{P}\right)$	$\mathbf{s}_i^{\mathrm{T}}$	$-\mathbf{s}_i^{\mathrm{T}}\tilde{\mathbf{s}}_j^{P}$
Spherical	3×12	$\mathbf{I}$	$-\tilde{\mathbf{s}}_i^{P}$	$-\mathbf{I}$	$\tilde{\mathbf{s}}_j^{P}$
Sph-sph	1×12	$-\mathbf{d}^{\mathrm{T}}$	$\mathbf{d}^{\mathrm{T}}\tilde{\mathbf{s}}_i^{P}$	$\mathbf{d}^{\mathrm{T}}$	$-\mathbf{d}^{\mathrm{T}}\tilde{\mathbf{s}}_j^{P}$

Table 2. Right-hand-side of acceleration constraints

Constraint	$\gamma = -\dot{\mathbf{D}}\mathbf{v}$
n1	$-\mathbf{s}_i^{\mathrm{T}}\tilde{\boldsymbol{\omega}}_j\dot{\mathbf{s}}_j - \mathbf{s}_j^{\mathrm{T}}\tilde{\boldsymbol{\omega}}_i\dot{\mathbf{s}}_i - 2\dot{\mathbf{s}}_j^{\mathrm{T}}\dot{\mathbf{s}}_i$
n2	$-\mathbf{d}^{\mathrm{T}}\tilde{\boldsymbol{\omega}}_i\dot{\mathbf{s}}_i - \mathbf{s}_i^{\mathrm{T}}(\tilde{\boldsymbol{\omega}}_i\dot{\mathbf{s}}_i^{P} - \tilde{\boldsymbol{\omega}}_j\dot{\mathbf{s}}_j^{P}) - 2\dot{\mathbf{d}}^{\mathrm{T}}\dot{\mathbf{s}}_i$
Spherical	$\tilde{\dot{\mathbf{s}}}_i^{P}\boldsymbol{\omega}_i - \tilde{\dot{\mathbf{s}}}_j^{P}\boldsymbol{\omega}_j$
Sph-sph	$2\,\mathbf{d}^{\mathrm{T}}(\tilde{\boldsymbol{\omega}}_i\dot{\mathbf{s}}_i^{P} - \tilde{\boldsymbol{\omega}}_j\dot{\mathbf{s}}_j^{P}) - 2\,\dot{\mathbf{d}}^{\mathrm{T}}\dot{\mathbf{d}}$

2.3   Newton-Euler equations

In the body-coordinate formulation, Newton-Euler equations provide the simplest description of the equations of motion. For a body with a mass m_i, if the sum of forces acting on the body is denoted as $\bar{f}_i$ as shown in Figure 6, the Newton equations of motion describing the translation of the mass center are written as:

$$m_i \ddot{\mathbf{r}}_i = \mathbf{f}_i . \tag{11}$$

The Euler's equations describing the rotation of the body are normally expressed in the body-attached $\xi_i - \eta_i - \zeta_i$ reference frame as:

$$\boldsymbol{J}_i \dot{\boldsymbol{\omega}}_i = \boldsymbol{n}_i - \tilde{\boldsymbol{\omega}}_i \boldsymbol{J}_i \boldsymbol{\omega}_i ,$$

where J_i is a 3×3 constant *rotational inertia* matrix. The Euler equations can also be described in the x–y–z frame as:

$$\mathbf{J}_i \dot{\boldsymbol{\omega}}_i = \mathbf{n}_i - \tilde{\boldsymbol{\omega}}_i \mathbf{J}_i \boldsymbol{\omega}_i , \tag{12}$$

where $\mathbf{J}_i = \mathbf{A}_i \, \boldsymbol{J}_i \, \mathbf{A}_i^{\mathrm{T}}$ is no longer a constant matrix.

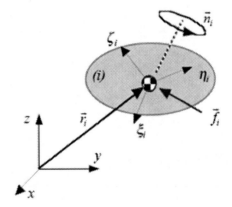

Figure 6. A force and a moment acting on a body

In order to be consistent with the constraint equations at velocity and acceleration levels, we choose the Euler equations from Eq. (12). Appending Eq. (12) to Eq. (11) provides six equations of motion for a free moving body:

$$\begin{bmatrix} m_i \mathbf{I} & \mathbf{0} \\ \mathbf{0} & \mathbf{J}_i \end{bmatrix} \begin{Bmatrix} \ddot{\mathbf{r}}_i \\ \dot{\boldsymbol{\omega}}_i \end{Bmatrix} = \begin{Bmatrix} \mathbf{f}_i \\ \mathbf{n}_i - \tilde{\boldsymbol{\omega}}_i \mathbf{J}_i \boldsymbol{\omega}_i \end{Bmatrix} , \text{ or, } \mathbf{M}_i \dot{\mathbf{v}}_i = \mathbf{g}_i . \tag{13}$$

If a kinematic joint connects two bodies, the *reaction force/momen* between the bodies must be included in the equations of motion. If we assume that bodies i and j are connected by some type of joint, and the Jacobian sub-matrices for the joint are described as $\mathbf{D}_i$ and $\mathbf{D}_j$, then Eq. (13) can be revised and written for both bodies as:

$$\begin{bmatrix} \mathbf{M}_i & \mathbf{0} \\ \mathbf{0} & \mathbf{M}_j \end{bmatrix} \begin{Bmatrix} \ddot{\mathbf{v}}_i \\ \dot{\mathbf{v}}_j \end{Bmatrix} = \begin{Bmatrix} \mathbf{g}_i \\ \mathbf{g}_j \end{Bmatrix} + \begin{bmatrix} \mathbf{D}_i^{\mathrm{T}} \\ \mathbf{D}_j^{\mathrm{T}} \end{bmatrix} \lambda, \tag{14}$$

where λ is an array of Lagrange multipliers containing as many elements as the number of constraints describing the joint.

2.4 Force elements

Typical forces that may act on a body are gravitational, spring, damper and actuator forces. These forces are either constants or position and velocity dependent. Computation of these forces is often a simple task.

As an example, assume a force element between two bodies is the linear spring-damper element shown in Figure 7. In a given configuration (known coordinates) and for known velocity, the force of this element can be determined and applied on the two bodies. A vector connecting the two attachment points is first computed as $\mathbf{d} = \mathbf{r}_j^P - \mathbf{r}_i^P$. The length of this vector is $l = (\mathbf{d}^T\mathbf{d})^{1/2}$. A unit vector along this axis is determined as $\mathbf{u} = \mathbf{d}/l$. The force of the spring, assuming that the undeformed length and stiffness are known, is computed as $f^{(s)} = k(l - l^{(0)})$. For the damping force we compute $\dot{l} = \mathbf{d}^T\dot{\mathbf{d}}/l$. Assuming that the damping coefficient is known, the damper force is computed as $f^{(d)} = c\,\dot{l}$. Now the force of the element on body i and body j is determined as $\mathbf{f}_i = (f^{(s)} + f^{(d)})\mathbf{u}$ and $\mathbf{f}_j = -(f^{(s)} + f^{(d)})\mathbf{u}$. Since these forces have moment arms, their moments must be included in the rotational equations as well; i.e., $\mathbf{n}_i = \tilde{\mathbf{s}}_i^P \mathbf{f}_i$ and $\mathbf{n}_j = \mathbf{s}_j^P \mathbf{f}_j$.

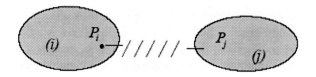

Figure 7. A spring-damper between two bodies

2.5 Complete set of equations of motion

Assume that a multibody system contains n_b rigid bodies. An array of coordinates, $\mathbf{c}$, is defined containing n_v coordinates. The number of coordinates is $6 \times n_b$ or $7 \times n_b$ depending whether we use Euler angles or Euler parameters. Arrays of velocities and accelerations are defined as $\mathbf{v}$ and $\dot{\mathbf{v}}$, each containing $n_e = 6 \times n_b$ elements.

The position constraints are expressed as n_c nonlinear algebraic equations:

$$\Phi(\mathbf{c}) = \mathbf{0}. \tag{15}$$

The n_c velocity constraints are written as:

$$\dot{\Phi} \equiv \mathbf{D}\mathbf{v} = \mathbf{0}. \tag{16}$$

Similarly, the n_c acceleration constraints are expressed as:

$$\ddot{\Phi} \equiv \mathbf{D}\dot{\mathbf{v}} + \dot{\mathbf{D}}\mathbf{v} = \mathbf{0}. \tag{17}$$

If all n_c constraints are independent, then the system has $n_{dof} = n_e - n_c$ degrees-of-freedom.

The Newton-Euler equations are expressed as n_e equations as

$$\mathbf{M}\dot{\mathbf{v}} - \mathbf{D}^T\lambda = \mathbf{g}, \tag{18}$$

where there are n_c multipliers in the array λ. We note that Eqs. (15)-(18) represent a set of mixed differential-algebraic equations.

3.   COORDINATE TRANSFORMATION

The equations of motion derived in the preceding section can be transformed to other forms. Meanwhile the number of equations can also be reduced. In this section we review a general and systematic procedure for such a transformation based on a simple velocity relation.

For a multibody system formulated with the body-coordinates, the array of velocities, $\mathbf{v}$, is assumed to contain $n_e = 6 \times n_b$ elements. Let us define another array of velocities, $\mathbf{w}$, containing n_w elements where $n_{dof} \leq n_w \leq n_e$. The array $\mathbf{w}$ can be a sub-set of $\mathbf{v}$ or its elements could be defined to be partially or completely different from $\mathbf{v}$. In a multibody system, at any given configuration, since the velocity relationships are linear in velocities, a velocity transformation can be established as:

$$\mathbf{v} = \mathbf{B}\,\mathbf{w}, \tag{19}$$

where **B** is called the *velocity transformation matrix* and it is a function of the coordinates. The product of the system Jacobian matrix and the velocity transformation matrix yields a new matrix **C** as:

$$\mathbf{DB} \Rightarrow \mathbf{C} . \tag{20}$$

We will discuss the characteristics of matrices **B** and **C** later in this section. The time derivative of Eq. (20) provides the acceleration transformation equations as:

$$\dot{\mathbf{v}} = \mathbf{B}\dot{\mathbf{w}} + \dot{\mathbf{B}}\mathbf{w} . \tag{21}$$

The kinematics constraints in Eqs. (15)-(17) are transformed into new sets. Specifically, the velocity constraints become:

$$\dot{\mathbf{\Phi}} \equiv \mathbf{DBw} = \mathbf{Cw} = \mathbf{0} . \tag{22}$$

Similarly, the n_c acceleration constraints are transformed as:

$$\ddot{\mathbf{\Phi}} \equiv \mathbf{D}(\mathbf{B}\dot{\mathbf{w}} + \dot{\mathbf{B}}\mathbf{w}) + \dot{\mathbf{D}}\mathbf{Bw} = \mathbf{C}\dot{\mathbf{w}} + \dot{\mathbf{C}}\mathbf{w} = \mathbf{0} . \tag{23}$$

The transformation is not performed explicitly at the coordinate level since (a) a new set of coordinates has not been defined yet, and (b) the constraints in Eq. (15) are nonlinear in the coordinates. The coordinate transformation is normally performed computationally in a recursive process.

The transformed set of Newton-Euler equations is obtained by substituting Eq. (21) into Eq. (18) and then pre-multiplying the result by $\mathbf{B}^T$:

$$\mathbf{B}^T \mathbf{M} (\mathbf{B}\dot{\mathbf{w}} + \dot{\mathbf{B}}\mathbf{w}) - \mathbf{B}^T \mathbf{D}^T \lambda = \mathbf{B}^T \mathbf{g} ,$$

or,

$$\mathbf{M}_{(w)} \dot{\mathbf{w}} - \mathbf{C}^T \lambda = \mathbf{g}_{(w)} , \tag{24}$$

where,

$$\mathbf{B}^T \mathbf{M} \mathbf{B} \equiv \mathbf{M}_{(w)}$$
$$\mathbf{B}^T (\mathbf{g} - \mathbf{M}\dot{\mathbf{B}}\mathbf{w}) \equiv \mathbf{g}_{(w)} \tag{25}$$

We can examine three cases for the transformed equations. We note that the **D** matrix is n_c x n_e and the **B** matrix is n_e x n_w:

Case 1: In this case $n_w = n_e$; i.e., we have as many elements in **w** as we do in **v**. The number of kinematic constraints and the number of Newton-Euler equations remain the same. The **B** matrix is square and the **C** matrix has the same dimensions as **D**.

Case 2: In this case $n_{dof} < n_w < n_e$; i.e., the transformed space has a smaller dimension but it is not the minimum dimension. The **B** matrix is rectangular (it has fewer columns that rows) and therefore **C** has fewer columns than **D**. Furthermore, **B** has the characteristics that its columns are orthogonal to $n_e - n_w$ rows of **D**. Therefore, **C** has $n_e - n_w$ fewer rows than **D**. This means that the transformed set has $n_e - n_w$ fewer constraints compared to the original set. This is also evident from the dimensions of the transformed mass matrix $\mathbf{M}_{(w)}$ and the generalized force array $\mathbf{g}_{(w)}$. Note that we still have a mixed set of differential-algebraic equations.

Case 3: In this case $n_{dof} = n_w < n_e$; i.e., the transformed space has a minimum dimension equal to the number of system degrees-of-freedom. All of the columns of **B** are orthogonal to all of the rows of **D** and, therefore, **C** and all of the constraints vanish. The equations of motion become:

$$\mathbf{M}_{(w)}\dot{\mathbf{w}} = \mathbf{g}_{(w)} .$$
(26)

These are $n_w = n_{dof}$ second-order differential equations without any constraints.

4.   POINT-COORDINATE FORMULATION

A multibody system can be represented as a collection of interconnected points [2]. If the transformation from body-coordinate formulation to pointcoordinate formulation is performed properly, the kinematics and inertial characteristics are preserved and, therefore, no approximations are involved. The point-coordinate representation eliminates the need to define rotational coordinates for a body. A similar formulation can be found in [3] describing a multibody system as a collection of points and vectors.

4.1   Describing a body as a collection of points

Two or more points can be used to describe a rigid body, as shown in Figure 8. These points may represent attachment points between bodies or a particular axis. We refer to such points as *primary* points. The coordinates of a primary point appear explicitly in the formulation of the constraints and equations of motion. The coordinates, velocity, and acceleration of a typical primary points are defined by their x-y-z components as $\mathbf{r}^i$, $\dot{\mathbf{r}}^i$, and $\ddot{\mathbf{r}}^i$.

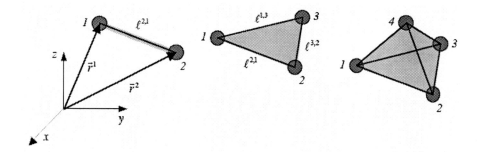

Figure 8. Rigid bodies described by two, three, and four primary points

For two primary points defined on a body a *length* constraint exists as:

$$\Phi^{(l,1)} \equiv \frac{1}{2}((\mathbf{r}^i - \mathbf{r}^j)^T(\mathbf{r}^i - \mathbf{r}^j) - l^{i,j^2}) = 0 . \tag{27}$$

For a body with 2 primary points we have 6 coordinates and one constraint; i.e., 5 DoF; 3 primary points provide 9 coordinates, 3 constraints, 6 DoF; and 4 primary points not being in the same plane yield 12 coordinates, 6 constraints, and 6 DoF.

Non-primary or *secondary* points can also be defined on a body. The coordinates of a secondary point are computed from the coordinates of the primary points. For a body defined by two primary points, the coordinates of the secondary point A, as shown in Figure 9a), can be computed as:

$$\mathbf{r}^A = (a^1\mathbf{r}^2 - a^2\mathbf{r}^1)/l^{1,2} , \tag{28}$$

where a^1 and a^2 are directional constants. For a body with three primary points as shown in Figure 9b), the coordinates of the secondary point A, which is not necessarily in the plane of the primary points, are computed as:

$$\mathbf{r}^A = \mathbf{r}^1 + a^1(\mathbf{r}^2 - \mathbf{r}^1) + a^2(\mathbf{r}^3 - \mathbf{r}^1) + a^3(\tilde{\mathbf{r}}^2 - \tilde{\mathbf{r}}^1)(\mathbf{r}^3 - \mathbf{r}^1) . \tag{29}$$

The three constants can be determined in several ways. For example if $\mathbf{r}^1$, $\mathbf{r}^2$, $\mathbf{r}^3$, and $\mathbf{r}^A$ are known initially, we compute $\mathbf{s}^{2,1} = \mathbf{r}^2 - \mathbf{r}^1$, $\mathbf{s}^{3,1} = \mathbf{r}^3 - \mathbf{r}^1$, $\mathbf{s}^A = \mathbf{r}^A - \mathbf{r}^1$, $\mathbf{s} = \tilde{\mathbf{s}}^{2,1}\mathbf{s}^{3,1}$. We then describe

$$\mathbf{s}^A = a^1\mathbf{s}^{2,1} + a^2\mathbf{s}^{3,1} + a^3\mathbf{s}, \ \mathbf{S} \equiv \begin{bmatrix} \mathbf{s}^{2,1} & \mathbf{s}^{3,1} & \mathbf{s} \end{bmatrix}, \ \mathbf{a} \equiv \{ a^1 \ \ a^2 \ \ a^3 \}^T$$

and $\mathbf{s}^A = \mathbf{Sa}$. Solving $\mathbf{a} = \mathbf{S}^{-1}\,\mathbf{s}^A$ yields the three constants.

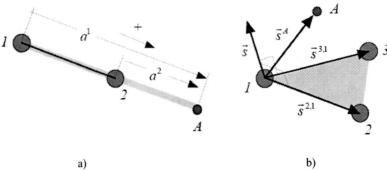

a)   b)

Figure 9. A secondary point defined on a) two and b) three point systems

4.2 Kinematic joints

Kinematic joints can be described between bodies by either allowing primary points to be shared between bodies, or describing simple conditions between vectors that connect primary points. If two primary points are shared, the two points require only one vector of coordinates. Figure 10 shows three types of joints: (a) a spherical joint where one point is shared; (b) a revolute joint with two shared points; and (c) a universal joint with one shared point and one constraint between two vectors; e.g., $\mathbf{s}_i^T\mathbf{s}_j = 0$. Other types of joints can be described in a similar manner.

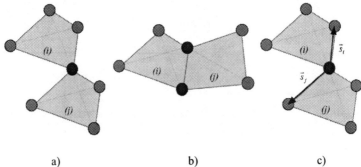

a)   b)   c)

Figure 10. Three types of joints between bodies defined by primary points

4.3 Force distribution

Forces and moments that act on a body must be properly distributed to the primary points. In general, if a force acts on point P on a body as shown in Figure 11a), and the body is described by n primary points, an equivalent set of forces acting on the primary points as shown in Figure 11b) must be determined. The equivalency equations are written as:

$$\sum_{j=1}^{n} \mathbf{f}^{j} = \mathbf{f}_{i}^{P}, \sum_{j=1}^{n} \tilde{\mathbf{s}}^{j} \mathbf{f}^{j} = \tilde{\mathbf{s}}_{i}^{P} \mathbf{f}_{i}^{P}$$

If we assume that all n forces are parallel to the original force, we substitute

$$\mathbf{f}_{i} = \alpha^{j} \mathbf{f}_{i}^{P} ; j = 1,..., n \tag{30}$$

in the above equations. Since the result must be applicable to any force acting on the body, after simplification we have:

$$\sum_{j=1}^{n} \alpha^{j} = 1, \ \sum_{j=1}^{n} \alpha^{j} \mathbf{s}^{j} = \mathbf{s}_{i}^{P} . \tag{31}$$

These equations are solved once for the constants. The unknown coefficients are a function of the position of point P and not a function of the magnitude or the direction of the applied force.

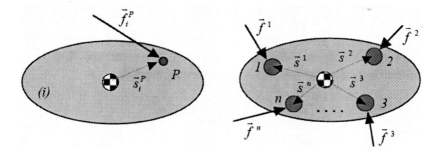

Figure 11. Force-moment equivalent systems.

Figure 12 shows two cases for two and four primary points. For two primary points, Eq. (31) finds the following form:

$$\begin{bmatrix} \varsigma^{1} & \varsigma^{2} \\ 1 & 1 \end{bmatrix} \begin{Bmatrix} \alpha^{1} \\ \alpha^{2} \end{Bmatrix} = \begin{Bmatrix} \varsigma^{P} \\ 1 \end{Bmatrix}$$

where there are only two constants to be determined. For four points, Eq. (31) becomes:

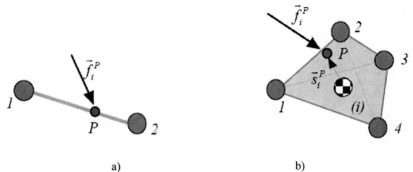

a)   b)

Figure 12. A force acting on bodies represented by primary points.

$$
\begin{bmatrix}
\xi^1 & \xi^2 & \xi^3 & \xi^4 \\
\eta^1 & \eta^2 & \eta^3 & \eta^4 \\
\zeta^1 & \zeta^2 & \zeta^3 & \zeta^4 \\
1 & 1 & 1 & 1
\end{bmatrix}
\begin{Bmatrix}
\alpha^1 \\
\alpha^2 \\
\alpha^3 \\
\alpha^4
\end{Bmatrix}
=
\begin{Bmatrix}
\xi^P \\
\eta^P \\
\zeta^P \\
1
\end{Bmatrix}.
$$

4.4   Mass distribution

Mass of a body must be distributed among the primary points in such a way that the inertial characteristics of the body are preserved. This requires the following conditions for a body described by n points:

$$
\sum_{j=1}^{n} \mathbf{m}^j = m_i \quad \text{(1 equation) (a)},
$$

$$
\sum_{j=1}^{n} \mathbf{s}^j \mathbf{m}^j = \mathbf{0} \quad \text{(3 equations) (b)}, \tag{32}
$$

$$
\sum_{j=1}^{n} \tilde{\mathbf{s}}^{j^{\mathrm{T}}} \tilde{\mathbf{s}}^j \mathbf{m}^j = \mathbf{J}_i \quad \text{(6 equations) (c)}.
$$

Note that Eq. (32)(c) is a matrix identity. Since the inertia matrix is symmetric, it yields 6 algebraic equations. Equation (32) represents 10 equations and therefore we can have up to 10 unknowns.

Figure 13 shows three cases: for a body with 4 primary points we temporarily define 6 secondary points; for 3 primary points we define 3 secondary points; and for 2 primary points we define 1 secondary point.

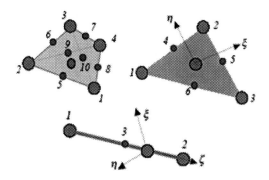

Figure 13. Body mass is distributed between primary and secondary points.

Secondary points are defined for convenience exactly equi-distance from two primary points. Note that for 3 and 2 primary point cases, the mass center must be respectively on the plane and on the axis of the body. For these cases we end up with 10, 6, and 3 equations respectively. These sets of equations can be solved for the point masses. The coordinates of the secondary points will be eliminated from the model (but not their masses) as it will be seen in the next sub-section.

4.5 Newton equations of motion and inertia matrix

In this section we eliminate the secondary points from the description of a body without any approximations. This process is described in detail for the case of 2 primary and 1 secondary points.

Assume that the mass center of the slender rod in Figure 13c) is at the geometric center of the rod where the secondary point is also defined. We further assume that the mass of these three points have already been computed as described in the preceding section. The position constraints between the three points are expressed as:

$$
\begin{aligned}
(\mathbf{r}^1 - \mathbf{r}^2)^{\mathrm{T}} (\mathbf{r}^1 - \mathbf{r}^2) - (l^{1,2})^2 = 0 \quad &\text{(a)} \\
\mathbf{r}^1 + \mathbf{r}^2 - 2\mathbf{r}^3 = \mathbf{0} \quad &\text{(b)}
\end{aligned}
\tag{33}
$$

The first time derivative of these equations provides the velocity constraints. The second time derivative is written in matrix form as:

$$
\begin{bmatrix}
(\mathbf{r}^1 - \mathbf{r}^2)^T & -(\mathbf{r}^1 - \mathbf{r}^2)^T & 0 \\
\mathbf{I} & \mathbf{I} & -2\mathbf{I}
\end{bmatrix}
\begin{bmatrix} \ddot{\mathbf{r}}^1 \\ \ddot{\mathbf{r}}^2 \\ \ddot{\mathbf{r}}^3 \end{bmatrix}
= \begin{Bmatrix} -(\dot{\mathbf{r}}^1 - \dot{\mathbf{r}}^2)^T (\dot{\mathbf{r}}^1 - \dot{\mathbf{r}}^2) \\ 0 \end{Bmatrix}.
\tag{34}
$$

Newton equations of motion with Lagrange multipliers are written as:

$$
\begin{bmatrix}
m^1\mathbf{I} & 0 & 0 \\
0 & m^2\mathbf{I} & 0 \\
0 & 0 & m^3\mathbf{I}
\end{bmatrix}
\begin{bmatrix} \ddot{\mathbf{r}}^1 \\ \ddot{\mathbf{r}}^2 \\ \ddot{\mathbf{r}}^3 \end{bmatrix}
-
\begin{bmatrix}
\mathbf{r}^1 - \mathbf{r}^2 & \mathbf{I} \\
\mathbf{r}^2 - \mathbf{r}^1 & \mathbf{I} \\
0 & -2\mathbf{I}
\end{bmatrix}
\begin{Bmatrix} \lambda^1 \\ \lambda^2 \end{Bmatrix}
= \begin{Bmatrix} \mathbf{f}^1 \\ \mathbf{f}^2 \\ 0 \end{Bmatrix}.
\tag{35}
$$

We eliminate the secondary point 3 from the equations by applying a *coordinate transformation* to obtain:

$$
\begin{bmatrix}
(m^1 + \dfrac{m^3}{4})\mathbf{I} & \dfrac{m^3}{4}\mathbf{I} \\[2mm]
\dfrac{m^3}{4}\mathbf{I} & (m^2 + \dfrac{m^3}{4})\mathbf{I}
\end{bmatrix}
\begin{Bmatrix} \ddot{\mathbf{r}}^1 \\ \ddot{\mathbf{r}}^2 \end{Bmatrix}
-
\begin{bmatrix} \mathbf{r}^1 - \mathbf{r}^2 \\ \mathbf{r}^2 - \mathbf{r}^1 \end{bmatrix} \lambda^1
= \begin{Bmatrix} \mathbf{f}^1 \\ \mathbf{f}^2 \end{Bmatrix}.
\tag{36}
$$

where the only comstraint left is Eq. (33)(a) and its first and second time derivates.

For this special case where the secondary point and the mass center are at the geometric center of the rod, we have $m^1 = m^2 = m/6$ and $m^3 = 2m/3$. After eliminating the secondary point from the equations of motion, the mass matrix becomes:

$$
\frac{m}{3}
\begin{bmatrix}
\mathbf{I} & \dfrac{1}{2}\mathbf{I} \\[2mm]
\dfrac{1}{2}\mathbf{I} & \mathbf{I}
\end{bmatrix}.
\tag{37}
$$

For three primary points, the equations of motion are found in a similar process to be:

$$
\begin{bmatrix}
m^{1,1}\mathbf{I} & m^{1,2}\mathbf{I} & m^{1,3}\mathbf{I} \\
m^{2,1}\mathbf{I} & m^{2,2}\mathbf{I} & m^{2,3}\mathbf{I} \\
m^{3,1}\mathbf{I} & m^{3,2}\mathbf{I} & m^{3,3}\mathbf{I}
\end{bmatrix}
\begin{Bmatrix} \ddot{\mathbf{r}}^1 \\ \ddot{\mathbf{r}}^2 \\ \ddot{\mathbf{r}}^3 \end{Bmatrix}
-
\begin{bmatrix}
\mathbf{r}^1 - \mathbf{r}^2 & 0 & \mathbf{r}^1 - \mathbf{r}^3 \\
\mathbf{r}^2 - \mathbf{r}^1 & \mathbf{r}^2 - \mathbf{r}^3 & 0 \\
0 & \mathbf{r}^3 - \mathbf{r}^2 & \mathbf{r}^3 - \mathbf{r}^1
\end{bmatrix}
\begin{Bmatrix} \lambda^1 \\ \lambda^2 \\ \lambda^3 \end{Bmatrix}
= \begin{Bmatrix} \mathbf{f}^1 \\ \mathbf{f}^2 \\ \mathbf{f}^3 \end{Bmatrix},
\tag{38}
$$

where

$$m^{1,1} = m^1 + \frac{m^4 + m^6}{4}, \quad m^{1,2} = m^{2,1} = \frac{m^4}{4}, \quad m^{1,3} = m^{3,1} = \frac{m^6}{4},$$

$$m^{2,2} = m^2 + \frac{m^4 + m^5}{4}, \quad m^{2,3} = m^{3,2} = \frac{m^5}{4}, \quad m^{3,3} = m^3 + \frac{m^5 + m^6}{4}.$$

For four primary points, the equations of motions are found to be:

$$
\begin{bmatrix}
m^{1,1}I & m^{1,2}I & m^{1,3}I & m^{1,4}I \\
m^{2,1}I & m^{2,2}I & m^{2,3}I & m^{2,4}I \\
m^{3,1}I & m^{3,2}I & m^{3,3}I & m^{3,4}I \\
m^{4,1}I & m^{4,2}I & m^{4,3}I & m^{4,4}I
\end{bmatrix}
\begin{Bmatrix}
\ddot{r}^1 \\
\ddot{r}^2 \\
\ddot{r}^3 \\
\ddot{r}^4
\end{Bmatrix} -
$$

$$
-\begin{bmatrix}
-s^{2,1} & -s^{3,1} & -s^{4,1} & 0 & 0 & 0 \\
s^{2,1} & 0 & 0 & -s^{3,2} & -s^{4,2} & 0 \\
0 & s^{3,1} & 0 & s^{3,2} & 0 & -s^{4,3} \\
0 & 0 & s^{4,1} & 0 & s^{4,2} & s^{4,3}
\end{bmatrix}
\begin{Bmatrix}
\lambda^1 \\
\lambda^2 \\
\lambda^3 \\
\lambda^4 \\
\lambda^5 \\
\lambda^6
\end{Bmatrix}
=
\begin{Bmatrix}
f^1 \\
f^2 \\
f^3 \\
f^4
\end{Bmatrix}, \quad (39)
$$

where

$$s^{i,j} = r^i - r^j, \quad m^{1,1} = m^1 + \frac{m^5 + m^8 + m^{10}}{4}, \quad m^{1,2} = m^{2,1} = \frac{m^5}{4},$$

$$m^{1,3} = m^{3,1} = \frac{m^{10}}{4}, \quad m^{1,4} = m^{4,1} = \frac{m^8}{4}, \quad m^{2,2} = m^2 + \frac{m^5 + m^6 + m^9}{4}$$

$$m^{2,3} = m^{3,2} = \frac{m^6}{4}, \quad m^{2,4} = m^{4,2} = \frac{m^9}{4}, \quad m^{3,3} = m^3 + \frac{m^6 + m^7 + m^{10}}{4},$$

$$m^{3,4} = m^{4,3} = \frac{m^7}{4}, \quad m^{4,4} = m^4 + \frac{m^7 + m^8 + m^9}{4}.$$

4.6   A multibody system

Assume that a multibody system is described by n_p primary points. An array of coordinates, **r**, is defined containing $n_v = 3 \times n_p$ coordinates. The position constraints can be described by n_c algebraic equations as:

$$\mathbf{\Phi}(\mathbf{r}) = \mathbf{0}.$$

(40)

The velocity and acceleration constraints are:

$$\dot{\mathbf{\Phi}} \equiv \mathbf{D}\dot{\mathbf{r}} = \mathbf{0}.$$

(41)

$$\ddot{\mathbf{\Phi}} \equiv \mathbf{D}\ddot{\mathbf{r}} + \dot{\mathbf{D}}\dot{\mathbf{r}} = \mathbf{0}.$$

(42)

The Newton's equations of motion with Lagrange multipliers are expressed as:

$$\mathbf{M}\ddot{\mathbf{r}} - \mathbf{D}^T \lambda = \mathbf{f}.$$

(43)

Special attention must be paid when we construct the inertia matrix and the force vector in Eq. (43). A shared primary point can receive its inertia and its vector of force from two bodies. We show this through a simple example. Assume that two rods, each described by two primary points, share one point as shown in Figure 14. The body and point indices are also shown in the Figure. Assume that the mass matrix for each rod is described as in Eq. (37). The equations of motion for this system of three primary points are expressed as (subscripts refer to the body indices):

$$
\begin{bmatrix}
\dfrac{m_1}{3}\mathbf{I} & \dfrac{m_1}{6}\mathbf{I} & 0 \\[2mm]
\dfrac{m_1}{6}\mathbf{I} & \dfrac{m_1 + m_2}{3}\mathbf{I} & \dfrac{m_2}{6}\mathbf{I} \\[2mm]
0 & \dfrac{m_2}{6}\mathbf{I} & \dfrac{m_2}{3}\mathbf{I}
\end{bmatrix}
\begin{Bmatrix}
\ddot{\mathbf{r}}^1 \\ \ddot{\mathbf{r}}^2 \\ \ddot{\mathbf{r}}^3
\end{Bmatrix}
-
\begin{bmatrix}
\mathbf{d}_1 & 0 \\
-\mathbf{d}_1 & \mathbf{d}_2 \\
0 & -\mathbf{d}_2
\end{bmatrix}
\begin{Bmatrix}
\lambda_1 \\ \lambda_2
\end{Bmatrix}
=
\begin{Bmatrix}
\mathbf{f}_1^1 \\ \mathbf{f}_1^2 + \mathbf{f}_2^2 \\ \mathbf{f}_2^3
\end{Bmatrix}.
$$

(44)

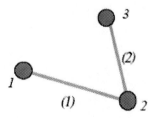

Figure 14. Two rods sharing a primary point.

5. JOINT-COORDINATE FORMULATION

In this formulation the equations of motion from the body coordinate formulation are transformed to a smaller set [4-6]. The generalized coordinates in this formulation are mostly relative coordinates associated with the kinematic joints that connect the bodies. This formulation is first described for *open-chain* systems (no closed kinematic loops) and then extended to *closed-chain* systems.

5.1 Body indices and joint velocities

In an open-chain system, the multibody is viewed as a *tree*. We start from the *ground* (or the root), then move to a body via *its* joint, then to another body and so on until we reach a *leaf* where no further bodies could be reached. A leaf represents the end of a *branch*. Bodies are numbered in any desired order, however numbering in ascending order provides easier book keeping. As shown in Figure 15, for a typical body *i*, the attached body closer to the root is referred to as body *i-1*. The joint between these two bodies is said to belong to body *i* and, therefore, all of its attributes carry the index of body *i*. Each body will be assigned a body-fixed reference frame with the origin at the mass center as in the body-coordinate formulation.

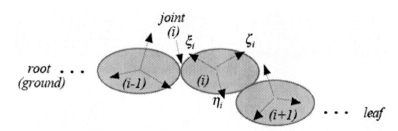

Figure 15. Body and joint indices.

In an open-chain system each kinematic joint is represented by a joint coordinate and an associated joint velocity. The number of components in a joint velocity is equal to the number of DoF of the joint. Figure 15 shows three typical kinematic joints: a revolute joint with 1 rotational velocity about the joint axis; a prismatic joint with 1 translational velocity along the joint axis; and a spherical joint with a relative angular velocity vector containing 3 components. If a body does not own an actual joint, it is assigned a floating joint with 6 DoF, $\mathbf{v}_i = \{\mathbf{r}_i^T \quad \boldsymbol{\omega}_i^T\}^T$. Single DoF joints can be combined to represent joints with more than 1 DoF. For example, as shown in Figure 16, a universal joint is represented by 2 revolute joints and a cylindrical joint is represented by one revolute joint and one prismatic joint.

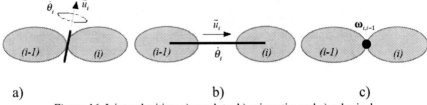

a)   b)   c)

Figure 16. Joint velocities: a) revolute; b) prismatic; and c) spherical.

For the system shown in Figure 18 with *b* bodies, whether being a single branch or multi-branch tree, an array of joint velocities is defined as:

$$\dot{\boldsymbol{\theta}} \equiv \left\{ \mathbf{v}_1^T \quad \dot{\theta}_2 \quad \cdots \quad \dot{\theta}_b \right\}^T \tag{45}$$

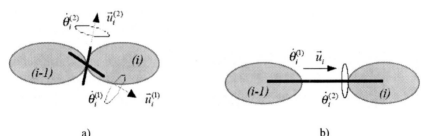

a)   b)

Figure 17. Joint velocities: a) universal; and b) cylindrical.

This array contains as many velocity components as the number of DoF. A useful definition is vector $\vec{d}_{i,j}$, shown in Figure 19, connecting the origin of body *i* to a point on joint *j*. This vector is computed as:

$$\mathbf{d}_{i,j} = \mathbf{r}_i - \mathbf{r}_j^P . \tag{46}$$

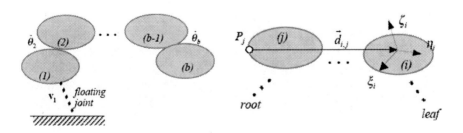

Figure 18. A single branch system.   *Figure 19.* Definition of vector d$_{ij}$

5.2   Velocity transformation

By deriving the velocity relationships, it can be shown that there is a transformation between *body velocities* and *joint velocities* as:

$$\mathbf{v} = \mathbf{B}\dot{\theta} \,. \tag{47}$$

Matrix **B** is called the *velocity transformation matrix* and it is orthogonal to the system Jacobian:

$$\mathbf{DB} = \mathbf{0} \,. \tag{48}$$

The structure of matrix **B** can be demonstrated through a simple example. Consider the open-chain single-branch system shown in Figure 20(a) containing 5 bodies, 1 floating joint, 2 revolute joints, 1 prismatic joint, and 1 spherical joint. The velocity transformation equation for this system, in symbolic form, is found to be:

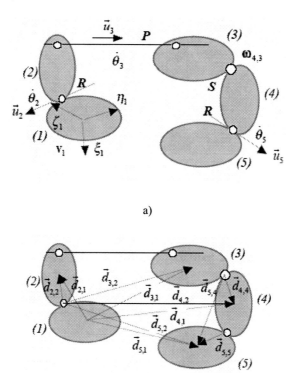

a)

b)

Figure 20. An example for constructing the velocity transformation matrix.

$$
\begin{Bmatrix} \mathbf{v}_1 \\ \mathbf{v}_2 \\ \mathbf{v}_3 \\ \mathbf{v}_4 \\ \mathbf{v}_5 \end{Bmatrix} = \begin{bmatrix} \mathbf{F}_{1,1} & \mathbf{0} & \mathbf{0} & \mathbf{0} & \mathbf{0} \\ \mathbf{F}_{2,1} & \mathbf{R}_{2,2} & \mathbf{0} & \mathbf{0} & \mathbf{0} \\ \mathbf{F}_{3,1} & \mathbf{R}_{3,2} & \mathbf{P}_{3,3} & \mathbf{0} & \mathbf{0} \\ \mathbf{F}_{4,1} & \mathbf{R}_{4,2} & \mathbf{P}_{4,3} & \mathbf{S}_{4,4} & \mathbf{0} \\ \mathbf{F}_{5,1} & \mathbf{R}_{5,2} & \mathbf{P}_{5,3} & \mathbf{S}_{5,4} & \mathbf{R}_{5,5} \end{bmatrix} \begin{Bmatrix} \mathbf{v}_1 \\ \dot{\theta}_2 \\ \dot{\theta}_3 \\ \omega_{4,3} \\ \dot{\theta}_5 \end{Bmatrix} .
$$

The elements in the velocity transformation matrix are called *block matrices*. Block matrices for four fundamental joints are described in Table 3. The description of the $\mathbf{d}_{i\,j}$, vectors that appear in the block-matrices for our example can be found in Figure 20b). We note that $\mathbf{d}_{1,1} = 0$. The main conclusion we draw from this example is that the velocity transformation matrix for any open-chain system can be constructed systematically from the topology of the tree structure and the block-matrices.

Table 3. Elementary block-matrices

Type	Size	ID	Entries	Time derivative
Floating	6x6	$\mathbf{F}_{i,j}$	$\begin{bmatrix} \mathbf{I} & -\tilde{\mathbf{d}}_{i,j} \\ \mathbf{0} & \mathbf{I} \end{bmatrix}$	$\begin{bmatrix} \mathbf{0} & -\dot{\tilde{\mathbf{d}}}_{i,j} \\ \mathbf{0} & \mathbf{0} \end{bmatrix}$
Revolute	6x1	$\mathbf{R}_{i,j}$	$\begin{bmatrix} -\tilde{\mathbf{d}}_{i,j}\mathbf{u}_j \\ \mathbf{u}_j \end{bmatrix}$	$\begin{bmatrix} -(\dot{\tilde{\mathbf{d}}}_{i,j} + \tilde{\mathbf{d}}_{i,j}\tilde{\omega}_j)\mathbf{u}_j \\ \tilde{\omega}_j\mathbf{u}_j \end{bmatrix}$
Prismatic	6x1	$\mathbf{P}_{i,j}$	$\begin{bmatrix} \mathbf{u}_j \\ \mathbf{0} \end{bmatrix}$	$\begin{bmatrix} \tilde{\omega}_j\mathbf{u}_j \\ \mathbf{0} \end{bmatrix}$
Spherical	6x3	$\mathbf{S}_{i,j}$	$\begin{bmatrix} -\tilde{\mathbf{d}}_{i,j} \\ \mathbf{I} \end{bmatrix}$	$\begin{bmatrix} -\dot{\tilde{\mathbf{d}}}_{i,j} \\ \mathbf{0} \end{bmatrix}$

5.3   Equations of motion for open-chain systems

The time derivative of Eq. (47) yields the acceleration transformation equation as:

$$
\dot{\mathbf{v}} = \mathbf{B}\ddot{\theta} + \dot{\mathbf{B}}\dot{\theta} . \tag{49}
$$

Substituting Eqs. (47) and (49) in the constraints of Eqs. (16) and (17), and then using Eq. (48) shows that all of the constraints will disappear when we transform the constraints to the joint coordinates. Substituting Eq. (49) into Eq. (18) and pre-multiplying the result by $\mathbf{B}^{\mathrm{T}}$, then using Eq. (48), yield the equations of motion as:

$$M\ddot{\theta} = f, \tag{50}$$

where

$$\begin{aligned} M &= \mathbf{B}^{\mathrm{T}}\mathbf{MB} \\ f &= \mathbf{B}^{\mathrm{T}}(\mathbf{g} - \mathbf{MB}\dot{\theta})^{\cdot} \end{aligned} \tag{51}$$

Equation (50) represents as many second-order differential equations as the number of DoF.

5.4   Closed-chain systems

A multibody system containing one or more closed chains can temporarily be transformed to an open-chain system, also known as a *reduced* system, by removing one kinematic joint from each loop. The removed joints are called *cut-joints* and they are denoted by the symbol "⊗". Figure 21 shows schematically a system before and after the cut-joint process. The reduced system is first modeled by joint coordinate formulation, and then the cut-joints are put back into the system. When a cut-joint is put back in a loop, the joint coordinates in that loop become dependent on one another. Therefore, for the cut-joints, the following kinematic constraints can be written:

$$\Phi^{\otimes}(\mathbf{c}) = \mathbf{0}, \tag{52}$$

$$\dot{\Phi}^{\otimes} \equiv \mathbf{D}^{\otimes}\mathbf{v} = \mathbf{D}^{\otimes}\mathbf{B}\dot{\theta} = \mathbf{C}\dot{\theta} = \mathbf{0}, \tag{53}$$

$$\ddot{\Phi}^{\otimes} \equiv \mathbf{D}^{\otimes}\dot{\mathbf{v}} + \dot{\mathbf{D}}^{\otimes}\mathbf{v} = \mathbf{C}\ddot{\theta} + \dot{\mathbf{C}}\dot{\theta} = \mathbf{0}, \tag{54}$$

where,

$$\mathbf{C} \equiv \mathbf{D}^{\otimes}\mathbf{B}, \tag{55}$$

$$\dot{\mathbf{C}} = \mathbf{D}^{\otimes}\dot{\mathbf{B}} + \dot{\mathbf{D}}^{\otimes}\mathbf{B}. \tag{56}$$

Since the position constraints are nonlinear in the coordinates, Eq. (52) is kept in terms of the body coordinates. For computational purposes, the term $\mathbf{C}\dot{\boldsymbol{\theta}}$ in Eq. (54) can be expressed as: $\mathbf{C}\dot{\boldsymbol{\theta}} = \mathbf{D}^{\otimes}\mathbf{B}\dot{\boldsymbol{\theta}} + \dot{\mathbf{D}}^{\otimes}\mathbf{v}.$

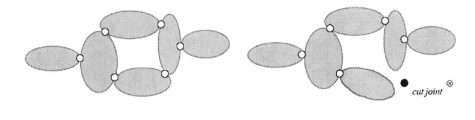

a)   b)

Figure 21. A closed-chain system and its reduced open-chain system.

5.5   Equations of motion for closed-chain systems

In the presence of constraints in a closed-chain system, the open-chain equations of motion; i.e., Eq. (50), is revised with the aid of Lagrange multipliers to obtain:

$$M\ddot{\boldsymbol{\theta}} - \mathbf{C}^{T\,\otimes}\boldsymbol{\lambda} = f . \tag{57}$$

Equations (52)-(57) provide mixed differential-algebraic equations of motion for a closed-chain system. The number of equations, however, is much smaller than that of the body-coordinate formulation.

5.6   Reduction to a minimal set

The equations of motion for a closed-chain system can be reduced to a minimal set by applying a second step of velocity transformation. For the multibody system of interest, we define a set of independent joint velocities, $\dot{\boldsymbol{\theta}}^{(i)}$. A closed-chain velocity transformation matrix $\mathbf{E}$ can be expressed as:

$$\dot{\boldsymbol{\theta}} = \mathbf{E}\dot{\boldsymbol{\theta}}^{(i)} . \tag{58}$$

Matrix $\mathbf{E}$ can be obtained from the $\mathbf{C}$ matrix as:

$$\dot{\boldsymbol{\Phi}}^{\otimes} \equiv \mathbf{C}\dot{\boldsymbol{\theta}} = \mathbf{C}^{(i)}\dot{\boldsymbol{\theta}}^{(i)} + \mathbf{C}^{(d)}\dot{\boldsymbol{\theta}}^{(d)} = \mathbf{0} ,$$

where $\dot{\boldsymbol{\theta}}^{(d)} = -\mathbf{C}^{(d)^{-1}}\mathbf{C}^{(i)}\dot{\boldsymbol{\theta}}^{(i)}$ contains dependent joint velocities. This yields:

$$\dot{\boldsymbol{\theta}} \equiv \begin{bmatrix} \dot{\boldsymbol{\theta}}^{(i)} \\ \dot{\boldsymbol{\theta}}^{(d)} \end{bmatrix} = \begin{bmatrix} \mathbf{I} \\ -\mathbf{C}^{(d)^{-1}} \mathbf{C}^{(i)} \end{bmatrix} \dot{\boldsymbol{\theta}}^{(i)}. \tag{59}$$

Therefore, the closed-chain velocity transformation matrix becomes:

$$\mathbf{E} = \begin{bmatrix} \mathbf{I} \\ -\mathbf{C}^{(d)^{-1}} \mathbf{C}^{(i)} \end{bmatrix}. \tag{60}$$

This matrix has the following characteristics:

$$\mathbf{CE} = 0. \tag{61}$$

In most practical applications, matrix $\mathbf{C}$ associated with a closed chain is very small in dimensions. Therefore, $\mathbf{E}$ can efficiently be evaluated in numerical form.

The acceleration transformation equation is written as:

$$\ddot{\boldsymbol{\theta}} = \mathbf{E}\ddot{\boldsymbol{\theta}}^{(i)} + \dot{\mathbf{E}}\dot{\boldsymbol{\theta}}^{(i)}. \tag{62}$$

It can be found that the term $\dot{\mathbf{E}}\dot{\boldsymbol{\theta}}^{(i)}$ can be computed as

$$\dot{\mathbf{E}}\dot{\boldsymbol{\theta}}^{(i)} = \begin{bmatrix} 0 \\ \mathbf{C}^{(d)^{-1}} \dot{\mathbf{C}}^{(i)}\dot{\boldsymbol{\theta}} \end{bmatrix}.$$

Substituting Eq. (62) into Eq. (57), pre-multiplying the result by $\mathbf{E}^{\mathrm{T}}$, and then employing Eq. (61) yield:

$$M^{(i)}\ddot{\boldsymbol{\theta}}^{(i)} = f^{(i)}, \tag{63}$$

where

$$\begin{aligned} M^{(i)} &= \mathbf{E}^{\mathrm{T}}M\mathbf{E} \\ f^{(i)} &= \mathbf{E}^{\mathrm{T}}(f - M\dot{\mathbf{E}}\dot{\boldsymbol{\theta}}^{(i)}) \end{aligned}. \tag{64}$$

Equation (63) represents a set of highly nonlinear second-order differential equations. There are no constraints or any associated Lagrange multipliers in these equations.

6.   RIGID-DEFORMABLE MULTIBODY SYSTEMS

A multibody system may contain both rigid and deformable bodies. Therefore in this section a brief overview of some fundamental issues associated with modeling deformable bodies in a multibody setting is provided. We first review the equations of motion for a single deformable body and discuss some of the corresponding issues. We then show the equations of motion for a multibody system containing both rigid and deformable bodies.

6.1   A deformable body

A deformable body undergoing large rigid-body motion can be modeled by finite-element methodology [7, 8]. The equations of motion, however, must be formulated properly in order to capture all the terms associated with the rigid-body motion. In this section, for notational simplification and without any loss of generality, we only consider nodal *translational* deflections. Assume that a deformable body is represented by n_{nodes} nodes. If the system is not constrained to the ground, it will have $n_{dof} = 3 \times n_{nodes}$ degrees-of-freedom. For such a body shown in Figure 22, if the $\xi - \eta - \zeta$ reference frame is attached to the ground, the equations of motion are described as:

$$M\ddot{\delta} = f - K\delta , \tag{65}$$

where all the entities are described in the $\xi - \eta - \zeta$ frame. In this equation, M, K, f, and δ are, respectively, the mass matrix, the stiffness matrix, the array of forces, and the array of deflections. As long as no boundary conditions are defined, the stiffness matrix K remains singular.

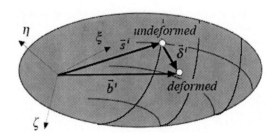

Figure 22. A deformable body and a non-moving reference frame.

If the non-moving $\xi - \eta - \zeta$ frame is positioned with respect to another non-moving *x-y-z* frame, as shown in Figure 23, Eq. (65) can be transformed using the rotational transformation matrix $\mathbf{A}$ to obtain:

$$\mathbf{M}\ddot{\boldsymbol{\delta}} = \mathbf{f} - \mathbf{K}\boldsymbol{\delta}, \tag{66}$$

where $\quad \mathbf{M} = \overline{\mathbf{A}} M \overline{\mathbf{A}}^{T}, \quad \mathbf{K} = \overline{\mathbf{A}} K \overline{\mathbf{A}}^{T}, \quad \mathbf{f} = \overline{\mathbf{A}} f, \quad \boldsymbol{\delta} = \overline{\mathbf{A}}^{T} \delta, \quad \ddot{\boldsymbol{\delta}} = \overline{\mathbf{A}}^{T} \ddot{\delta}, \quad$ and
$\overline{\mathbf{A}} = block - diag.[\mathbf{A} \quad \cdots \quad \mathbf{A}]$. For a properly constructed mass matrix, $\mathbf{M} = M$.

If the body and its body-fixed frame are allowed to move with respect to the non-moving *x-y-z* frame, the equations of motion are revised as:

$$\mathbf{M}\ddot{\mathbf{d}} = \mathbf{f} - \mathbf{K}\boldsymbol{\delta}, \tag{67}$$

where $\ddot{\mathbf{d}}$ is the array of nodal accelerations as viewed in the *x-y-z* frame.

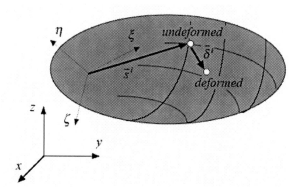

Figure 23. A deformable body and two non-moving reference frames.

6.2   Nodal kinematics

The position of node *i* in the *x-y-z* frame, as shown in Figure 24, can be written as:

$$\mathbf{d}^{i} = \mathbf{r} + \mathbf{b}^{i} = \mathbf{r} + \mathbf{s}^{i} + \boldsymbol{\delta}^{i}. \tag{68}$$

The first and second time derivatives of Eq. (68) yield the velocity and acceleration as:

$$\dot{\mathbf{d}}^{i} = \dot{\mathbf{r}} - \tilde{\mathbf{b}}^{i}\boldsymbol{\omega} + \dot{\boldsymbol{\delta}}^{i}, \tag{69}$$

$$\ddot{\mathbf{d}}^{i} = \ddot{\mathbf{r}} - \tilde{\mathbf{b}}^{i}\dot{\boldsymbol{\omega}} + \ddot{\boldsymbol{\delta}}^{i} + \mathbf{w}^{i}, \tag{70}$$

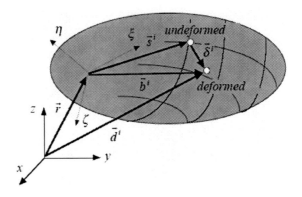

Figure 24. A deformable body moving in space

where $\dot{\boldsymbol{\omega}}$ and $\boldsymbol{\omega}$ are the angular velocity and acceleration of $\xi - \eta - \zeta$ frame, and

$$\mathbf{w}^i = \tilde{\boldsymbol{\omega}}\tilde{\boldsymbol{\omega}}\mathbf{b}^i + 2\tilde{\boldsymbol{\omega}}\dot{\boldsymbol{\delta}}^i \tag{71}$$

contains quadratic velocity terms. Equations (68)-(70) can be written for all the nodes as:

$$\mathbf{d} = \hat{\mathbf{I}}\mathbf{r} + \mathbf{b}, \tag{72}$$

$$\dot{\mathbf{d}} = \hat{\mathbf{I}}\dot{\mathbf{r}} - \hat{\tilde{\mathbf{b}}}\boldsymbol{\omega} + \dot{\boldsymbol{\delta}} = \begin{bmatrix} \hat{\mathbf{I}} & -\hat{\tilde{\mathbf{b}}} & \bar{\mathbf{I}} \end{bmatrix} \begin{Bmatrix} \dot{\mathbf{r}} \\ \boldsymbol{\omega} \\ \dot{\boldsymbol{\delta}} \end{Bmatrix}, \tag{73}$$

$$\ddot{\mathbf{d}} = \hat{\mathbf{I}}\ddot{\mathbf{r}} - \hat{\tilde{\mathbf{b}}}\dot{\boldsymbol{\omega}} + \ddot{\boldsymbol{\delta}} + \mathbf{w} = \begin{bmatrix} \hat{\mathbf{I}} & -\hat{\tilde{\mathbf{b}}} & \bar{\mathbf{I}} \end{bmatrix} \begin{Bmatrix} \ddot{\mathbf{r}} \\ \dot{\boldsymbol{\omega}} \\ \ddot{\boldsymbol{\delta}} \end{Bmatrix} + \bar{\tilde{\boldsymbol{\omega}}}\bar{\tilde{\boldsymbol{\omega}}}\mathbf{b} + 2\bar{\tilde{\boldsymbol{\omega}}}\dot{\boldsymbol{\delta}}, \tag{74}$$

where

$$\hat{\mathbf{I}} = \begin{bmatrix} \mathbf{I} \\ \vdots \\ \mathbf{I} \end{bmatrix}, \quad \bar{\mathbf{I}} = \begin{bmatrix} \mathbf{I} & \cdots & 0 \\ \vdots & \ddots & \vdots \\ 0 & \cdots & \mathbf{I} \end{bmatrix}, \quad \hat{\tilde{\mathbf{b}}} = \begin{bmatrix} \tilde{\mathbf{b}}^1 \\ \vdots \\ \tilde{\mathbf{b}}^{n_{nodes}} \end{bmatrix}, \quad \bar{\tilde{\boldsymbol{\omega}}} = \begin{bmatrix} \tilde{\boldsymbol{\omega}} & \cdots & 0 \\ \vdots & \ddots & \vdots \\ 0 & \cdots & \tilde{\boldsymbol{\omega}} \end{bmatrix}.$$

6.3   Equations of motion

Substituting Eq. (74) into Eq. (67) yields:

$$\mathbf{M}\begin{bmatrix} \hat{\mathbf{I}} & -\hat{\tilde{\mathbf{b}}} & \hat{\mathbf{I}} \end{bmatrix}\begin{Bmatrix} \ddot{\mathbf{r}} \\ \dot{\omega} \\ \ddot{\delta} \end{Bmatrix} = \mathbf{f} - \mathbf{Mw} - \mathbf{K}\delta. \tag{75}$$

These equations are not solvable for the accelerations since there are 6 more accelerations than equations. Equation (75) can be transformed to another form by pre-multiplying it by the transpose of the velocity transformation matrix in Eq. (73) to get:

$$\begin{bmatrix} m\mathbf{I} & -\hat{\mathbf{I}}^T\mathbf{M}\hat{\tilde{\mathbf{b}}} & \hat{\mathbf{I}}^T\mathbf{M} \\ -\hat{\tilde{\mathbf{b}}}^T\mathbf{M}\hat{\mathbf{I}} & \mathbf{J} & -\hat{\tilde{\mathbf{b}}}^T\mathbf{M} \\ \mathbf{M}\hat{\mathbf{I}} & -\mathbf{M}\hat{\tilde{\mathbf{b}}} & \mathbf{M} \end{bmatrix}\begin{bmatrix} \ddot{\mathbf{r}} \\ \dot{\omega} \\ \ddot{\delta} \end{bmatrix} = \begin{Bmatrix} \hat{\mathbf{I}}^T(\mathbf{f}-\mathbf{Mw}) \\ -\hat{\tilde{\mathbf{b}}}^T(\mathbf{f}-\mathbf{Mw}) \\ \mathbf{f}-\mathbf{Mw}-\mathbf{K}\delta \end{Bmatrix}, \tag{76}$$

where

$$\hat{\mathbf{I}}^T\mathbf{M}\hat{\mathbf{I}} = m\mathbf{I},\ \hat{\tilde{\mathbf{b}}}^T\mathbf{M}\hat{\tilde{\mathbf{b}}} = \mathbf{J},\ \hat{\mathbf{I}}^T\mathbf{K}\delta = 0,\ \hat{\tilde{\mathbf{b}}}^T\mathbf{K}\delta = 0. \tag{77}$$

Equation (76) is not solvable for the accelerations even though there are as many equations as accelerations. The reason is that we have not yet defined any boundary conditions on how $\xi - \eta - \zeta$ frame is attached to the nodes. Equations (75) and (76) can be expressed in terms of their $\xi - \eta - \zeta$ components if desired, except for $\ddot{\mathbf{r}}$.

6.4   Boundary conditions

The most common practice in the finite element modeling for boundary conditions is to properly set six of the nodal deflections to zero. We can assume that the origin of the $\xi - \eta - \zeta$ frame stays at node o, node j stays on the ξ-axis, and node k remains on the $\xi - \eta$ plane. These conditions are expressed as:

$$\delta^0_{(\xi)} = \delta^0_{(\eta)} = \delta^0_{(\zeta)} = 0,\ \delta^j_{(\eta)} = \delta^j_{(\zeta)} = 0,\ \delta^k_{(\zeta)} = 0 \tag{78}$$

These are called *nodal-fixed* conditions. If these conditions are introduced in the equations of motion, either in Eq. (75) or Eq. (76), the equations become solvable for the unknown accelerations.

Even though the nodal-fixed conditions are popular in structural and deformable multibody dynamics, other forms of boundary conditions exist.

The *mean-axis* conditions are based on minimization of kinetic energy due to deformation [9]. These conditions are expressed at the velocity level as:

$$\hat{\mathbf{I}}^T \mathbf{M}\dot{\boldsymbol{\delta}} = \mathbf{0}, \ \hat{\tilde{\mathbf{b}}}^T \mathbf{M}\dot{\boldsymbol{\delta}} = \mathbf{0}. \tag{79}$$

These are three translational and three rotational algebraic equations. The time derivative of these conditions provides acceleration conditions. Depending on the characteristics of the mass matrix, the acceleration conditions can be expressed as:

$$\hat{\mathbf{I}}^T \mathbf{M}\ddot{\boldsymbol{\delta}} = \mathbf{0}, \ \hat{\tilde{\mathbf{b}}}^T \mathbf{M}\ddot{\boldsymbol{\delta}} = \mathbf{0}. \tag{80}$$

The *translational* position condition can be expressed in several forms as:

$$\hat{\mathbf{I}}^T \mathbf{M}\boldsymbol{\delta} = \mathbf{0}, \ \hat{\mathbf{I}}^T \mathbf{M}\mathbf{b} = \mathbf{0}, \ \hat{\mathbf{I}}^T \mathbf{M}\hat{\tilde{\mathbf{b}}} = \hat{\tilde{\mathbf{b}}}^T \mathbf{M}\hat{\mathbf{I}} = \mathbf{0}. \tag{81}$$

Interpretation of the conditions at position level is that if at the undeformed state the origin of the reference frame is positioned at the body mass center, it will remain at the mass center as the body deforms.

Substituting the mean-axis conditions into Eq. (76) yields:

$$\begin{bmatrix} m\mathbf{I} & \mathbf{0} & \mathbf{0} \\ \mathbf{0} & \mathbf{J} & \mathbf{0} \\ \mathbf{M}\hat{\mathbf{I}} & -\mathbf{M}\hat{\tilde{\mathbf{b}}} & \mathbf{M} \end{bmatrix} \begin{Bmatrix} \ddot{\mathbf{r}} \\ \dot{\boldsymbol{\omega}} \\ \ddot{\boldsymbol{\delta}} \end{Bmatrix} = \begin{Bmatrix} \hat{\mathbf{I}}^T (\mathbf{f} - \mathbf{Mw}) \\ -\hat{\tilde{\mathbf{b}}}^T (\mathbf{f} - \mathbf{Mw}) \\ \mathbf{f} - \mathbf{K}\boldsymbol{\delta} - \mathbf{Mw} \end{Bmatrix}, \tag{82}$$

where further simplifications can be performed; e.g., $\hat{\mathbf{I}}^T \mathbf{Mw} = \mathbf{0}$.

6.5 Boundary and unconstrained nodes

In a multibody system, some of the nodes of a deformable body may be connected to other bodies through kinematic constraints. We distinguish these nodes from the free nodes by designating the following superscripts to entities associated with a node: *b* for boundary; *u* for unconstrained (free). As an example, the stiffness and mass matrices associated with a deformable body can be partitioned as:

$$\mathbf{K} = \begin{bmatrix} \mathbf{K}^{bb} & \mathbf{K}^{bu} \\ \mathbf{K}^{ub} & \mathbf{K}^{uu} \end{bmatrix} = \begin{bmatrix} \mathbf{K}^{b,} \\ \mathbf{K}^{u,} \end{bmatrix} = \begin{bmatrix} \mathbf{K}^{,b} & \mathbf{K}^{,u} \end{bmatrix},$$

$$\mathbf{M} = \begin{bmatrix} \mathbf{M}^{bb} & \mathbf{M}^{bu} \\ \mathbf{M}^{ub} & \mathbf{M}^{uu} \end{bmatrix} = \begin{bmatrix} \mathbf{M}^{b,} \\ \mathbf{M}^{u,} \end{bmatrix} = \begin{bmatrix} \mathbf{M}^{\cdot b} & \mathbf{M}^{\cdot u} \end{bmatrix}.$$

6.6 A simple multibody system

In this section we derive the equations of motion for rigid-deformable multibody systems. Without any loss of generality, we show the equations for a simple system containing one rigid and one deformable body connected by a spherical joint as shown in Figure 25. It is further assumed that arbitrary forces and moments also act on the system. In this Figure, subscript r refers to entities associated with the rigid body.

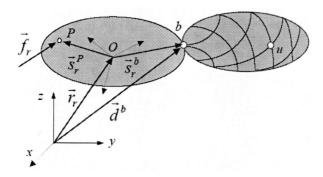

Figure 25. A simple rigid-deformable multibody system.

The position constraints for the spherical joint can be described as:

$$\mathbf{r}_r + \mathbf{s}_r^b - \mathbf{d}^b = \mathbf{0} .$$ (83)

The velocity and acceleration constraints are:

$$\dot{\mathbf{r}}_r - \tilde{\mathbf{s}}_r^b \boldsymbol{\omega}_r - \dot{\mathbf{d}}^b = \mathbf{0} ,$$ (84)

$$\ddot{\mathbf{r}}_r - \tilde{\mathbf{s}}_r^b \dot{\boldsymbol{\omega}}_r - \ddot{\mathbf{d}}^b + \tilde{\boldsymbol{\omega}}_r \tilde{\boldsymbol{\omega}}_r \mathbf{s}_r^b = \mathbf{0} .$$ (85)

Using Eq. (70), the acceleration constraints can also be written as:

$$\ddot{\mathbf{r}}_r - \tilde{\mathbf{s}}_r^b \dot{\boldsymbol{\omega}}_r - \ddot{\mathbf{r}} + \tilde{\mathbf{b}}^b \dot{\boldsymbol{\omega}} - \ddot{\boldsymbol{\delta}}^b = \boldsymbol{\gamma}^b ,$$ (86)

where

$$\gamma^b = -\tilde{\boldsymbol{\omega}}_r \tilde{\boldsymbol{\omega}}_r \mathbf{s}_r^b + \mathbf{w}^b . \tag{87}$$

The equations of motion for the system can be written in a variety of forms. First form uses Eqs. (13), (67), (85), and Lagrange multipliers:

$$
\begin{bmatrix}
\mathbf{M}_r & 0 & 0 & 0 & \mathbf{I} \\
0 & \mathbf{J}_r & 0 & 0 & -\tilde{\mathbf{s}}_r^{bT} \\
0 & 0 & \mathbf{M}^{bb} & \mathbf{M}^{bu} & -\mathbf{I} \\
0 & 0 & \mathbf{M}^{ub} & \mathbf{M}^{uu} & 0 \\
\mathbf{I} & -\tilde{\mathbf{s}}_r^{b} & -\mathbf{I} & 0 & 0
\end{bmatrix}
\begin{Bmatrix}
\ddot{\mathbf{r}}_r \\
\dot{\boldsymbol{\omega}}_r \\
\ddot{\mathbf{d}}^b \\
\ddot{\mathbf{d}}^u \\
\boldsymbol{\lambda}
\end{Bmatrix}
=
\begin{Bmatrix}
\mathbf{f}_r \\
\mathbf{n}_r \\
\mathbf{f}^b - \mathbf{K}^{b,}\boldsymbol{\delta} \\
\mathbf{f}^u - \mathbf{K}^{u,}\boldsymbol{\delta} \\
-\tilde{\boldsymbol{\omega}}_r \tilde{\boldsymbol{\omega}}_r \mathbf{s}_r^b
\end{Bmatrix} . \tag{88}
$$

In a second form, we use Eqs. (13), (82), (86), and then we define $\mathbf{f}_0 = \mathbf{I}^T(\mathbf{f} - \mathbf{M}\mathbf{w})$, $\mathbf{n}_0 = -\mathbf{b}^T(\mathbf{f} - \mathbf{M}\mathbf{w})$ and $\mathbf{f}_t^j \equiv \mathbf{f}^j - \mathbf{M}^j \mathbf{w} - \mathbf{K}^{j,}\boldsymbol{\delta}$, $j = b,\ u$, to get:

$$
\begin{bmatrix}
\mathbf{M}_r & 0 & 0 & 0 & 0 & 0 & \mathbf{I} \\
0 & \mathbf{J}_r & 0 & 0 & 0 & 0 & -\tilde{\mathbf{s}}_r^{bT} \\
0 & 0 & m\mathbf{I} & 0 & 0 & 0 & -\mathbf{I} \\
0 & 0 & 0 & \mathbf{J} & 0 & 0 & \tilde{\mathbf{b}}^{bT} \\
0 & 0 & \mathbf{M}^{b,}\mathbf{I} & -\mathbf{M}^{b,}\hat{\tilde{\mathbf{b}}} & \mathbf{M}^{bb} & \mathbf{M}^{bu} & -\mathbf{I} \\
0 & 0 & \mathbf{M}^{u,}\hat{\mathbf{I}} & -\mathbf{M}^{u,}\hat{\tilde{\mathbf{b}}} & \mathbf{M}^{ub} & \mathbf{M}^{uu} & 0 \\
\hat{\mathbf{I}} & -\tilde{\mathbf{s}}_r^{b} & -\mathbf{I} & \tilde{\mathbf{b}}^{b} & -\mathbf{I} & 0 & 0
\end{bmatrix}
\begin{Bmatrix}
\ddot{\mathbf{r}}_r \\
\dot{\boldsymbol{\omega}}_r \\
\ddot{\mathbf{r}} \\
\dot{\boldsymbol{\omega}} \\
\ddot{\boldsymbol{\delta}}^b \\
\ddot{\boldsymbol{\delta}}^u \\
\boldsymbol{\lambda}
\end{Bmatrix}
=
\begin{Bmatrix}
\mathbf{f}_r \\
\mathbf{n}_r \\
\mathbf{f}_0 \\
\mathbf{n}_0 \\
\mathbf{f}_t^b \\
\mathbf{f}_t^u \\
\boldsymbol{\gamma}^b
\end{Bmatrix} . \tag{89}
$$

Other forms of these equations can be derived if we use the nodal-fixed conditions instead of the mean-axis conditions. For more complex multibody system, we could transform the equations to the joint coordinate formulation. In order to reduce the number of degrees-of-freedom associated with the deformable body, it is a common practice to transform the equations associated with the deformable body to modal coordinates and, then, truncate the number of modes. This issue requires specific attention to how the modal data is extracted and how the modes are truncated [7]. Due to space limitation, this issue will not be discussed in this paper.

7.   SUMMARY

In this paper we have reviewed three different formulations for rigid multibody dynamics. The body-coordinate formulation is based on the constrained Newton-Euler equations and results into a large set of differential-algebraic equations. The point-coordinate formulation employs the constrained Newton equations and, therefore, eliminates the use of rotational coordinates. This formulation also yields a large set of differential-algebraic equations. The joint-coordinate formulation uses relative coordinates and velocities to yield a much smaller set of equations. These equations are obtained through a systematic process by transforming the body-coordinate formulation to the joint space. If the dimension of the joint space is equal to the number of degrees-of-freedom, the transformed equations of motion will be second-order differential equations. Each formulation has its own advantages and disadvantages and, therefore, we must choose a formulation depending on our objectives. We have also reviewed briefly how a deformable body can be modeled within a multibody system. We have discussed how a moving reference frame could be defined to follow the nodes of a deformable body. Due to space limitation, this discussion is brief and only introductory. Any of the formulations discussed in this paper can be implemented either in a special-purpose or a general-purpose computer program. In a special-purpose program we can even mix different formulations in order to take advantage of certain capabilities of each formulation.

REFERENCES

1. P. E. NIKRAVESH, *Computer-Aided Analysis of Mechanical Systems* (Prentice-Hall, 1988).
2. P. E. NIKRAVESH, H. A. AFFIFI, *Construction of the Equations of Motion for Multibody Dynamics Using Point and Joint Coordinates* (Computer-Aided Analysis of Rigid and Flexible Mechanical Systems, Kluwer academic publishers, NATO ASI Series E: Applied Sciences - Vol. 268, 1994, pp. 31-60).
3. J. G. DE JALON, E. BAYO, *Kinematic and Dynamic Simulations of Multibody Systems* (Springer-Verlag, 1994).
4. W. JERKOVSKY, *The Structure of Multibody Dynamics Equations* (J. Guidance and Control, Vol. 1, No. 3, 1978, pp. 173-182).
5. S. S. KIM, M. J. VANDERPLOEG, *A General and Efficient Method for Dynamic Analysis of Mechanical Systems Using Velocity Transformation* (ASME J. Mech., Trans., and Auto. in Design, Vol. 108, 1986, pp 176-182).
6. P. E. NIKRAVESH, G. GIM, *Systematic Construction of the Equations of Motion for Multibody Systems Containing Closed Kinematic Loops* (ASME J. of Mechanical Design, Vol. 115, No. 1, 1993, pp 143-149).
7. P. E. NIKRAVESH, Y. LIN, *Body reference Frames in Deformable Multibody Systems* (Int. J. of Multiscale Computational Engineering, Vol. 1, Issues 2 & 3, 2003, pp 201-217).
8. P. E. NIKRAVESH, *Model Reduction Techniques in Flexible Multibody Dynamics* (Virtual Nonlinear Multibody Systems, Kluwer academic publishers, 2003, pp 83-102).

9. O. P. AGRAWAL, A. A. SHABANA, *Application of Deformable-Body Mean Axis To Flexible Multibody System Dynamics* (Computer Methods in Applied Mechanics and Engineering 56, 1986, pp 217-245).

DYNAMIC MODELS IN MULTI-BODY SYSTEMS

A product life cycle key technology

D. Talabă and Cs. Antonya

Transilvania University of Brasov, Romania

Abstract: Over the last decades, time simulation of mechanisms is performed using special computer programs that implement general formulations, known in the literature as Multi-Body Systems (MBS) formulations. The name is related to the model structure used for the mechanism, which is often regarded as a collection of bodies inter-connected through joints. This paper aims to present an overview of the multi-body formulations as well as a new formulation based on a model in which the mechanism is regarded as a collection of particles rather than bodies. A comparative study is provided, in which the mechanism mobility calculation and interpretation is presented in each case as well as a simulation case study for which the main formulations have been tested. Advantages and disadvantages are discussed for the formulations utilized.

Key words: multi-body systems, multi-particle systems, mechanism, dynamics.

1.   THE MBS SIMULATION IN THE CONTEXT OF PRODUCT LIFE CYCLE APPLICATIONS

The mechanical systems dynamic simulation and control is nowadays the subject of intensive studies using various theories and computer technologies. A very popular technology for this purpose is usually referred in the technical literature as "multi-body system simulation technology", according to one of the more usual representations - as a collection of interconnected bodies.

In recent years, multi-body systems (MBS) technology evolved in powerful computer analysis and control software, which became widely used in industry, research and development areas, complementing the Finite

D. Talabă and T. Roche (eds.), Product Engineering, 227–252.

Element Method for the simulation of the systems with large displacements in space. The need to perform evaluation and verification in all product life cycle stages has increased the emphasis given to simulation activities in research and industry. The MBS commercially available codes include nowadays a wide range of facilities allowing simulation of sophisticated experiments with virtual prototypes of mechanical systems.

The product life cycle related software includes a full range of tools, i.e. for Design (CAD), Analysis (CAE), Testing (CAT), Manufacturing, Maintenance and Recycling (CAM, CAPP). These tools are usually developed as independent software modules, generally taking input data from the same geometric model database generated by the CAD module.

Although, the kinematic simulation is now a standard capability included in all major CAD packages, the physical aspects of objects are still necessary in the database in order to manage the *dynamic* interaction between objects in an integrated manner. As MBS is the science dealing with the physical behavior of the objects, it is a natural evolution towards more and more MBS features included as standard facilities of CAD systems, such as to provide *native virtual prototyping capabilities*. The geometric model tends thus to become a virtual prototype while the more comprehensive database will enable a qualitative shift in the entire range of the product life cycle software, creating the basis for a new generation of software (Figure 1).

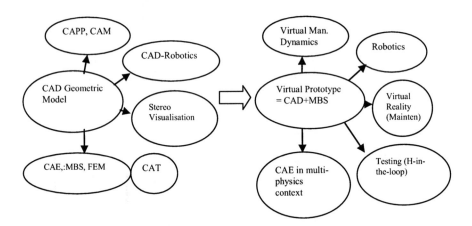

Figure 1. Evolution of the product life cycle software which will be enabled through the integration the MBS capabilities with the CAD packages

Given the importance of the MBS simulation in the context of the product life cycle, this paper aims to present the main approaches for dynamic simulation of articulated systems from the *mechanical model* perspective and within this classification, a model developed by the authors, based on a multi-particle representation.

In the dynamic formulation, according to the literature, three main approaches can be identified, namely "reference point coordinates" [9, 4, 8], "relative coordinates" [15] and "natural coordinates" [3], formulations. Trying to associate *mechanical models* to the three formulations we arrive at two mechanical models that are supporting the mathematical formulations. The first type is based on a *multi-body* representation, in which the mechanism is regarded as a system of independent bodies *inter-constrained* through joints. The second type of formulation is based on a representation that regards the mechanism as a collection of *mobile kinematic chains inter-constrained* through joints.

Out of these models, a new one was proposed by the authors in which the system is regarded as a collection of inter-constrained particles [13].

A brief presentation of this model is provided together with a comparative study, with respect to the various models. Advantages and disadvantages are discussed for the formulations utilized.

2. THE MULTI-BODY SYSTEM MODEL

According to this model, the mechanism is represented as a *collection of bodies,* the motion of which is subject to a set of absolute and relative constraints. Many authors have utilized this model since 1977 [9, 4, 8]. The mobility of the mechanism is obtained by cumulating all body's degrees of freedom considered as free bodies, from which the number of joint constraints is subtracted (Gruebler's formula):

$$M = S \cdot n - \sum i \cdot C_i , \qquad (1)$$

in which S is the motion dimension of the space (S=6 for the spatial mechanisms and S=3 for planar mechanisms), n is the number of the mechanism mobile bodies and C_i – the number of joints of class i (the class of a joint is given by the number of constraints introduced).

In order to represent the motion of the mechanism each body is associated with a Body Reference Frame (BRF) characterized by six (three for the planar mechanisms) generalized coordinates (most often the origin coordinates and the orientation angle with respect to the Global Reference Frame – GRF), the mechanism position being thus characterized by $S \cdot n = 3 \cdot 5 = 15$ generalized coordinates.

The total number of generalized coordinates is $6 \cdot n$ in space and $3n$ for planar mechanisms. For the spatial case, the generalized coordinate's vector is:

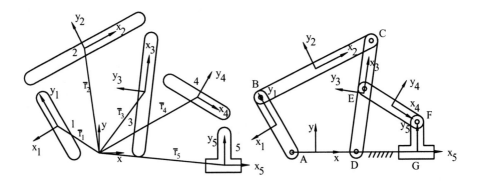

Figure 2. The multi-body system model

$$[q] = [x_1 \ y_1 \ z_1 \ \varphi_1 \ \psi_1 \ \theta_1 \ x_2 \ y_2 \ z_2 \ \ldots \ x_n \ y_n \ z_n \ \varphi_n \ \psi_n \ \theta_n]^T$$

or

$$[q] = [q_1 \quad q_2 \ \ldots \ q_{6n}]^T. \tag{2}$$

Not all the coordinates are independent because of the geometrical constraints introduced by the joints. Each constraint is represented by a geometric condition written mathematically as an algebraic equation linking the generalized coordinates of the adjacent bodies. For example a tri-mobile joint (f=3, c=3) introduces three algebraic equations, a mono-mobile joint five algebraic equations etc. In total, the number of equations for all joints is $\Sigma i \cdot C_i$, where C_i is the numbers of joints of class i.

The motion of the mechanism is kinematically determined when each independent generalized coordinate corresponds to a driving motion expressed by an algebraic equation. Thus, the number of independent generalized coordinates (i.e. that can not be calculated from the constraint equations) equals the mechanism mobility,

$$N_{qi} = M = 6 \cdot n - \Sigma \, i \cdot C_i \, , \tag{3}$$

therefore the formulation for the mechanism position includes a number of $M + \Sigma \, i \cdot C_i$ equations. The velocity and acceleration equations are obtained generally by differentiation with respect to time of the position equations yielding kinematic relations with the general expression :

$$\Phi(q,t) = 0, \quad \dot{\Phi}(q,t) = 0, \quad \ddot{\Phi}(q,t) = 0. \tag{4}$$

For the dynamic formulation, most usually a set of S differential equations is written for each body that is S·n equations for the entire mechanism with the general form

$$m\ddot{q} - J^T \lambda = Q_{ex},$$ (5)

where J is the constraints Jacobian, λ the Lagrange multipliers vector and Q_{ex} the generalized external forces. As the number of unknowns (the generalized accelerations and lagrange multipliers exceed the number of equations, in order to solve the dynamic equations by numerical integration, one has to consider the kinematic equations also, constituting the differential algebraic system (DAE) with the typical general form

$$\begin{cases} \Phi(q,t) = 0 \\ m\ddot{q} - J^T \lambda = Q_{ex} \end{cases}.$$ (6)

For illustration, the planar mechanism given in figure 2 have 5 bodies and 7 joints with two geometric constraints each, that is M = 3·5-7·2 = 1. Thus, a number of 7·2=14 constraint equations (two for each constraint) that express the algebraic conditions for the revolute joints A...F and the translational joint G, respectively can be formulated as:

$$\bar{r}_{A1} = 0, \ \bar{r}_{B1} - \bar{r}_{B2} = 0, \ \bar{r}_{C2} - \bar{r}_{C3} = 0, \ \bar{r}_{D3} = 0, \ \bar{r}_{E1} - \bar{r}_{E2} = 0,$$ (7a)

$$\bar{r}_{E3} - \bar{r}_{E4} = 0, \ \bar{r}_{F4} - \bar{r}_{F5} = 0, \ r_G = 0, \ \varphi_5 = 0, \ \bar{r}_{Xi} = \bar{r}_i + A_i \bar{r}_{Xi}^i$$ (7b)

where in general $\bar{r}_{Xi}$ represents the position vector of the point X on the body i $\bar{r}_i$ is the position vector of the frame "i", $\bar{r}_{Xi}^i$ is the position of the point in the local frame i and

$$A_i = \begin{bmatrix} \cos\varphi_i & -\sin\varphi_i \\ \sin\varphi_i & \cos\varphi_i \end{bmatrix}, \ i=1...4$$

are the transformation matrices of the bodies 1...4.

Taking into account that $[H] = \begin{bmatrix} 0 & -1 \\ 1 & 0 \end{bmatrix}$,

$$\dot{A}_i = \dot{\varphi}_i \begin{bmatrix} -\sin\varphi_i & -\cos\varphi_i \\ \cos\varphi_i & -\sin\varphi_i \end{bmatrix} = \dot{\varphi}_i[H][A_i],$$

$$\ddot{A}_i = \ddot{\varphi}_i[H][A_i] - \dot{\varphi}_i^2[A_i],$$

and differentiating twice the position equations (7a) and (7b), one can obtain successively the velocity and acceleration equations of the form

$$[J][\dot{q}] = 0 \text{ and } [J][\ddot{q}] = [\psi], \tag{8}$$

where J - the constraints jacobian matrix has 14 lines (2 lines for each joint) and 15 columns corresponding to the 15 generalized coordinates and $[\psi]$ has 14 lines:

$$[J] = \begin{bmatrix} [J_{1A}] & 0 & 0 & 0 & 0 \\ [J_{1B}] & [J_{2A}] & 0 & 0 & 0 \\ 0 & [J_{2C}] & [J_{3C}] & 0 & 0 \\ 0 & 0 & [J_{3D}] & 0 & 0 \\ 0 & 0 & [J_{3E}] & [J_{4E}] & 0 \\ 0 & 0 & 0 & [J_{4F}] & [J_{5F}] \\ 0 & 0 & 0 & 0 & [J_{5G}] \end{bmatrix}, \ [\psi] = \begin{bmatrix} \dot{\varphi}_1^2[A_1] \\ \dot{\varphi}_1^2[A_1] + \dot{\varphi}_2^2[A_2] \\ \dot{\varphi}_2^2[A_2] + \dot{\varphi}_3^2[A_3] \\ \dot{\varphi}_3^2[A_3] \\ \dot{\varphi}_3^2[A_3] + \dot{\varphi}_4^2[A_4] \\ \dot{\varphi}_4^2[A_4] + \dot{\varphi}_5^2[A_5] \\ \dot{\varphi}_5^2[A_5] \end{bmatrix},$$

where $[J_{iX}]$ are 2×3 jacobian components given by joint X on the body i of the form:

$$[J_{iX}] = [I] + [r_X^i][H][A_i].$$

For each of the 5 bodies one could write 3 equilibrium equations of the form

$$\begin{bmatrix} m_i & 0 & 0 \\ 0 & m_i & 0 \\ 0 & 0 & I_i \end{bmatrix} \begin{bmatrix} \ddot{x}_i \\ \ddot{y}_i \\ \ddot{\varphi}_i \end{bmatrix} - \begin{bmatrix} R_i^x \\ R_i^y \\ R_i^T \end{bmatrix} = \begin{bmatrix} Q_{exi}^x \\ Q_{exi}^y \\ Q_{exi}^T \end{bmatrix}, \ i=1\ldots5,$$

in which Q_{exi} represents the generalized forces acting on the body i and $[R_i]$ are the reaction forces that are dependent on the constraints. These can be expressed using the Lagrange multipliers and for the five bodies, the 3·5=15 dynamic equations could be written as

$$[m][\ddot{q}] - [J]^T[\lambda] = [Q_{ex}].$$ (9)

The constraint accelerations and dynamic equations constitute a set of *29 differential algebraic equations (DAE)* with 29 unknowns: 15 generalized coordinates and 14 Lagrange multipliers

$$\begin{bmatrix} [m] & [J]^T \\ [J] & 0 \end{bmatrix} \begin{bmatrix} \ddot{q} \\ \lambda \end{bmatrix} = \begin{bmatrix} Q_{ex} \\ \psi \end{bmatrix}.$$

The system is usually solved in a first stage as a linear system in $\ddot{q}$ and λ, then integrated for q.

3. THE KINEMATIC CHAIN MODEL

According to this model, the mechanism is represented by a *chain* of bodies and joints with the role of transmitting and transforming the motion. The kinematic chain may be serial (open loop) or parallel (closed loop) and its structure is usually represented by a graph (Figure 3), which enable an easy identification of the independent loops. This model [15, 11], is very popular especially in robotics and control systems, because facilitates the direct control of the joints motion.

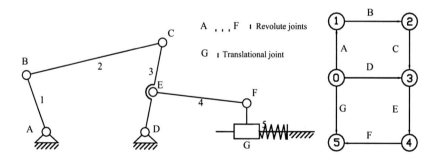

Figure 3. Parallel mechanism and the corresponding graph

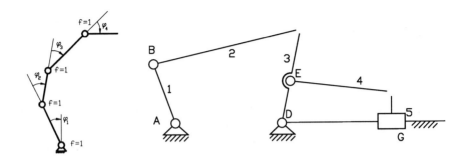

a) A serial mechanism: b) Mobility of a mechanism with closed loops after opening
        M = Σf$_i$ = 4   and replacing two joints with equivalent constraints:
                                         M = Σf$_i$-Σr$_i$ = 5-2x2 = 1

Figure 4. The kinematic chain model

In case of the serial mechanisms, the terminal body is cumulating the degrees of freedom of the preceding joints (Figure 4, a), the structure mobility relationship being thus

$$M = \sum f_i .$$ (10)

The parallel structures are regarded as a set of *inter-constrained kinematic chains*. Thus the mobility is calculated in two steps first the mechanism is converted into a serial structure (Figure 4, b) in order to allow the use of the relation (1). This is achieved by opening a joint in each closed loop and replacing it with equivalent constraints. In this way the mobility relation becomes:

$$M = \sum f_i - \sum r_i$$

in which $\sum_i r_i$ is the number constraints corresponding to the opened joints. The constraints are modeled by algebraic equations, in the same way as for the multi-body model. Together with the driving motions equations, these constraints provide a number of algebraic equations, which, in case of M drivers, equals the d.o.f. allowed by the joints ($\sum f_i$), according to the relation:

$$\sum f_i = M + \sum r_i$$ (11)

that is, the kinematic motion of the mechanism modelled as a kinematic chain is characterized by a set of $\sum f_i$ kinematic equations.

For the dynamic analysis, the Lagrange equations allow writing a differential equation for each generalized coordinate. In this way, the differential algebraic equation (DAE) system is obtained as:

$$\begin{cases} \left[\dfrac{d}{dt}\dfrac{\partial L}{\partial \dot{q}}\right]-\left[\dfrac{\partial L}{\partial q}\right]-[J]^{T}[\lambda]=[Q_{ex}] \\ [[J][\ddot{q}]=[\psi] \end{cases} \tag{12}$$

and contains n_q equations with $n_q+\Sigma r_i$ unknowns: n_q generalized coordinates (the joint angles) and Σr_i Lagrange multipliers.

For the sample mechanism in Figure 5, closure loop constraint equations impose the coincidence of points C_2-C_3 and F_4-F_5 and the corresponding equations are:

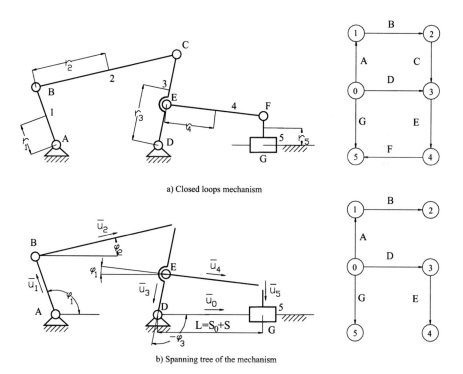

a) Closed loops mechanism

b) Spanning tree of the mechanism

Figure 5. The kinematic model

$$\begin{array}{l} l_1[u_1]+l_2[u_2]=l_0[u_0]-l_3[u_3] \\ -d_3[u_3]+l_4[u_4]=L[u_0]-l_5[u_5] \end{array}, \quad [u_k]=\begin{bmatrix}\cos\varphi_k \\ \sin\varphi_k\end{bmatrix}, \text{k=1...4,} \qquad (13)$$

with $[u_0]=\begin{bmatrix}1 \\ 0\end{bmatrix}$, $[u_5]=\begin{bmatrix}0 \\ -1\end{bmatrix}$, where

$[q]=[\varphi_1 \quad \varphi_2 \quad \varphi_3 \quad \varphi_4 \quad L]^T$ are the system generalized coordinates. Consequently, the generalized velocities and accelerations are $[\dot{q}]=[\dot{\varphi}_1 \quad \dot{\varphi}_2 \quad \dot{\varphi}_3 \quad \dot{\varphi}_4 \quad \dot{L}]^T$ and $[\ddot{q}]=[\ddot{\varphi}_1 \quad \ddot{\varphi}_2 \quad \ddot{\varphi}_3 \quad \ddot{\varphi}_4 \quad \ddot{L}]^T$.

Through successive differentiation of constraint equations one can easily obtain the velocity and acceleration equations as

$$[J][\dot{q}]=0 \text{ and } [[J][\ddot{q}]=[\psi], \qquad (14)$$

where

$$[J]=\begin{bmatrix} -l_1\sin\varphi_1 & -l_2\sin\varphi_2 & -l_3\sin\varphi_3 & 0 & 0 \\ l_1\cos\varphi_1 & l_2\cos\varphi_2 & l_3\cos\varphi_3 & 0 & 0 \\ 0 & 0 & d_3\sin\varphi_3 & -l_4\sin\varphi_4 & 1 \\ 0 & 0 & -d_3\cos\varphi_3 & l_4\cos\varphi_4 & 0 \end{bmatrix}, \qquad (15)$$

$$[\psi]=\begin{bmatrix} l_1\sin\varphi_1+l_2\sin\varphi_2+l_3\sin\varphi_3 \\ l_1\cos\varphi_1+l_2\cos\varphi_2+l_3\cos\varphi_3 \\ -d_3\sin\varphi_3+l_4\sin\varphi_4 \\ -d_3\cos\varphi_3+l_4\cos\varphi_4 \end{bmatrix} \qquad (16)$$

The Lagrange equations written for each generalized coordinate provide 5 differential equations that can be expressed together in the matrix form given in (9). Since the potential energy of the system is considered null (assuming that the gravity is acting in a plane perpendicular to the motion), the lagrangian is: $L=E_c-E_p= E_c$, that is

$$L = E_{c1} + E_{c2} + E_{c3} + E_{c4} + E_{c5} \text{ and } E_{ci} = \frac{1}{2}I_i\dot{q}_i^2 + \frac{1}{2}m_i v_i^2,$$

where v_i are the velocities of the mass centers fo each body (Figure 6):

$v_i = r_i \omega_i$, for the i=1 and i=3 (elements 1 and 3),

$$\overline{v}_i = \overline{l}_{i-1} \times \overline{\omega}_{i-1} + \overline{r}_i \times \overline{\omega}_i,$$

$$v_i = \sqrt{(l_{i-1}\omega_{i-1})^2 + (r_i\omega_i)^2 + 2l_{i-1}r_i\omega_{i-1}\omega_i \cos(\varphi_{i-1} - \varphi_i)}$$

for the elements 2 and 4 and $v_5 = \dot{q}_5$ is the generalized velocity in the translational joint G.

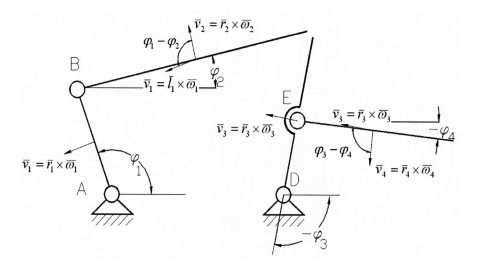

Figure 6. Velocities of the center of mass of the bodies 1...4

Carrying out the necessary differentiations, the matriceal form of Lagrange equations takes the general form

$$\left[\frac{d}{dt}\frac{\partial L}{\partial \dot{q}}\right] - \left[\frac{\partial L}{\partial q}\right] - [J]^T[\lambda] = [Q_{ex}]$$

that can be written generically as

$[m][\ddot{q}] - [J]^T[\lambda] = [Q_{ex}]$, where the mass matrix is not constant,

$$[m] = \begin{bmatrix} I_1 + m_1 r_1^2 + m_2 l_1^2 & m_2 l_1 r_2 \cos(\varphi_2 - \varphi_1) & 0 & 0 & 0 \\ m_2 l_1 r_2 \cos(\varphi_2 - \varphi_1) & I2 + m_2 r_2^2 & 0 & 0 & 0 \\ 0 & 0 & I_3 + m_3 r_3^2 + m_4 d_3^2 & m_4 d_3 r_4 \cos(\varphi_4 - \varphi_3) & 0 \\ 0 & 0 & m_4 d_3 r_4 \cos(\varphi_4 - \varphi_3) & I_4 + m_4 r_4^2 & 0 \\ 0 & 0 & 0 & 0 & m_5 \end{bmatrix}$$

and the generalized forces include also the velocity dependent terms from the LHS of Lagrange equations, out of the components coresponding to the external forces:

$$[Q_{ex}] = \begin{bmatrix} m_2 l_1 r_2 \omega_2 (\omega_2 - \omega_1) \sin(\varphi_2 - \varphi_1) - m_2 l_1 r_2 \omega_1 \omega_2 \sin(\varphi_2 - \varphi_1) + M_1 \\ m_2 l_1 r_2 \omega_1 (\omega_2 - \omega_1) \sin(\varphi_2 - \varphi_1) + m_2 l_1 r_2 \omega_1 \omega_2 \sin(\varphi_2 - \varphi_1) \\ m_4 d_3 r_4 \omega_4 (\omega_4 - \omega_3) \sin(\varphi_4 - \varphi_3) - m_4 d_3 r_4 \omega_3 \omega_4 \sin(\varphi_4 - \varphi_3) \\ m_4 d_3 r_4 \omega_3 (\omega_4 - \omega_3) \sin(\varphi_4 - \varphi_3) + m_4 d_3 r_4 \omega_3 \omega_4 \sin(\varphi_4 - \varphi_3) \\ -k_{arc}(L - L_0) \end{bmatrix}.$$

With these notations, the differential algebraic equation system can be expressed as

$$\begin{bmatrix} [m] & [J]^T \\ [J] & 0 \end{bmatrix} \begin{bmatrix} \ddot{q} \\ \lambda \end{bmatrix} = \begin{bmatrix} Q_{ex} \\ \psi \end{bmatrix}. \tag{17}$$

4. THE NATURAL COODINATES MODEL

According to this model [6], the system is represented as a *chain* of low level geometric entities like *basic points* and *unit vectors* and interconnecting constraints. The generalized coordinates are the basic point coordinates and the unit vectors components. These entities are usually chosen such as to facilitate the definition of joints. In order to define an open kinematic chain, one must impose just the rigid body constraints between the various points and unit vectors without needing to define any other constraints. For the closed loop chains, joint constraints must be added, corresponding to the loop closing joint. The method allows very simple formulations for the constraint equations, which expresses constant distance conditions between the vectors and points and very simple algebraic conditions for the closing loop constraints (point coincidence, unit vector coincidence etc). Each basic point or a unit vector is introducing three natural coordinates. For example two basic points of a body are introducing

6 generalized coordinates from which only 5 are independent because of the constant distance between the two points.

An element with two basic points and a unit vector introduces 9 natural coordinates, 6 of which are independent etc.

For the planar mechanism in Figure 6, all elements are modeled by basic points: body 1, 2, 4 with 2 points each and body 3 and 5 with 3 points (points B, C, E and F are shared between the adjacent bodies). Thus, for the 7 mobile basic points in plane, the number of the natural coordinates is 14.

There are 9 constant distance constraints between the basic points and 4 constraints introduced by the closing loop joints D and GH that is structure mobility can be calculated as

$$M = 2n_{p,u} - \sum d_k - \sum r_i = 2 \times 7 - 9 - 4 = 1, \tag{18}$$

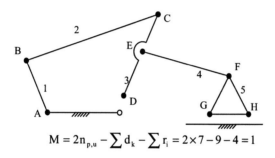

$$M = 2n_{p,u} - \sum d_k - \sum r_i = 2 \times 7 - 9 - 4 = 1$$

Figure 7. Natural coordinates model

The kinematic model of the mechanism (Figure 7), includes 9 constant distance equations and other 4 scalar equations for the closing loop joints, in total 13 algebraic equations:

$$\begin{cases} (x_A - x_B)^2 + (y_A - y_B)^2 = l_1^2 \\ (x_B - x_C)^2 + (y_B - y_C)^2 = l_2^2 \\ (x_C - x_D)^2 + (y_C - y_D)^2 = l_3^2 \\ (x_C - x_E)^2 + (y_C - y_E)^2 = l_{CE}^2 \\ (x_D - x_E)^2 + (y_D - y_E)^2 = l_{DE}^2, \quad \begin{cases} x_D = l_0 \\ y_D = 0 \\ y_G = 0 \\ y_H = 0 \end{cases} \\ (x_E - x_F)^2 + (y_E - y_F)^2 = l_4^2 \\ (x_F - x_G)^2 + (y_F - y_G)^2 = l_{FG}^2 \\ (x_F - x_H)^2 + (y_F - y_H)^2 = l_{FH}^2 \\ (x_G - x_H)^2 + (y_G - y_H)^2 = l_{GH}^2 \end{cases} \tag{19}$$

The velocity and acceleration equations take the same matriceal form (14) as in the previous model, in which the jacobian is a 13×14 sparse matrix, with very simple expressions (only non-null elements are given below):

$$J_{1,1...1,2} = \{-x_A + x_B \quad -y_A + y_B\},$$

$$J_{2,1...2,4} = \{x_B - x_C \quad y_B - y_C \quad -x_B + x_C \quad -y_B + y_C\},$$

$$J_{3,3...3,6} = \{x_C - x_D \quad y_C - y_D \quad -x_C + x_D \quad -y_C + y_D\},$$

$$J_{4,3 \ 4,4 \ 4,7 \ 4,8} = \{x_C - x_E \quad y_C - y_E \quad -x_C + x_E \quad -y_C + y_E\},$$

$$J_{5,5...5,8} = \{x_D - x_E \quad y_D - y_E \quad -x_D + x_E \quad -y_D + y_E\}, \qquad (20)$$

$$J_{6,7...6,10} = \{x_E - x_F \quad y_E - y_F \quad -x_E + x_F \quad -y_E + y_F\},$$

$$J_{7,9...7,12} = \{x_F - x_G \quad y_F - y_G \quad -x_F + x_G \quad -y_F + y_G\},$$

$$J_{8,9 \ 8,10 \ 8,13 \ 8,14} = \{x_F - x_H \quad y_F - y_H \quad -x_F + x_H \quad -y_F + y_H\},$$

$$J_{9,11...9,14} = \{x_G - x_H \quad y_G - y_H \quad -x_G + x_H \quad -y_G + y_H\},$$

$$J_{10,7} = 1, \quad J_{11,8} = 1, \quad J_{12,12} = 1, \quad J_{13,14} = 1.$$

and

$$[\psi] = -[\dot{J}][\dot{q}] = \begin{bmatrix} \dot{x}_B(\dot{x}_A - \dot{x}_B) + \dot{y}_B(\dot{y}_A - \dot{y}_B) \\ -\dot{x}_B(\dot{x}_B - \dot{x}_C) - \dot{y}_B(\dot{y}_B - \dot{y}_C) + \dot{x}_C(\dot{x}_B - \dot{x}_C) + \dot{y}_C(\dot{y}_B - \dot{y}_C) \\ -\dot{x}_C(\dot{x}_C - \dot{x}_D) - \dot{y}_C(\dot{y}_C - \dot{y}_D) + \dot{x}_D(\dot{x}_C - \dot{x}_D) + \dot{y}_D(\dot{y}_C - \dot{y}_D) \\ -\dot{x}_C(\dot{x}_C - \dot{x}_E) - \dot{y}_C(\dot{y}_C - \dot{y}_E) + \dot{x}_E(\dot{x}_C - \dot{x}_E) + \dot{y}_E(\dot{y}_C - \dot{y}_E) \\ -\dot{x}_D(\dot{x}_D - \dot{x}_E) - \dot{y}_D(\dot{y}_D - \dot{y}_E) + \dot{x}_E(\dot{x}_D - \dot{x}_E) + \dot{y}_E(\dot{y}_D - \dot{y}_E) \\ -\dot{x}_E(\dot{x}_E - \dot{x}_F) - \dot{y}_E(\dot{y}_E - \dot{y}_F) + \dot{x}_F(\dot{x}_E - \dot{x}_F) + \dot{y}_F(\dot{y}_E - \dot{y}_F) \\ -\dot{x}_F(\dot{x}_F - \dot{x}_G) - \dot{y}_F(\dot{y}_F - \dot{y}_G) + \dot{x}_G(\dot{x}_F - \dot{x}_G) + \dot{y}_G(\dot{y}_F - \dot{y}_G) \\ -\dot{x}_F(\dot{x}_F - \dot{x}_H) - \dot{y}_F(\dot{y}_F - \dot{y}_H) + \dot{x}_H(\dot{x}_F - \dot{x}_H) + \dot{y}_H(\dot{y}_F - \dot{y}_H) \\ -\dot{x}_G(\dot{x}_G - \dot{x}_H) - \dot{y}_G(\dot{y}_G - \dot{y}_H) + \dot{x}_H(\dot{x}_G - \dot{x}_H) + \dot{y}_H(\dot{y}_G - \dot{y}_H) \\ 0 \\ 0 \\ 0 \\ 0 \end{bmatrix} \qquad (21)$$

For the dynamic formulation, the mass matrix and external force terms have to be computed in relation with the set of generalized coordinates. These matrices are computed *for each body* with respect to the associated basic points and unit vectors. For this purpose the virtual power principle could be utilized. To keep the relations simple, for the considered mechanism, we adopt the 2 point mass matrix for each body [6], that is:

$$[m_i] = \begin{bmatrix} \dfrac{m_i}{3} & 0 & \dfrac{m_i}{6} & 0 \\ 0 & \dfrac{m_i}{3} & 0 & \dfrac{m_i}{6} \\ \dfrac{m_i}{6} & 0 & \dfrac{m_i}{3} & 0 \\ 0 & \dfrac{m_i}{6} & 0 & \dfrac{m_i}{3} \end{bmatrix}, k=1...5. \tag{22}$$

For the mass matrix of the bodies 3 and 5, only the pairs of points (B, D) respectively (G, H) will be considered. In the next stage, the dynamic equations could be written using Lagrange formulation, in a similar manner as for the kinematic chain model, taking the form (15). For this purpose, the system lagrangian is

$$L = E_c - E_p = E_c = \sum_{i=1}^{5} E_{ci} = \frac{1}{2} [\dot{q}_i]^T [m_i][\dot{q}_i] \tag{23}$$

and after differentiations, the terms from Lagrange equations become:

$$\left[\frac{d}{dt}\frac{\partial L}{\partial \dot{q}}\right] = \begin{bmatrix} m_1\ddot{x}_A/6 + m_1\ddot{x}_B/3 + m_2\ddot{x}_B/3 + m_2\ddot{x}_C/6 \\ m_1\ddot{y}_A/6 + m_1\ddot{y}_B/3 + m_2\ddot{y}_B/3 + m_2\ddot{y}_C/6 \\ m_2\ddot{x}_B/6 + m_2\ddot{x}_C/3 + m_3\ddot{x}_C/3 + m_3\ddot{x}_D/6 \\ m_2\ddot{y}_B/6 + m_2\ddot{y}_C/3 + m_3\ddot{y}_C/3 + m_3\ddot{y}_D/6 \\ m_3\ddot{x}_C/6 + m_3\ddot{x}_D/3 \\ m_3\ddot{y}_C/6 + m_3\ddot{y}_D/3 \\ m_4\ddot{x}_E/3 + m_4\ddot{x}_F/6 \\ m_4\ddot{y}_E/3 + m_4\ddot{y}_F/6 \\ m_4\ddot{x}_E/6 + m_4\ddot{x}_F/3 \\ m_4\ddot{y}_E/6 + m_4\ddot{y}_F/3 \\ m_5\ddot{x}_G/3 + m_5\ddot{x}_H/6 \\ m_5\ddot{y}_G/3 + m_5\ddot{y}_H/6 \\ m_5\ddot{x}_G/6 + m_5\ddot{x}_H/3 \\ m_5\ddot{y}_G/6 + m_5\ddot{y}_H/3 \end{bmatrix}, \left[\frac{\partial L}{\partial q}\right] = 0. \tag{24}$$

The number of dynamic equations is equal to the number of natural coordinates i.e. for the sample mechanism - 14 differential equations are obtained. Considering the 13 algebraic constraints, the DAE system has thus 27 equations with 27 unknowns: the 14 generalized coordinates and 13 Lagrange multipliers corresponding to the 13 constraints defined.

5.   THE MULTI-PARTICLE SYSTEM (MPS) MODEL

This model considers the mechanism as a *collection of particles* subject to a set of absolute and relative constraints. The main difference with respect to the natural coordinate's formulation is that each body is totally replaced by a set of mass points equivalent from the inertial viewpoint. Thus, the mechanism representation includes a *particle based model* for the rigid body and *point contact models* for the joints.

The *body model* consists in a set of particles separated by constant distances, each particle being associated with a concentrated mass according to the inertial equivalence with the real object. Once the particles location is established in the body frame, the concentrated masses can be easily obtained from the inertial equivalence conditions. These must ensure that for the particle system representing the body, the centroid position, the cumulated mass of the particles and the axial/centrifugal moments of inertia are the same as for the original body:

$$m_1 y_1^2 + m_1 z_1^2 + m_2 y_2^2 + m_2 z_2^2 + m_3 y_3^2 + m_3 z_3^2 + m_4 y_4^2 + m_4 z_4^2 = J_{xx}$$

$$m_1 y_1^2 + m_1 z_1^2 + m_2 y_2^2 + m_2 z_2^2 + m_3 y_3^2 + m_3 z_3^2 + m_4 y_4^2 + m_4 z_4^2 = J_{xx}$$

$$m_1 x_1^2 + m_1 y_1^2 + m_2 x_2^2 + m_2 y_2^2 + m_3 x_3^2 + m_3 y_3^2 + m_4 x_4^2 + m_4 y_4^2 = J_{xx} , \quad (25a)$$

$$m_1 x_1 y_1 + m_2 x_2 y_2 + m_3 x_3 y_3 + m_4 x_4 y_4 = J_{xy}$$

$$m_1 y_1 z_1 + m_2 y_2 z_2 + m_3 y_3 z_3 + m_4 y_4 z_4 = J_{yz}$$

$$m_1 x_1 z_1 + m_2 x_2 z_2 + m_3 x_3 z_3 + m_4 x_4 z_4 = J_{xz}$$

$$m_1 x_1 + m_2 x_2 + m_3 x_3 + m_4 x_4 = 0$$

$$m_1 y_1 + m_2 y_2 + m_3 y_3 + m_4 y_4 = 0, \qquad m_1 + m_2 + m_3 + m_4 = m . (25b)$$

$$m_1 z_1 + m_2 z_2 + m_3 z_3 + m_4 z_4 = 0$$

According to these equations, 10 unknowns can be computed for each set of particle associated to a body that is at least 3 mass points (characterized by 12 parameters) are necessary to fully represent the inertial properties of a body. Higher number of points is usually considered and in this case the remaining unknown parameters must be numerically adopted by the user. For the planar mechanisms, only 6 equivalence equations could be formulated and therefore a planar body could be fully represented by minimum 2 mass points.

Regarding the joint MBS model, we note that in the 3D space, a particle has 3 degrees of freedom ($f=3$), therefore maximum three types of constraints can be imposed (Figure 8).

a) Coincidence with a point (or another particle) → $f = 0$, $c = 3$.
b) Contact with a 3D curve → $f = 1$, $c = 2$.
c) Contact with a 3D surface → $f = 2$, $c = 1$.

The full body model includes the modeling particles and a set of constant distance constraints between them, which represent the ideal rigid conditions, according to the usual definition of the rigid body. For a body represented by 4 particles involving $4 \times 3 = 12$ generalized coordinates, a number of 6 distance constraints have to be imposed resulting finally only 6 independent coordinates (Figure 8):

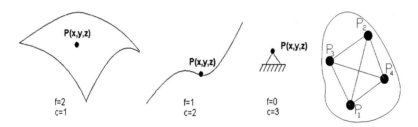

Figure 8. Particle constraints

$$\begin{cases} (x_{P_1} - x_{P_2})^2 + (y_{P_1} - y_{P_2})^2 + (z_{P_1} - z_{P_2})^2 = P_1P_2^{\,2} \\ (x_{P_1} - x_{P_3})^2 + (y_{P_1} - y_{P_3})^2 + (z_{P_1} - z_{P_3})^2 = P_1P_3^{\,2} \\ (x_{P_2} - x_{P_3})^2 + (y_{P_2} - y_{P_3})^2 + (z_{P_2} - z_{P_3})^2 = P_2P_3^{\,2} \\ (x_{P_1} - x_{P_4})^2 + (y_{P_1} - y_{P_4})^2 + (z_{P_1} - z_{P_4})^2 = P_1P_4^{\,2} \\ (x_{P_2} - x_{P_4})^2 + (y_{P_2} - y_{P_4})^2 + (z_{P_2} - z_{P_4})^2 = P_2P_4^{\,2} \\ (x_{P_3} - x_{P_4})^2 + (y_{P_3} - y_{P_4})^2 + (z_{P_3} - z_{P_4})^2 = P_3P_4^{\,2} \end{cases} \qquad (26)$$

The *joint model* is defined as combination of constraints between the particles composing the two adjacent bodies. Each joint could be represented by a set of constraints defined between the particles of the two adjacent bodies. The point type contact model allows the definition of practically any type of joints. The models of the most usual joints are detailed in Table 1. With these models defined for body and joint, a new criterion can be formulated for the mechanism mobility as:

$$M = S \cdot p - \Sigma c_i, \qquad (27)$$

in which p is the number of the particles included in the model, S is the space dimension (S=3 for 3D space and S=2 for 2D space) and c_i is the number of constraints. The generalized coordinates vector has the form:

$$[q] = [x1 \quad y1 \quad z1 \quad x2 \quad y2 \quad z2 \quad \dots \quad xp \quad yp \quad zp]^T, \qquad (28)$$

Table 1. The usual joints representation for the multi-particle system (MPS) model

Joint type	Particle model	Constraints	Constraints Equation
Spherical joint		$P_1 \equiv Q_1$ $c = 3$ $f = 3$	$x_{P1} = x_{Q1},$ $y_{P1} = y_{Q1},$ $z_{P1} = z_{Q1}.$
Cylindri-cal joint		$P_1 \in Q_1 Q_2$ axis $P_2 \in Q_1 Q_2$ axis $c = 4$ $f = 2$	$\dfrac{x_{P_1} - x_{Q_1}}{x_{Q_2} - x_{Q_1}} = \dfrac{y_{P_1} - y_{Q_1}}{y_{Q_2} - y_{Q_1}} = \dfrac{z_{P_1} - z_{Q_1}}{z_{Q_2} - z_{Q_1}},$ $\dfrac{x_{P_2} - x_{Q_1}}{x_{Q_2} - x_{Q_1}} = \dfrac{y_{P_2} - y_{Q_1}}{y_{Q_2} - y_{Q_1}} = \dfrac{z_{P_2} - z_{Q_1}}{z_{Q_2} - z_{Q_1}}.$
Transla-tion joint		$P_1 \in Q_1 Q_2$ axis $P_2 \in Q_1 Q_2$ axis $P_3 \in Q_1 Q_2 Q_3$ plane $c = 5$ $f = 1$	Idem and $\begin{vmatrix} x_{P_3} & y_{P_3} & z_{P_3} & 1 \\ x_{Q_1} & y_{Q_1} & z_{Q_1} & 1 \\ x_{Q_2} & y_{Q_2} & z_{Q_2} & 1 \\ x_{Q_3} & y_{Q_3} & z_{Q_3} & 1 \end{vmatrix} = 0.$
Revolute joint		$P_1 \equiv Q_1$ $P_2 \in Q_1 Q_2$ axis $c = 5$ $f = 1$	$x_{P1} = x_{Q1}, \ y_{P1} = y_{Q1}, \ z_{P1} = z_{Q1},$ $\dfrac{x_{P_2} - x_{Q_1}}{x_{Q_2} - x_{Q_1}} = \dfrac{y_{P_2} - y_{Q_1}}{y_{Q_2} - y_{Q_1}} = \dfrac{z_{P_2} - z_{Q_1}}{z_{Q_2} - z_{Q_1}}.$
Plane joint		$P_1 \in Q_1 Q_2 Q_3$ plane $P_2 \in Q_1 Q_2 Q_3$ plane $P_3 \in Q_1 Q_2 Q_3$ plane $c = 3$ $f = 3$	$\begin{vmatrix} x_{P_i} & y_{P_i} & z_{P_i} & 1 \\ x_{Q_1} & y_{Q_1} & z_{Q_1} & 1 \\ x_{Q_2} & y_{Q_2} & z_{Q_2} & 1 \\ x_{Q_3} & y_{Q_3} & z_{Q_3} & 1 \end{vmatrix} = 0,$ $i = 1, 2, 3.$

The vector [q] can be obtained by numerically solving the system of M + Σc_i algebraic equations corresponding to the M driving motions and Σc_i joint constraints. It must be noted that for the MPS model the joint equations can take a very limited number of forms. As far as the usual joints from the Table 1 are involved – only four type of equations:
- Distance equation

$$(x_{P_1} - x_{P_2})^2 + (y_{P_1} - y_{P_2})^2 + (z_{P_1} - z_{P_2})^2 = P_1 P_2^2. \tag{29}$$

- Coincidence equation

$$x_{P1} = x_{Q1}, y_{P1} = y_{Q1}, z_{P1} = z_{Q1}. \tag{30}$$

- Co-linearity equations

$$\frac{x_{P_1} - x_{Q_1}}{x_{Q_2} - x_{Q_1}} = \frac{y_{P_1} - y_{Q_1}}{y_{Q_2} - y_{Q_1}} = \frac{z_{P_1} - z_{Q_1}}{z_{Q_2} - z_{Q_1}}. \tag{31}$$

- Co-planarity equation

$$\begin{vmatrix} x_{P_3} & y_{P_3} & z_{P_3} & 1 \\ x_{Q_1} & y_{Q_1} & z_{Q_1} & 1 \\ x_{Q_2} & y_{Q_2} & z_{Q_2} & 1 \\ x_{Q_3} & y_{Q_3} & z_{Q_3} & 1 \end{vmatrix} = 0. \tag{32}$$

In the next step, through successive differentiation, velocity and acceleration equations can be easily derived yielding the set of kinematic equations of the general form (8).

For dynamic simulation, the equations have the same general form as for MBS model - relation (9), in which the mass matrix is

$$m = \text{diag}[\quad m_1 \quad m_1 \quad m_1 \quad m_2 \quad m_2 \quad m_2 \quad m_3 \quad m_3 \quad m_{31} \quad \ldots \quad m_p \quad m_p \quad m_p]. \tag{33}$$

The Σc_i Lagrange multipliers include the joint reaction forces (including no torques) and also the internal reaction forces between the particles of the bodies.

For the sample mechanism modeled as in Figure 9, the number of particles per body is 2, except bodies 3 and 5, which are defined with three particles each. The total number of mobile particles is p = 12 (A_1, B_1, B_2, C_2, C_3, D_3, E_3, E_4, F_4, F_5, G_5, H_5), that is $S \cdot p = 2 \times 12 = 24$ generalized coordinates (two Cartesian coordinates for each particle):

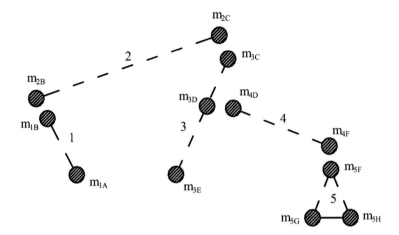

$$\text{Particle set : dof} = 2\sum p_i = 2 \times 12 = 24$$

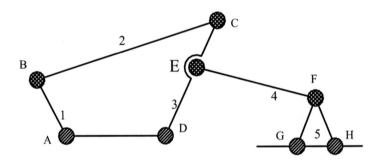

$$\text{Mechanism : dof} = 2\sum p_i - \sum r_i = 24 - (9 + 7 \times 2) = 1$$

Figure 9. The multi-particle model

$$q = [x_{A1} \; y_{A1} \; z_{A1} \; x_{B1} \; y_{B1} \; z_{B1} \; x_{B2} \; y_{B2} \; z_{B2} \dots x_{H5} \; y_{H5} \; z_{H5}]^T. \qquad (34)$$

As constraints, there are 9 rigid body constant distances (AB, BC, CD, DE, CE, EF, FG, FH, GH) and 14 joint constraints, yielding $\sum c_i = 23$, that is $M = S \cdot p - \sum c_i = 24 - 23 = 1$. The constraint equations set is very similar to the one written in natural coordinates case, the only difference being the joint constraints, which are written in this case for each joint:

$$\begin{cases} (x_{A_1} - x_{B_1})^2 + (y_{A_1} - y_{B_1})^2 = l_1^2 \\ (x_{B_2} - x_{C_2})^2 + (y_{B_2} - y_{C_2})^2 = l_2^2 \\ (x_{C_3} - x_{D_3})^2 + (y_{C_3} - y_{D_3})^2 = l_3^2 \\ (x_{C_3} - x_{E_3})^2 + (y_{C_3} - y_{E_3})^2 = l_{CE}^2 \\ (x_{D_3} - x_{E_3})^2 + (y_{D_3} - y_{E_3})^2 = l_{DE}^2 \\ (x_{E_4} - x_{F_4})^2 + (y_{E_4} - y_{F_4})^2 = l_4^2 \\ (x_{F_5} - x_{G_5})^2 + (y_{F_5} - y_{G_5})^2 = l_{FG}^2 \\ (x_{F_5} - x_{H_5})^2 + (y_{F_5} - y_{H_5})^2 = l_{FH}^2 \\ (x_{G_5} - x_{H_5})^2 + (y_{G_5} - y_{H_5})^2 = l_{GH}^2 \end{cases} , \begin{cases} x_{A_1} - x_{A_0} = 0, & y_{A_1} - y_{A_0} = 0 \\ x_{B_2} - x_{B_1} = 0, & y_{B_2} - y_{B_1} = 0 \\ x_{C_3} - x_{C_2} = 0, & y_{C_3} - y_{C_2} = 0 \\ x_{D_3} - x_{D_0} = 0, & y_{D_3} - y_{D_0} = 0 \\ x_{E_4} - x_{E_3} = 0, & y_{E_4} - y_{E_3} = 0 \\ x_{F_5} - x_{F_4} = 0, & y_{F_5} - y_{F_4} = 0 \\ y_{H_5} = 0, & y_{G_5} = 0 \end{cases} \cdot (35)$$

The velocity and acceleration equations are derived by differentiation of the position equations that is the 23×24 jacobian matrix is very similar to the one derived for natural coordinates, the only difference consisting in some more non-null elements due to the joint constraints.

$$\begin{cases} (x_{A_1} - x_{B_1})(\dot{x}_{A_1} - \dot{x}_{B_1}) + (y_{A_1} - y_{B_1})(\dot{y}_{A_1} - \dot{y}_{B_1}) = 0 \\ (x_{B_2} - x_{C_2})(\dot{x}_{B_2} - \dot{x}_{C_2}) + (y_{B_2} - y_{C_2})(\dot{y}_{B_2} - \dot{y}_{C_2}) = 0 \\ (x_{C_3} - x_{D_3})(\dot{x}_{C_3} - \dot{x}_{D_3}) + (y_{C_3} - y_{D_3})(\dot{y}_{C_3} - \dot{y}_{D_3}) = 0 \\ (x_{C_3} - x_{E_3})(\dot{x}_{C_3} - \dot{x}_{E_3}) + (y_{C_3} - y_{E_3})(\dot{y}_{C_3} - \dot{y}_{E_3}) = 0 \\ (x_{D_3} - x_{E_3})(\dot{x}_{D_3} - \dot{x}_{E_3}) + (y_{D_3} - y_{E_3})(\dot{y}_{D_3} - \dot{y}_{E_3}) = 0 , \\ (x_{E_4} - x_{F_4})(\dot{x}_{E_4} - \dot{x}_{F_4}) + (y_{E_4} - y_{F_4})(\dot{y}_{E_4} - \dot{y}_{F_4}) = 0 \\ (x_{F_5} - x_{G_5})(\dot{x}_{F_5} - \dot{x}_{G_5}) + (y_{F_5} - y_{G_5})(\dot{y}_{F_5} - \dot{y}_{G_5}) = 0 \\ (x_{F_5} - x_{H_5})(\dot{x}_{F_5} - \dot{x}_{H_5}) + (y_{F_5} - y_{H_5})(\dot{y}_{F_5} - \dot{y}_{H_5}) = 0 \\ \qquad (x_{G_5} - x_{H_5}) + (y_{G_5} - y_{H_5}) = 0 \end{cases} \quad (36a)$$

$$\begin{cases} \dot{x}_{A_1} = 0, & \dot{y}_{A_1} = 0 \\ \dot{x}_{B_2} - \dot{x}_{B_1} = 0, & \dot{y}_{B_2} - \dot{y}_{B_1} = 0 \\ \dot{x}_{C_3} - \dot{x}_{C_2} = 0, & \dot{y}_{C_3} - \dot{y}_{C_2} = 0 \\ \dot{x}_{D_3} - \dot{x}_{D_0} = 0, & \dot{y}_{D_3} - \dot{y}_{D_0} = 0 . \\ \dot{x}_{E_4} - \dot{x}_{E_3} = 0, & \dot{y}_{E_4} - \dot{y}_{E_3} = 0 \\ \dot{x}_{F_5} - \dot{x}_{F_4} = 0, & \dot{y}_{F_5} - \dot{y}_{F_4} = 0 \\ \dot{y}_{H_5} = 0, & \dot{y}_{G_5} = 0 \end{cases} \quad (36b)$$

For the dynamic analysis, one has to take into account the particles are acted by external, reaction and inertia forces. The general matrix form of the differential equations is given also by (12), in which the mass matrix is a 24×24 diagonal matrix,

$$m = diag[\; m_{A_1}\;\; m_{A_1}\;\; m_{A_1}\;\; m_{B_1}\;\; m_{B_1}\;\; m_{B_1}\;\; m_{B_2}\;\; m_{B_2}\;\; m_{B_2}\;\; \cdots \;\; m_{H_5}\;\; m_{H_5}\;\; m_{H_5}]$$

the Jacobian is a 23×24 matrix corresponding to the joint constraints and constant distances and the Lagrange multiplier vector λ has also 23 components. The DAE system has 47 equations with 47 unknown: 24 generalized coordinates and 23 Lagrange multipliers.

6.   COMPARATIVE SIMULATIONS

The sample mechanism illustrated for the mathematical formulation for each model was comparatively simulated. For this purpose, identical conditions have been set up in terms of computing power (P4, 2,4 GHz), programming language numerical methods and numerical data of the mechanism subject to simulations.

The integration method used is the Adams-Moulton, Adams-Bashfort predictor-corrector method. At time step i the predicted value of the generalized position and velocities can be computed as:

$$v_{(i+1)_p} = v_i + \frac{\Delta t}{12}(23a_{pre.} - 16a_{i-1} + 5a_{i-2}), \tag{37}$$

$$x_{(i+1)_p} = x_i + \frac{\Delta t}{12}(23v_{pre.} - 16v_{i-1} + 5v_{i-2}). \tag{38}$$

The corrected velocities and positions are

$$v_{i+1} = v_i + \frac{\Delta t}{24}(9a_{cor.} + 19a_{pre.} - 5a_{i-1} + a_{i-2}), \tag{39}$$

$$x_{i+1} = x_i + \frac{\Delta t}{24}(9v_{cor.} + 19v_{pre.} - 5v_{i-1} + v_{i-2}).$$

Constraints stabilization method employed was the one proposed by Baumgarte. According to this, the modified acceleration equation can be written as:

$$\Phi_q \ddot{q} = \psi - \alpha(\Phi_q \dot{q} + \Phi_t) - \beta\Phi \tag{40}$$

and for α and β greater than zero the equation is stable. According to (7), the two coefficients were selected function on the integration step size.

$$\alpha = \frac{\hat{\alpha}}{\Delta t}, \tag{41}$$

$$\beta = \frac{\hat{\beta}}{\Delta t^2}, \tag{42}$$

where the values selected were $\hat{\alpha} = 0.3$ and $\hat{\beta} = 0.02$.

The numerical data of the mechanism considered was the same for the simulations based on the various models, including the integration step size, as follows:

l_1=100 mm; l_2=164.9242 mm; l_3=145.6022 mm; d_3=83.2036 mm; l_4= 217.2415 mm; l_5=70 mm; r_1=l_1/2; r2=l_2/2; r_3=d_3; r_4=l_4/2; r_5=l_5/2; φ_{10}=Pi/2; φ_{20}=14°; φ_{30}=254.055°; φ_{40}=316.332°: m_1=m_2=m_4= 0.4 kg, m_3=0.5 kg, m_5=0.45 kg; I_1:=1000+0.4 l_1^2/2; I_2:= 2720+0.4 l_2^2/2; I_3:=1822.5+0.4 d_3^2/2; ω_1=ω_2=ω_3=ω_4=v_5=0.

The mechanism is considered operated by a torque at the level of joint A, M_1=-100000 Nmm and with the slider restricted by a spring with k_{arc}=12 N/mm; L_{arc0}=120 mm; L_{arc}=120 mm:

The mechanism has been modelled and a simulation for the same input data was performed also using ADAMS software. In al cases (the 4 models and the ADAMS model) the results obtained are identical, which validates the numerical results. The time history of slider displacement is presented in the Figure 10 and a comparison of the four models is presented in Table 2.

Table 2 Comparison between the four models

Model	Constraint equations	Generalized coordinates	Total DAE	DAE Matrix sparsity	Run-time [sec]
MBS	14	15	29	90.36%	31.34
Relative coordinates	4	5	9	62.96%	13.66
Natural coordinates	13	14	27	86.83%	29.09
MPS	23	24	47	94.56%	53.77

According to the results, some remarkable issues can be noted:
a) The number of equations of the relative coordinate model is significantly lower (only 9 equations for the considered case), and therefore this model is the most efficient.

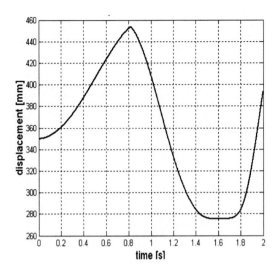

Figure 10. Displacement of the slider

b) The more number of equations, the sparser the DAE system is.
c) The model efficiency is lower for the models with larger number of differential algebraic equations.
d) For the models with larger number of DAE equations the sparsity is very high (94.5% for the MPS model). This is important for the mechanical systems with large number of differential equations, for which the sparse matrix techniques could bring significant computational efficiency. For example, in the case of a civil engineering structure (Figure 11), modelled with 745 particles [14], the use of sparse matrix techniques reduced the runtime from over 20 hours to 20.77 min. This demonstrates that in case of large mechanical systems, the MPS model could be very efficient through the exploitation of the DAE matrix high sparsity.

7.   CONCLUSIONS

A model based classification has been outlined. Two types of models have been identified: reference point based models (MBS and MPS) and joint coordinate based model (relative coordinate and natural coordinates models). This classification allows a clear background of each formulation including specific formulas for mobility computation and finally a proper evaluation of the potential for each of them.

The model proposed by the authors (MPS), despite of the larger number of equations, provides several features with relevance to the non-linear multibody simulation:

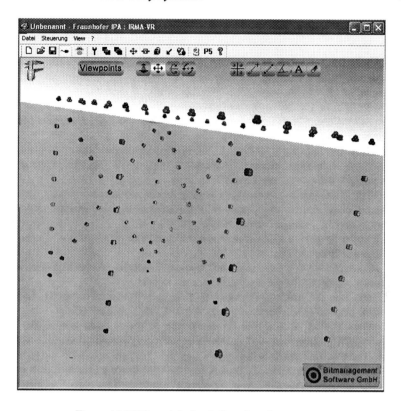

Figure 11. MPS model of a civil engineering structure

- The representation of forces and inertial mass properties is significantly simplified.
- The constraints and the corresponding algebraic equations are of small variety. This is simplifying both constraint and Jacobian matrix formulation.
- The MPS model allows the extension for the treatment of flexible multibody systems by replacing the distance equations with the flexibility principles for each body [13]. However for this purpose, proper identification methods are missing for the moment.
- For systems with large number of bodies, the MPS model becomes more efficient making use of sparse matrix computational techniques.

AKNOWLEDGEMENTS

This research was carried out thanks to the generous support offered by the European Commission under the FP5-Growth scheme, through the Project ADEPT, contract no. G1MA-CT-2002-04038/ADEPT.

REFERENCES

1. P. ALEXANDRU, I. VIŞA, D. TALABA, *Utilization of the Cartesian coordinates for the linkages study.* (in Romanian, The Romanian Symposium on Mechanisms and Machine Theory MTM '88, Cluj-Napoca, vol. I, 1988, pp. 1-10).
2. C. ANTONYA, D. TALABA, *Dynamic simulation of civil engineering structures in virtual reality environment.* (The International Conference on Multi-Body Dynamics MBD-MBS '04, Loughborough, 2004, pp.137-145).
3. J. GARCIA. DE JALON, J. UNDA, A. AVELLO, *Natural coordinates for the computer analysis of multi-body systems.* (Computer Methods in Applied Mechanics and Engineering, vol. 56, 1986, pp. 309-327).
4. E. J. HAUG, *Elements and methods of computational dynamics.* (in Haug E.J. (eds) NATO Advanced Study Institute on "*Computer Aided Analysis and optimization of mechanical systems dynamics*", Springer Verlag, 1984).
5. J. E. HAUG, *Computers Aided Kinematics and Dynamics of Mechanical System.* (vol. I. Ed. Allyn and Bacon, 1989).
6. J. G. JALON, E. BAYO, *Kinematic and Dynamic Simulation of Multibody Systems – The real time challenge.* (Springer-Verlag, New York, 1994).
7. S. T. LIN, J. N. HUANG, *Stabilization of Baumgarte's Method Using the Runge-Kutta Approach.* (Journal of Mechanical Design, Vol. 124, 2002, pp. 633 – 641).
8. P. E. NIKRAVESH, I. S. CHUNG, *Application of Euler parameters to the dynamic analysis of Three Dimensional Constrained Mechanical Systems.* (Journal of Mechanical Design, vol 104, 1982, pp. 785-791).
9. N. ORLANDEA, M. A. CHACE, D. A. CALAHAN, *A Sparsity Oriented Approach to the Dynamics Analysis and Design of Mechanical Systems.* (Journal of Engineering for Industry, August, 1977, pp 773-784).
10. R. ROBERSON, R. SCHWERTASSEK, *Dynamics of Multibody Systems.* (Springer Verlag, New York, 1988).
11. W. O. SCHIEHLEN, E. J. KREUZER, *Symbolic Computer Derivation of Equation of Motion.* (Dynamics of multi-body systems, Springer Verlag Berlin, Heildelberg, 1978, pp. 290-305).
12. W. O. SCHIEHLEN, *Multibody Systems Handbook.* (Springer Verlag, Berlin-New York, 1990).
13. D. TALABA, *A particle model for mechanical system simulation.* (Proceedings of NATO Advanced Study Institute „Nonlinear Virtual Multi-body Systems", Prague, 2002).
14. D. TALABA, C. ANTONYA, *The multi-particle system (mps) model as a tool for simulation of mechanisms with rigid and elastic bodies.* (The International Conference on Multi-Body Dynamics MBD-MBS '04, Loughborough, 2004, pp.111-119).
15. J. WITTEMBURG, *Dynamics of system of rigid bodies.* (Teubner, Stuttgart, 1977).

REAL-TIME MBS FORMULATIONS: TOWARDS VIRTUAL ENGINEERING

J. Cuadrado, M. Gonzalez, R. Gutierrez and M.A. Naya
Laboratory of Mechanical Engineering, University of La Coruña, Ferrol, Spain

Abstract: This paper presents the research conducted during the last years by the Laboratory of Mechanical Engineering on real-time formulations for the dynamics of multi-body systems, a topic of great relevance for the development of new virtual reality applications. The work carried out by our group has been focused on: a) the development of real-time formulations capable of performing very fast calculations of the dynamics of complex rigid-flexible multi-body systems; b) the experimental validation of the motions, deformations and forces obtained through the application of the above-mentioned formulations, so as to verify that reality is being reasonably well imitated; c) the study of the aptitude of such formulations for becoming part of a virtual reality environment, in connection with user-interface devices. In the paper, a real-time formulation developed by the authors is described, along with some examples of rigid-flexible multi-body systems simulated with such formulation. Moreover, results of the experimental validation of one of the examples are shown. Finally, a simulator based on the proposed formulation is presented.

Key words: multi-body dynamics, real-time formulations, virtual reality, product life-cycle, experimental validation, simulator.

1.   INTRODUCTION

Entities in the real world can be considered from two points of view: graphical and physical. If a virtual environment is to be created in which just the graphical aspect of objects is accounted for, only the computer graphics discipline is needed. In such a virtual world, the user will be able to see, and perhaps to hear, but never to touch or to be touched. Although these capabilities can be enough for some applications, it is clear that they are far from replicating reality. On the other hand, a virtual environment can be

D. Talabă and T. Roche (eds.), Product Engineering, 253–272.

conceived in which objects possess physical properties, besides the graphical ones, of course. Then, solids will experiment motion and/or deformation when forces act upon them, and the user will be allowed to interact with its environment and to feel through the sense of touch too. This virtual world is much closer to reality than the previously described one.

In order to develop environments with such types of capabilities, computational methods for the dynamics of rigid-flexible multi-body systems will be demanded which encompass the two following characteristics:

a) adequate consideration of mechanical phenomena, like flexibility, contact, impact, friction, damping, etc.;

b) very fast calculation of all the kinematic and dynamic magnitudes concerning the virtual bodies in the simulation and the user itself.

Applications for physics-based virtual environments can be very diverse, but undoubtedly some of them are fully related to the product life-cycle: design (CAD), analysis (CAE), testing (CAT), manufacturing (CAM and CAPP), maintenance and end-of-the-product. More specifically, the practical implementation of some eco-design concepts, like modularity, disassembling, reusability, recycling, new materials, low energy consumption, noise reduction, low cost, integration of intelligent systems, security, etc., can benefit from such applications.

Having the mentioned objectives in mind, the work carried out by our group has been focused on: a) the development of real-time formulations capable of performing very fast calculations of the dynamics of complex rigid-flexible multi-body systems; b) the experimental validation of the motions, deformations and forces obtained through the application of the abovementioned formulations, so as to verify that reality is being reasonably well imitated; c) the study of the aptitude of such formulations for becoming part of a virtual reality environment, in connection with user-interface devices.

The paper has been organized as follows: Section 2 describes a real-time formulation developed by the authors for the dynamics of rigid-flexible multi-body systems; Section 3 shows some examples of rigid-flexible multi-body systems simulated with such formulation, paying attention to the achieved efficiency, robustness and accuracy; Section 4 reports the results of the experimental validation of one of the examples; Section 5 presents a simulator based on the proposed formulation; and, finally, Section 6 summarizes the conclusions of the work.

2. THE REAL-TIME FORMULATION

The three main elements (modeling, dynamic equations and numerical integrator) of the proposed approach to solve the dynamics of rigid-flexible multi-body systems are described in what follows. For the modeling, fully-Cartesian coordinates, also known as *natural coordinates*, are used. Both rigid and flexible links are modeled with this type of coordinates in a total compatible form. Further explanation about these coordinates can be found in [1]. The dynamic equations are stated through an improved version of the index-3 augmented Lagrangian formulation with mass-orthogonal projections given in [2]. For the integration scheme, the unconditionally-stable implicit single-step trapezoidal rule has been chosen [3].

2.1 Modeling

The modeling of rigid bodies follows the general rules given in [1], and won't be described here, since it can be considered as a well established technique.

The modeling of flexible bodies is carried out by a floating frame of reference approach with component mode synthesis for small elastic displacements. The global motion of each flexible body is described as a superposition of the large-amplitude motion of a moving frame, rigidly attached to a certain point of the body, and the small elastic displacements of the body with respect to an undeformed configuration, taken as reference. Any deformed configuration of the body is expressed as a linear combination of static and dynamic modes, in the sense of the mode synthesis approach with fixed boundaries. The static modes depend on the natural coordinates defined at the joints of the body, while the number of internal, dynamic modes should be decided by the analyst. A detailed description of the kinematics of this method can be found in [4].

However, the approach proposed in the abovementioned reference for the dynamics has not been adopted, as long as it showed to be too involved, particularly in what refers to the form of the inertia terms. Therefore, the co-rotational approximation suggested in [5] has been considered in the way proposed by [6] when natural coordinates are used for the modeling. This modified approach provides the same level of accuracy while notably simplifying the formulation of the mass matrix and velocity dependent forces vector. In what follows, the approach is briefly described.

Figure 1 shows a general flexible body. Point r_o and orthogonal unit vectors **a**, **b** and **c** serve to define the body local reference frame, and are always needed, so that they must compulsorily be considered as problem variables. Additional points and unit vectors (further problem variables) may be defined at the joints of the body (they will produce the static modes) in the typical way featured by the natural coordinates. The position of any point of the body can then be expressed as

$$\mathbf{r} = \mathbf{r}_o + \mathbf{A}(\overline{\mathbf{r}}_u + \overline{\mathbf{u}}),$$ (1)

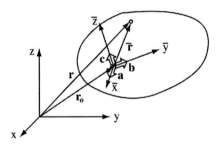

Figure 1. General flexible body.

where $\mathbf{r}_o$ is the position of the origin of the local reference frame, $\mathbf{A}$ the rotation matrix which columns are the three orthogonal unit vectors $\mathbf{a}$, $\mathbf{b}$ and $\mathbf{c}$ already mentioned, $\overline{\mathbf{r}}_u$ the position of the point in local coordinates for the undeformed configuration of the body, and $\overline{\mathbf{u}}$ the elastic displacement suffered by the point, also expressed in local coordinates. If the finite element method is used, the elastic displacement can be written as

$$\overline{\mathbf{u}} = \mathbf{N}\overline{\mathbf{u}}^*,$$ (2)

where $\mathbf{N}$ stands for the interpolation matrix which contains the interpolation functions, and $\overline{\mathbf{u}}^*$ are the elastic displacements of the nodes. Substituting Eq. (2) into Eq. (1) yields

$$\mathbf{r} = \mathbf{r}_o + \mathbf{A}(\overline{\mathbf{r}}_u + \mathbf{N}\overline{\mathbf{u}}^*).$$ (3)

At this point, the approach called *co-rotational* [5] is introduced. It implies the following interpolation for the velocity of any point of the flexible body

$$\dot{\mathbf{r}} = \mathbf{N}\mathbf{v}^*,$$ (4)

where $\mathbf{v}^*$ are the velocities of the nodes. Obviously, expression given by Eq. (4) for the velocities is not consistent with expression given by Eq. (3) for the positions, as long as the former should be the derivative of the latter.

Based on Eq. (4), the kinetic energy of the flexible body can be written in the form

$$T = \frac{1}{2}\int_{v} \dot{\mathbf{r}}^{\mathrm{T}} \dot{\mathbf{r}} dm = \frac{1}{2}\int_{v} \mathbf{v}^{*\mathrm{T}} \mathbf{N}^{\mathrm{T}} \mathbf{N} \mathbf{v}^{*} dm = \frac{1}{2}\mathbf{v}^{*\mathrm{T}} \mathbf{M}_{\mathrm{FEM}} \mathbf{v}^{*}, \qquad (5)$$

where $\mathbf{M}_{\mathrm{FEM}}$ is the mass matrix of the finite element method. In order to express the kinetic energy in terms of the problem variables, velocities of the nodes, $\mathbf{v}^{*}$, should be related with the problem variables. Differentiating Eq. (3) and evaluating the result at the nodes gives

$$\mathbf{v}^{*} = \dot{\mathbf{r}}_{o} + \dot{\mathbf{A}}(\bar{\mathbf{r}}_{u} + \bar{\mathbf{u}}^{*}) + \mathbf{A}\dot{\bar{\mathbf{u}}}^{*} \qquad (6)$$

and considering that unit vectors $\mathbf{a}$, $\mathbf{b}$ and $\mathbf{c}$ are the columns of rotation matrix $\mathbf{A}$,

$$\mathbf{v}^{*} = \left\{ \begin{matrix} \mathbf{v}_1 \\ \vdots \\ \vdots \\ \vdots \\ \mathbf{v}_p \end{matrix} \right\} = \begin{bmatrix} \mathbf{I} & a_{11}\mathbf{I} & a_{12}\mathbf{I} & a_{13}\mathbf{I} & \mathbf{A} & \cdots & \cdots & \cdots & \mathbf{0} \\ \vdots & & & & & & & & \vdots \\ \vdots & & & & & & & & \vdots \\ \vdots & & & & & & & & \vdots \\ \mathbf{I} & a_{p1}\mathbf{I} & a_{p2}\mathbf{I} & a_{p3}\mathbf{I} & \mathbf{0} & \cdots & \cdots & \cdots & \mathbf{A} \end{bmatrix} \left\{ \begin{matrix} \dot{\mathbf{r}}_o \\ \dot{\mathbf{a}} \\ \dot{\mathbf{b}} \\ \dot{\mathbf{c}} \\ \dot{\bar{\mathbf{u}}}_1 \\ \vdots \\ \vdots \\ \dot{\bar{\mathbf{u}}}_p \end{matrix} \right\}, \qquad (7)$$

where p is the number of nodes of the body, $\mathbf{I}$ the 3 by 3 identity matrix, and

$$a_{11} = \bar{r}_{ux1} + \bar{u}_{x1} \quad ; \quad a_{12} = \bar{r}_{uy1} + \bar{u}_{y1} \quad ; \quad a_{13} = \bar{r}_{uz1} + \bar{u}_{z1} \quad ;$$

$$a_{p1} = \bar{r}_{uxp} + \bar{u}_{xp} \quad ; \quad a_{p2} = \bar{r}_{uyp} + \bar{u}_{yp} \quad ; \quad a_{p3} = \bar{r}_{uzp} + \bar{u}_{zp} . \qquad (8)$$

The elastic displacements can be expressed as superposition of static and dynamic modes,

$$\bar{\mathbf{u}}^{*} = \sum_{i=1}^{n_s} \bar{\boldsymbol{\Omega}}_i \eta_i + \sum_{j=1}^{n_d} \bar{\boldsymbol{\Psi}}_j \xi_j, \qquad (9)$$

where n_s and n_d are the number of static and dynamic modes respectively, $\overline{\Omega}_i$ are the static modes, η_i are their amplitudes, $\overline{\Psi}_j$ are the dynamic modes, and ξ_j are their amplitudes. Either the amplitudes of the static modes as well as those of the dynamic modes are considered as problem variables. Therefore, the problem variables for a general flexible body when using the proposed modeling technique are: the origin of the local reference frame, $\mathbf{r}_o$, the thee orthogonal unit vectors, $\mathbf{a}$, $\mathbf{b}$, and $\mathbf{c}$, which define de local axes, and the amplitudes of static and dynamic modes, η_i and ξ_j, respectively. Substituting Eq. (9) into Eq. (7) leads to

$$
\mathbf{v}^* =
\begin{bmatrix}
\mathbf{I} & b_{11}\mathbf{I} & b_{12}\mathbf{I} & b_{13}\mathbf{I} & A\overline{\Omega}_1^1 & \cdots & A\overline{\Omega}_{n_s}^1 & A\overline{\Psi}_1^1 & \cdots & A\overline{\Psi}_{n_d}^1 \\
\vdots & & & & \vdots & & \vdots & \vdots & & \vdots \\
\vdots & & & & \vdots & & \vdots & \vdots & & \vdots \\
\vdots & & & & \vdots & & \vdots & \vdots & & \vdots \\
\mathbf{I} & b_{p1}\mathbf{I} & b_{p2}\mathbf{I} & b_{p3}\mathbf{I} & A\overline{\Omega}_1^p & \cdots & A\overline{\Omega}_{n_s}^p & A\overline{\Psi}_1^p & \cdots & A\overline{\Psi}_{n_d}^p
\end{bmatrix}
\begin{Bmatrix}
\dot{\mathbf{r}}_o \\
\dot{\mathbf{a}} \\
\dot{\mathbf{b}} \\
\dot{\mathbf{c}} \\
\dot{\eta}_1 \\
\vdots \\
\dot{\eta}_{n_s} \\
\dot{\xi}_1 \\
\vdots \\
\dot{\xi}_{n_d}
\end{Bmatrix} =
$$

$$
= \mathbf{B}\dot{\mathbf{q}}, \tag{10}
$$

where $\overline{\Omega}_i^r$ is a 3 by 1 vector containing static mode i evaluated at node r, $\overline{\Psi}_j^s$ is a 3 by 1 vector containing dynamic mode j evaluated at node s, and

$$
b_{11} = \overline{r}_{ux1} + \sum_{i=1}^{n_s} \overline{\Omega}_{ix1}\eta_i + \sum_{j=1}^{n_d} \overline{\Psi}_{jx1}\xi_j ; \quad b_{p1} = \overline{r}_{uxp} + \sum_{i=1}^{n_s} \overline{\Omega}_{ixp}\eta_i + \sum_{j=1}^{n_d} \overline{\Psi}_{jxp}\xi_j ;
$$

$$
b_{12} = \overline{r}_{uy1} + \sum_{i=1}^{n_s} \overline{\Omega}_{iy1}\eta_i + \sum_{j=1}^{n_d} \overline{\Psi}_{jy1}\xi_j \quad b_{p2} = \overline{r}_{uyp} + \sum_{i=1}^{n_s} \overline{\Omega}_{iyp}\eta_i + \sum_{j=1}^{n_d} \overline{\Psi}_{jyp}\xi_j ;
$$

$$
b_{13} = \overline{r}_{uz1} + \sum_{i=1}^{n_s} \overline{\Omega}_{iz1}\eta_i + \sum_{j=1}^{n_d} \overline{\Psi}_{jz1}\xi_j \quad b_{p3} = \overline{r}_{uzp} + \sum_{i=1}^{n_s} \overline{\Omega}_{izp}\eta_i + \sum_{j=1}^{n_d} \overline{\Psi}_{jzp}\xi_j ; \tag{11}
$$

The vector appearing in the right-hand-side of Eq. (10), $\dot{\mathbf{q}}$, is the vector of the derivatives (velocities) of the body variables $\mathbf{q}$, and then, substituting the value of $\mathbf{v}^*$ given by this equation into Eq. (5), the kinetic energy of the body can be written as

$$T = \frac{1}{2}\dot{\mathbf{q}}^{\mathrm{T}}\mathbf{B}^{\mathrm{T}}\mathbf{M}_{\mathrm{FEM}}\mathbf{B}\dot{\mathbf{q}} = \frac{1}{2}\dot{\mathbf{q}}^{\mathrm{T}}\mathbf{M}\dot{\mathbf{q}}, \tag{12}$$

which means that the mass matrix of the general flexible body is given by the expression

$$\mathbf{M} = \mathbf{B}^{\mathrm{T}}\mathbf{M}_{\mathrm{FEM}}\mathbf{B}. \tag{13}$$

Application of Lagrange equations with the kinetic energy of the flexible body given by Eq. (12) leads to calculate the vector of velocity dependent inertia forces

$$\mathbf{Q}_{\mathrm{v}} = -\mathbf{B}^{\mathrm{T}}\mathbf{M}_{\mathrm{FEM}}\dot{\mathbf{B}}\dot{\mathbf{q}}. \tag{14}$$

Therefore, it can be seen that, unlike their predecessors in [4], the final expressions for the inertia terms obtained through the abovementioned approach, given by Eqs. (13) and (14), are compact and relatively easy to evaluate.

2.2   Dynamic equations and numerical integrator

Once the modeling of a general flexible body has been explained, the approach to formulate the dynamic equations of the whole multi-body system as well as the numerical procedure to integrate them along the time is now addressed. Note that, as previously mentioned, either rigid and flexible links can be combined in a totally compatible form when using the proposed formulation, so that the actual mechanism can present both types of bodies.

The equations of motion of the whole multi-body system are given by an index-3 augmented Lagrangian formulation in the form

$$\mathbf{M}\ddot{\mathbf{q}} + \mathbf{\Phi}_{\mathbf{q}}^{\mathrm{T}}\alpha\mathbf{\Phi} + \mathbf{\Phi}_{\mathbf{q}}^{\mathrm{T}}\lambda^* = \mathbf{Q}, \tag{15}$$

where $\mathbf{M}$ is the mass matrix, $\ddot{\mathbf{q}}$ are the accelerations, $\mathbf{\Phi}_{\mathbf{q}}$ the Jacobian matrix of the constraint equations, α the penalty factor, $\mathbf{\Phi}$ the constraints vector, λ^* the Lagrange multipliers and $\mathbf{Q}$ the vector of applied and velocity dependent inertia forces. The main sources of the constraints, grouped into vector $\mathbf{\Phi}$, are: the rigid body conditions among the variables of each rigid body; the rigid body conditions among the variables of the local reference frame of each flexible body, and among such variables and those defining static modes whose deformation is to be neglected; the relative motion conditions imposed by each joint among the variables of the two neighbor bodies connected by such joint. The Lagrange multipliers are obtained from the following iteration process (given by sub-index i, while sub-index n stands for time-step),

$$\lambda_{i+1}^* = \lambda_i^* + \alpha\Phi_{i+1}, \ i=0, 1, 2,\dots,$$ (16)

where the value of λ_0^* is taken equal to the λ^* obtained in the previous time-step.

As integration scheme, the implicit single-step trapezoidal rule has been adopted. The corresponding difference equations in velocities and accelerations are:

$$\dot{q}_{n+1} = \frac{2}{\Delta t}q_{n+1} + \hat{\dot{q}}_n \text{ with } \hat{\dot{q}}_n = -\left(\frac{2}{\Delta t}q_n + \dot{q}_n\right),$$ (17)

$$\ddot{q}_{n+1} = \frac{4}{\Delta t^2}q_{n+1} + \hat{\ddot{q}}_n \text{ with } \hat{\ddot{q}}_n = -\left(\frac{4}{\Delta t^2}q_n + \frac{4}{\Delta t}\dot{q}_n + \ddot{q}_n\right).$$ (18)

Dynamic equilibrium can be established at time-step $n+1$ by introducing the difference equations (17) and (18) into the equations of motion (15), leading to

$$\frac{4}{\Delta t^2}Mq_{n+1} + \Phi_{q_{n+1}}^T\left(\alpha\Phi_{n+1} + \lambda_{n+1}\right) - Q_{n+1} + M\hat{\ddot{q}}_n = 0.$$ (19)

For numerical reasons, the scaling of Eq. (19) by a factor of $\Delta t^2/4$ seems to be convenient, thus yielding

$$Mq_{n+1} + \frac{\Delta t^2}{4}\Phi_{q_{n+1}}^T\left(\alpha\Phi_{n+1} + \lambda_{n+1}\right) - \frac{\Delta t^2}{4}Q_{n+1} + \frac{\Delta t^2}{4}M\hat{\ddot{q}}_n = 0 \ (20)$$

or, symbolically $f(q_{n+1}) = 0$.

In order to obtain the solution of this nonlinear system, the widely used iterative Newton-Raphson method may be applied

$$\left[\frac{\partial \mathbf{f}(\mathbf{q})}{\partial \mathbf{q}}\right]_i \Delta \mathbf{q}_{i+1} = -\left[\mathbf{f}(\mathbf{q})\right]_i \qquad (21)$$

being the residual vector

$$\left[\mathbf{f}(\mathbf{q})\right] = \frac{\Delta t^2}{4}\left(\mathbf{M}\ddot{\mathbf{q}} + \mathbf{\Phi}_\mathbf{q}^\mathrm{T}\alpha\mathbf{\Phi} + \mathbf{\Phi}_\mathbf{q}^\mathrm{T}\lambda^* - \mathbf{Q}\right) \qquad (22)$$

and the approximated tangent matrix

$$\left[\frac{\partial \mathbf{f}(\mathbf{q})}{\partial \mathbf{q}}\right] = \mathbf{M} + \frac{\Delta t}{2}\mathbf{C} + \frac{\Delta t^2}{4}\left(\mathbf{\Phi}_\mathbf{q}^\mathrm{T}\alpha\mathbf{\Phi}_\mathbf{q} + \mathbf{K}\right), \qquad (23)$$

where $\mathbf{C}$ and $\mathbf{K}$ represent the contribution of damping and elastic forces of the system provided they exist.

A closer look at the tangent matrix reveals that ill-conditioning may appear when the time-step becomes small. It may be seen in Eq. (23) that $\mathbf{K}$ and the constraint terms are multiplied by Δt^2, $\mathbf{C}$ by Δt and $\mathbf{M}$ is not affected by the step size. As a consequence when Δt reaches small values, large round-off errors will occur. In fact, it has been demonstrated in [7] that for an index-3 differential-algebraic equation, the tangent matrix has a condition number of order $1/\Delta t^3$. Consequently, the method is bound to have round-off errors for step sizes smaller than 10^{-5}, which lets a sufficient range for solving practical problems.

The procedure explained above yields a set of positions $\mathbf{q}_{n+1}$ that not only satisfies the equations of motion (19), but also the constraint condition $\mathbf{\Phi} = 0$. However, it is not expected that the corresponding sets of velocities and accelerations satisfy $\dot{\mathbf{\Phi}} = 0$ and $\ddot{\mathbf{\Phi}} = 0$, because these conditions have not been imposed in the solution process. To overcome this difficulty, mass-damping-stiffness-orthogonal projections in velocities and accelerations are performed. It can be seen that the projections leading matrix is the same tangent matrix appearing in Eq. (23). Therefore, triangularization is avoided and projections in velocities and accelerations are carried out with just forward reductions and back substitutions.

If $\dot{\mathbf{q}}^*$ and $\ddot{\mathbf{q}}^*$ are the velocities and accelerations obtained after convergence has been achieved in the Newton-Raphson iteration, their cleaned counterparts $\dot{\mathbf{q}}$ and $\ddot{\mathbf{q}}$ are calculated from

$$\left[\mathbf{M} + \frac{\Delta t}{2}\mathbf{C} + \frac{\Delta t^2}{4}(\mathbf{\Phi}_q^T\alpha\mathbf{\Phi}_q + \mathbf{K})\right]\dot{\mathbf{q}} =$$

$$\left[\mathbf{M} + \frac{\Delta t}{2}\mathbf{C} + \frac{\Delta t^2}{4}\mathbf{K} \right]\dot{\mathbf{q}}^* - \frac{\Delta t^2}{4}\boldsymbol{\Phi}_q^T \alpha \boldsymbol{\Phi}_t \tag{24}$$

for the velocities, and

$$\left[\mathbf{M} + \frac{\Delta t}{2}\mathbf{C} + \frac{\Delta t^2}{4}(\boldsymbol{\Phi}_q^T \alpha \boldsymbol{\Phi}_q + \mathbf{K}) \right]\ddot{\mathbf{q}} =$$

$$\left[\mathbf{M} + \frac{\Delta t}{2}\mathbf{C} + \frac{\Delta t^2}{4}\mathbf{K} \right]\ddot{\mathbf{q}}^* - \frac{\Delta t^2}{4}\boldsymbol{\Phi}_q^T \alpha \left(\dot{\boldsymbol{\Phi}}_q \dot{q} + \dot{\boldsymbol{\Phi}}_t \right) \tag{25}$$

for the accelerations.

Sparse matrix technology can be implemented to solve all the linear sets of equations that arise when applying this method, provided the size of the system is large enough for such technique to be worthwhile. The fact that the leading matrix is symmetric and positive-definite can be taken into account as well. The products $\mathbf{B}^T \mathbf{M}_{FEM} \mathbf{B}$ and $\mathbf{B}^T \mathbf{M}_{FEM} \dot{\mathbf{B}} \dot{\mathbf{q}}$ for each flexible body and $\boldsymbol{\Phi}_q^T \boldsymbol{\Phi}_q$ for the whole system, can also be optimized by considering the sparse character of the matrices.

3.   EXAMPLES

In order to show the behavior of the described formulation, the following examples of rigid and flexible multi-body systems have been solved. In all cases, the simulations have been run on common personal computers. Based on the obtained results, the performance of the method is evaluated in terms of efficiency, robustness and accuracy.

3.1   Double four-bar planar mechanism

The first example is a one degree-of-freedom assembly of two four-bar linkages, illustrated in Figure 2, which has been proposed [8] as an example to test the performance in cases where the mechanism undergoes singular configurations. When the mechanism reaches a horizontal position, the number of degrees of freedom instantaneously increases from 1 to 3. All the links have a uniformly distributed unit mass and a unit length. The gravity force acts in the negative vertical direction. At the beginning, the links pinned to the ground are in vertical position and upwards, receiving a unit clockwise angular velocity. The simulation lasts for 10 s, during which the mechanism goes through the singular position ten times.

Table 1 shows the obtained results for different time-steps (*Δt*). The error is measured as the maximum deviation, in J, suffered by the system energy.

As it can be seen in Table 1, the formulation shows a robust behavior when facing singular configurations, and large time-steps can be taken with acceptable accuracy. Regarding the efficiency, the example has been implemented in the computing environment Matlab. Although real-time performance is achieved, the reported CPU-times would be drastically reduced if the same programs were written in Fortran or C languages.

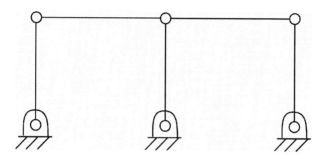

Figure 2. Double four-bar mechanism.

Table 1. Error and CPU-time for the double four-bar mechanism.

Δt (s)	Error (J)	CPU-time (s)
0.01	-0.21	0.86
0.03	-0.83	0.37
0.05	-9.03	0.26
0.1	wrong results	wrong results

3.2   Serial robot

The PUMA robot, designed by Unimation & Co. and shown in Figure 3, is an example of a 6 degrees-of-freedom serial manipulator. It has been often used by different authors [9] to illustrate methods and procedures in several areas of robotics. In this work, the robot has been taken as an example of multibody system undergoing changing configurations.

Starting from rest, torques at the six hinges of the robot are provided so that, in a time of 2 s, it arrives at a new position in the space, again in rest conditions. Once the new position has been reached, a point of the hand is attached to the ground, so that the robot loses 3 degrees-of-freedom. In this new configuration, torques are applied to the three rotational pairs of the hand, and a second maneuver, which lasts 4 s and ends with rest conditions, is performed. Finally, the robot is released from its attachment and returned in 2 s to the initial position by torques acting at the six revolute joints, once more finishing the maneuver at rest conditions. Therefore, the total simulation time is 8 s.

Table 2 illustrates the results obtained when applying the proposed method. The error has been calculated as the distance, in mm, between the hand positions at initial and final times (which, ideally, should be coincident).

From the results reported in Table 2, it is obvious that the formulation deals well with changing configurations. As in the previous example, large time-steps can be taken with acceptable accuracy. Although the example has been implemented in the computing environment Matlab, real-time performance is comfortably achieved.

Table 2. Error and CPU-time for the PUMA robot.

Δt (s)	Error (mm)	CPU-time (s)
0.01	0.53	6.25
0.05	3.4	1.92
0.1	no convergence	no convergence

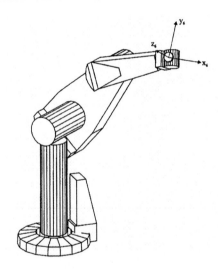

Figure 3. PUMA robot.

3.3   Four-bar mechanism with assembly defect

The third example is a four-bar mechanism, shown in Figure 4. The angular velocity ω of the left crank is kinematically guided at a constant rate of 1 rad/s. Gravity effects are neglected. Bar lengths are: AB=0.12 m; BC=0.24 m; CD=0.12 m; AD=0.24 m.

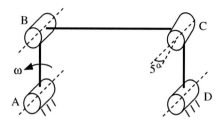

Figure 4. Four-bar mechanism with assembly defect.

Typically, the axes of rotation of the four revolute joints are orthogonal to the plane of the mechanism, and then it moves as a planar mechanism. However, in this case, the axis of rotation of joint C is at a 5 degree angle with respect to the plane normal to simulate an assembly defect in the mechanism. If the bars were rigid, no motion would be possible as the mechanism would lock. For elastic bars, motion becomes possible, but generates large internal forces. Physical properties of the bars are: mass density 3000 Kg/m^3, modulus of elasticity 7×10^{10} N/m^2, Poisson modulus 0.33. The section of the bars is circular with a diameter of 5 mm. For this example, the objective is to simulate 10 seconds of motion.

In order to evaluate the error, the example has also been solved through the nonlinear module of a finite element commercial code. Figures 5 and 6 show a good agreement between the torsion and bending moments at the middle section of the coupler. This time, the program was implemented in Fortran language. Using a time-step of 0.01 s, the analysis needed a CPU-time of 8 s, thus reaching real-time performance.

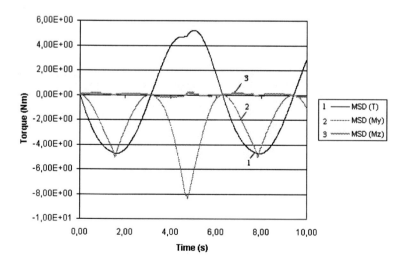

Figure 5. Torsion and bending moments at the middle section of the coupler: the proposed method

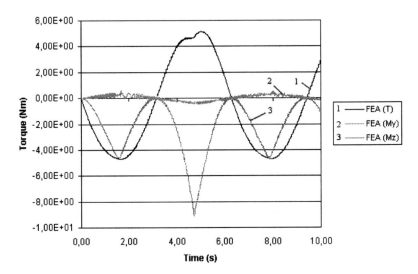

Figure 6. Torsion and bending moments at the middle section of the coupler: Finite element commercial code

3.4   Prototype car

A prototype car, which can be seen in Figure 7, has been manufactured in steel tubing and parts from old cars have been cannibalized and conveniently adapted (i.e. engine, suspension and steering systems, wheels, etc.).

The total mass of the car is 597 kg. Front shock-absorbers are characterized by a stiffness constant of 12360 N/m and a free length of 0.750 m, while their rear counterparts have values of 10595 N/m and 0.916 m for the same magnitudes. Damping of these front and rear elements has been set to 1000 Ns/m. Radii of the tires are 0.27 and 0.29 m for the front and rear tires respectively, which show a vertical stiffness of 60000 N/m.

The example has been used in two different ways corresponding to different purposes.

On one hand, the computer model of the car has been generated, considering the chassis as a flexible body, so that the stresses suffered by such body along the motion could be calculated, and compared with their experimentally measured counterparts. Therefore, the first purpose was the experimental validation of the proposed formulation.

On the other hand, a second computer model of the car has been implemented, but this time all the system bodies have been taken as rigid. This model has served as the calculation engine of a driving simulator, so as to verify whether the proposed formulation was reliable when integrated in an interactive real-time application.

A detailed description of each study is given in the following Sections.

Figure 7. The prototype car.

4.   EXPERIMENTAL VALIDATION

A maneuver has been defined in order to compare the resulting stresses at the chassis coming from measurement and simulation: starting from rest, the car accelerates and travels direct until reaching an approximate speed of 5 m/s, then drops down a curb of 8 cm height, and finally brakes until complete stop is attained. In this way, the bending modes of the chassis are excited. The described maneuver is illustrated in Figure 8.

To obtain the motion of the car, four tri-axial accelerometers have been attached to the chassis at the four corners of its base. Accelerations provided by the sensors in the motion direction have been integrated twice in order to obtain the position and velocity of the chassis at all times. On the other hand, four Wheatstone bridges have been installed at four points of the chassis to measure bending stresses. They are half-bridges, with an upper and a lower band each, to measure the normal bending stresses on the steel tubes surfaces. The sixteen captured signals –twelve accelerations and four strains– were recorded by a laptop PC after passing through a data-acquisition board.

The actual maneuver of 15 s has been simulated by implementation of the proposed formulation, with a fixed time-step of 0.01 s, the same sample period used for data acquisition during the experimental test, where a low-pass filter was applied with a cutoff frequency of 10 Hz to avoid noise from engine vibration.

Figure 8. The maneuver.

4.1   Accuracy

The comparison between measured and calculated bending moments at a certain point of the chassis is shown in Figure 9. At the view of the results, three time intervals can be distinguished.

The first one is the [0,5] interval. During this period of time, the car is at rest, but some peaks can be detected at about 2 s. They are explained because, at that time, the gear was set on *drive* position. This kind of engine loads is not considered by the model since engine torque is derived from acceleration information. Therefore, if the car is at rest, no engine torque is introduced in the model. This fact justifies the discrepancies shown between measurements and calculations during this time interval.

The second time interval is [5,10]. During this five seconds, the car accelerates, goes over the curb, and brakes until reaching the rest again. Good correlation is found during the accelerating and braking phases. The high peaks due to the falling action are not perfectly reproduced, although their intensities are reasonably captured. An explanation for such peak discrepancy can be the appearance of percussional forces in the actual prototype, due to the presence of clearances in some joints which have not been modeled.

The third and final time interval is [10,15], when the vehicle is again at rest. Some amount of error in the measurements can be detected in the figures, since the bending moments should return to zero, given that the initial and final positions of the car are exactly the same. However, the calculations come back correctly to zero.

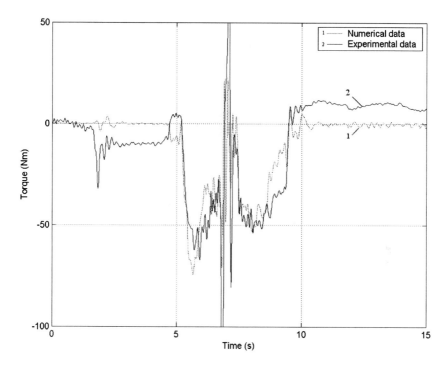

Figure 9. Comparison of bending moments.

4.2   Efficiency and robustness

The proposed formulation has shown an efficient behavior in the analysis of this case-study. The method was implemented on Fortran language, and the CPU time spent to run this 15-second-simulation was 48 s, that is, only a factor of 3 to real-time performance. The average number of Newton-Raphson iterations to attain convergence into each time-step of 0.01 s was equal to 2, needing a maximum of 3 iterations at the more difficult points.

Some important figures should be taken into account to adequately appraise the performance of the method. The number of problem variables is 293, related through 239 constraints, which means that 2 x 293 = 586 variables (positions and velocities) are integrated along the time, and that the Φ_q matrix size is (239 x 293), which in turn determines the cost of the product $\Phi_q^T \Phi_q$. The number of degrees of freedom of the chassis finite element model is 1266, while the number of static plus dynamic modes is 65, so that the M_{FEM} matrix size is (1266 x 1266) and the B matrix size is (1266 x 77). Therefore, both products $B^T M_{FEM} B$ and $B^T M_{FEM} \dot{B} \dot{q}$ are also rather time-consuming. Undoubtedly, as many readers can think when considering the adopted modeling, an effort could be done in order to reduce all these figures, since a special emphasis on keeping the problem size as small as possible has not been carried out. Consequently, perhaps the efficiency could be slightly enhanced, but probably the improvement would not be significant and, what is more important, it would not contribute to better show the power of the method, which was the objective of this work.

On the other hand, and this also affects to robustness consideration, the performed maneuver is particularly violent, since the curb descent produces high values of vertical acceleration. Moreover, the kinematic guidance of the forward motion of the prototype shows a curly shape which supposes an additional difficulty for the numerical integration.

5.   DRIVING SIMULATOR

The computational model of the car has been used to create a driving simulator, shown in Figure 10, by combining the already mentioned vehicle dynamics code, along with realistic graphical output and game-type driving peripherals (steering wheel and pedals).

The proposed formulation has shown to be efficient enough to enable the interactive driving of the car, and robust enough to bear very long driving sessions (over 1000 s), not free of violent maneuvers like jumps from elevated positions to the ground or sudden turns.

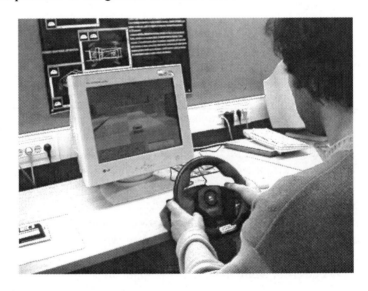

Figure 10. Driving simulator.

6.   CONCLUSIONS

Based on the results, the following conclusions may be established:

a) A real-time formulation has been developed capable of performing very fast calculations of the dynamics of complex rigid-flexible multi-body systems.

b) The modeling is carried out with natural coordinates, making use of the floating frame of reference approach with static and dynamic modes for the flexible bodies, along with the co-rotational approach to simplify the inertial terms.

c) The equations of motion are stated through an index-3 augmented Lagrangian scheme, which is combined with the implicit, single-step, trapezoidal rule to perform the numerical integration. Mass-damping-stiffness-orthogonal projections in velocities and accelerations are needed at each time-step to provide stability.

d) The formulation is global and, therefore, very easy to implement. This fact represents an advantage with respect to topological formulations, often used for real-time purposes.

e) The formulation provides good levels of efficiency, robustness and accuracy, as demonstrated through the solved examples. It can handle singular positions, changing configurations, stiff systems and large models.

f) The formulation has been experimentally validated through the comparison between calculated and measured stresses on a component of a complex and realistic multi-body system.

g) The formulation has shown to be valid for becoming part of a virtual reality environment, in connection with user-interface devices, since it has successfully served as the calculation engine for a driving simulator.

h) Therefore, it can be affirmed that the presented formulation is a good candidate for future virtual engineering applications.

REFERENCES

1. J. Garcia de Jalon, E. Bayo, *Kinematic and Dynamic Simulation of Multibody Systems –The Real-Time Challenge* (Springer-Verlag, 1994).
2. E. Bayo, R. Ledesma, *Augmented Lagrangian and Mass-Orthogonal Projection Methods for Constrained Multibody Dynamics, Nonlinear Dynamics* (Vol. 9, 1996, pp. 113-130).
3. E. Eich-Soellner, C. Führer, *Numerical Methods in Multibody Dynamics* (B.G. Teubner Stuttgart, 1998).
4. J. Cuadrado, J. Cardenal, J. Garcia de Jalon, *Flexible Mechanisms through Natural Coordinates and Component Synthesis: An Approach Fully Compatible with the Rigid Case* (Int. Journal for Numerical Methods in Engineering, Vol. 39, No. 20, 1996, pp. 3535-3551).

5. M. Geradin, A. Cardona, *Flexible Multibody Dynamics -A Finite Element Approach* (John Wiley and Sons, 2001).
6. A. Avello, *Simulación Dinámica Interactiva de Mecanismos Flexibles con Pequeñas Deformaciones* (Ph. D. Thesis, Universidad de Navarra, Spain, 1995).
7. K. E. Brenan, S. L. Campbell, L. R. Petzold, *The Numerical Solution of Initial Value Problems in Differential-Algebraic Equations* (Elsevier, 1989).
8. E. Bayo, A. Avello, *Singularity Free Augmented Lagrangian Algorithms for Constrained Multibody Dynamics, Nonlinear Dynamics* (Vol. 5, 1994, pp. 209-231).
9. J. Angeles, O. Ma *Dynamic Simulation of n-axis Serial Robotic Manipulators using a Natural Orthogonal Complement* (Int. J. of Robotic Research, Vol. 7, No. 5, 1988, pp. 32-47).

MULTI-BODY DYNAMICS: AN EVOLUTION FROM CONSTRAINED DYNAMICS TO A MULTI-PSYSICS INTERACTIVE FRAMEWORK

M. Teodorescu, S. Theodossiades and H. Rahnejat
Wolfson School of Mechanical and Manufacturing Engineering, Loughborough University, Loughborough, U.K.

Abstract: This paper provides an overview of a multi-physics approach to dynamics analysis across the physics of scale, typically existing in many machines and mechanisms. The interactive nature of phenomena calls for such holistic approaches within objective system engineering applications. The paper provides a number of powertrain investigations of significant concern in industry, and shows the validity of the proposed approach in their investigation. The emphasis is put upon the hierarchical and interactive nature of the problems and physics of scale from sub-micrometer to large displacements, including surface interactions, lubricated concentrated contacts, structural vibration, acoustic radiation and constrained inertial dynamics.

Key words: multi-body systems, dynamics.

1.   INTRODUCTION

In recent years greater attention has progressively been put on the interactions of various phenomena in the dynamic analysis of systems [1]. These phenomena may follow a different strand of fundamental physics, and thus served by a particular evolved discipline, such as tribology, thermodynamics and structural integrity. However, the core discipline remains the physics of motion. As the investigation tools have improved due to technological advances, the long recognised important interactive nature of these phenomena can be taken into account. Within the discipline of dynamics, the motion of particles and bodies can be studied with regard to interaction of matter at all physical scales from nano-scale through to macro-scale systems. The methodology for dealing with such problems had already

D. Talabă and T. Roche (eds.), Product Engineering, 273–298.
© 2004 *Springer. Printed in the Netherlands.*

existed to a large extent from the mid 17th and early 18th Centuries. These include the principle of virtual work as manifested in Lagrangian dynamics, which forms the basis of multi-body analysis, and elasticity from principle of superposition through to modern finite element analysis [2]. Other interactive phenomena contribute largely in the form of kinetics of various kind, governed by physics of processes such as tribology (friction, wear and lubrication) and thermodynamics (for example interaction of matter by chemical reactions). This means that the domain of analysis may neither be confined to any particular scale nor indeed to any given discipline. Thus, the opportunity has arisen for investigation of phenomena within a multi-scale, multi-physics framework. In short, one can delve into a system dynamic problem in a hierarchical manner in which an observed effect at the higher echelon of the framework may be traced back to the underlying causes at lower levels, down to the minutia of microcosm, where the interactions can no longer be described as multi-body dynamics but a case of many-body systems.

The trend towards this approach has already commenced, thanks to the refinement of numerical techniques, sensor technology, and measurement and monitoring techniques. This, however, is not to advocate that all analyses should be carried out to this level of detail, but simply to recognise that these opportunities have now emerged. This paper provides two examples of this approach. The first example deals with a drivetrain noise and vibration concern, referred to as clonk. The second example deals with a tribo-dynamic problem in valve train systems.

Nomenclature

C_k constraint function

D damping energy dissipation function

E^* the reduced elastic modulus of impacting solids:

$E_{1,2}$ elastic modulus

F impact force

F_{ξ_j} generalised applied forces projected on ξ

G Green's function

L the Lagrangian

L_b bearing length

R the reduced radius of the a sphere impacting a flat

R_j radious of the journal

T	kinetic energy
V	Potential energy
V_z	velocity of approach: $(v2 - v1)\, t=0$
Z	damping Ratios Matrix
h	oil film thickness
m	the equivalent mass impacting a rigid flat semi-infinite planenumber of independent modal co-ordinates
c	journal clearance
$m_{1,2}$	impact sphere masses
n	number of constraints
p	oil film pressure
q	elastic modal co-ordinates
r_i	point of interest
r_Γ	is a point at the boundary Γ
u	linear nodal deformation vector
x, y, z	translational degrees of freedom
$\gamma_{1,2}$	Poisson ration
δ	local elastic Hertzian depression
ε	the eccentricity between the centre of the journal and that of the bearing bush
η_0	atmospheric dynamic viscosity of the oil
λ_κ	lagrange multipliers
$v_{1,2}$	velocity of approach at start of impact
ξ_j	generalised co-ordinates
ϕ	deformation shape function
ψ, θ, ϕ	Euler angles

2. HIGH FREQUENCY STRUCTURAL-ACOUSTIC BEHAVIOUR OF POWERTRAIN SYSTEMS

2.1   Defining the problem

The vehicular drivetrain is an ill-configured very lightly damped system, which has evolved in an almost empirical manner. As a consequence the dynamics of the system is adversely affected by a plethora of unwanted dynamic problems across the physics of scale. These include coupled natural rigid body torsional and longitudinal motions at low frequency, referred to as shuffle and shunt respectively [3]. Shuffle is the lowest natural frequency of the powertrain system, typically in the range 3-10 Hz, whilst shunt is the fore and aft motion of the vehicle in tip-in and back-out at the same frequency. These can be easily noted when the vehicle goes from coast to drive condition with throttle tip-in action or conversely by throttle tip-out action. Other rigid body oscillations also exist as the result of engine order vibration caused by inherent inertial unbalance, excited by the combustion process. The amplitudes of vibration are large relative to structural vibration at higher frequencies, and are usually attenuated by provisions to limit the unbalance, such as the introduction of counterbalance masses to eliminate the first engine order (i.e. the primary torsional response of the crankshaft system), or by counter-rotating shafts with respect to the crankshaft system in order to reduce the second engine order response of a 4-cylinder, 4-stroke engine (i.e. the Lanchester balancer shaft) [4].

All transmissions, differentials and spline joints in powertrain systems are regarded as lash zones, where clearly functioning of meshing pairs calls for a certain amount of lash, the existence of which also leads to unwanted impact. Under sudden application of throttle, as described above, these lash zones can act as impulsive pairs, through which the impact energy can propagate as structural waves which in lightly damped structures can coincide with acoustic modes of cavities or the surrounding fluid media (such as air) and radiate [5]. Such hollow cavities are in abundance in the drivetrain systems, such as the transmission bell housing and the hollow driveshaft tubes. There are a host of structural modes for each of these elements, which increase in number, when made in the form of thin-walled structures, something which is common place nowadays in order to reduce the inertial imbalance, and thus improve the fuel efficiency of the system. One drawback, however, has been that the gain in inertial balance can be more than off-set by multitude of structural modal responses. The net result is a plethora of higher frequency, lower amplitude structural vibration that can lead to noise propagation, a phenomenon which is onomatopoeically referred to as clonk [6].

Up to recently there has been a dearth of analysis of the clonk problem, which is an endemic concern in the automotive sector, affecting many ranges of vehicles from small saloons to large trucks. The problem has been studied very much along the traditional disciplinary divides, such as inertial dynamics at the system level, transmission error and impulsive actions, and structural and acoustic behaviour of resonators such as the hollow driveshaft tubes, all at the sub-system or component levels. The implications of this approach have been that palliative rather than root cause solutions to the problem have been favoured. The plethora of inter-related noise, vibration and harshness (NVH) concerns has meant that palliation of one has often run contrary to the resolution of others, with the outcome that much of these problems have remained for the best part of two decades.

2.2   Identifying the underlying physics

Since impulsive action in the lash zones itself requires an initiator, it is clear that driver behaviour, road input and engine all play a role. It has emerged that abrupt actuation of the clutch, sudden throttle demands as in tip-in and tip-out, shift of the transmission and conditions at the contact patch all play a role in inertial dynamics of the powertrain system, which in turn can react in the lash zones in the form of impact of mating pairs, causing transfer of momentum into impulsive action upon structural elements, thus propagation of waves, which can coincide with acoustic modes of thin walled cavities, thus leading to clonk conditions. This rather simple explanation indicates the clear connection between powertrain shuffle, vehicle shunt, modal behaviour of structures and the emanating clonk conditions. It is interesting to note the profound nature of impulse-momentum relationship as the long established mechanism in physics of motion, simple in concept and yet manifested in all manner of complex phenomena, such as in clonk. The vehicular powertrain system is just an example of any other elastic solid such as a ball that may be subject to an impact, responding structurally that coincide with acoustic wave propagation. This very simple system is the basis of the well known Newton trolley (see Figure 1).

When the strain energy stored in such a solid sphere: $\frac{2}{5}k\delta^{5/2}$ is below the required energy for propagating wave energy (i.e. below the natural modes of the solid), then the deformation can be considered as localised and the emitted noise is regarded as accelerative of short finite duration, termed as the impact time. These were realised and formulated by Hertz in 1882. Now if the sphere was made hollow, the required energy to excite its natural modal behaviour can be reached more readily, and the accelerative noise, typically of duration of tenths to a couple of milliseconds, is followed by a much longer decaying ringing noise of several to tens of milliseconds. The ringing noise is as the result of coincidence of natural structural modes of the elastic solid with the acoustic modes of the contained volume, termed as breathing modes. In other words the hollow solids act as good resonators. After all this forms the basis of loudspeakers. The structural wave propagation was described by St Venant, and the propagation in adjacent fluids in the form of pressure waves by Von Helmholtz. How effective is a resonator depends on the material composition, modal density of the structure, geometry and fluid medium. Therefore, the principles that apply to a hollow sphere, equally well apply to any other solid structure, including the powertrain system. The differences lie in the detail.

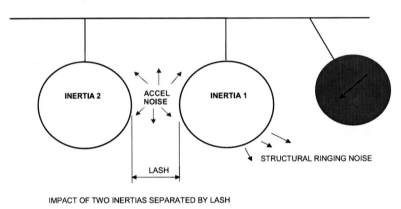

Figure 1. Newton trolley

2.3   Adhering to the principle of causality

Thus, the problem may be viewed from its component level behaviour right up to their system level interactions, with many facets of physical behaviour. A multi-physics analysis will mimic this reality, by incorporating salient features of components as well as their system level interactions. In this approach the system integration is effected by the traditional multi-body dynamic formulation, based on constrained Lagrangian dynamics. In any problem, it is natural to assume the position of an observer, thus determining the actual perceived phenomenon, this being the observed effect. In the case of problems encountered in industry, the observer is termed the *customer*, and the untoward *effect* is referred to as a *concern*. Therefore, in the case of problem at hand (i.e. clonk), the *effect* or *concern* is the radiated noise, being the central issue of investigation. In the multi-physics approach, there is a chain of causes and effects which culminate in the eventual concern, as described above. The extent to which one traces backwards along this chain should be based upon the practicability of investigation. In industry, this process is carried down to the "*underlying causes*" which may be affected by some practical and cost effective remedial action. This approach determines the boundary of the problem at hand.

Therefore, in the case of the clonk problem the central issue is *acoustic radiation* from system resonators (i.e. the hollow driveshaft tubes), and the "*underlying cause*" is the impulsive action through lash zones. The physics of the problem to be included are: inertial and flexible body dynamics, mechanism of impact and elasto-acoustic coupling. The boundary of the system is the powertrain system from the flywheel to the rear axle, with interfaces to the extraneous world at the flywheel to the power generating source (i.e. the engine), the driver (at throttle and clutch systems), the vehicle at the driveshaft centre bearings, rear axle, engine and transmission mounts, and the road at the rear road wheels' contact patches.

3.   COMPONENT LEVEL INVESTIGATION: ACOUSTIC BEHAVIOUR OF DRIVE SHAFT TUBES

In the case of thin walled hollow tubes that may be characterised as finite length cylindrical shells their acoustic properties in a light medium such as air, may be obtained by either a statistical approach or in a deterministic manner. In the latter approach the velocity distributions of each mode is required, which is affected by the boundary conditions. Therefore, the various modes are individually considered. This is a rather time consuming but accurate for the determination of the acoustic radiation efficiency of each mode. Specifically for shell structures, the influence of the curvature on sound radiation derives from its effect on the flexural wave characteristics, which are mainly in low wave numbers. Cylinder curvature increases the flexural wave phase velocities through the mechanism of mid-plane strain with a consequent increase in radiation efficiency. Simultaneously, a reduction in the density of natural frequencies is observed, compared to the acoustic efficiency of flat panels. The parameter that indicates the frequency range for which curvature effects are important is the so-called ring frequency f_r, which is defined as the frequency when the wavelength of extensional (axial) waves in the shell is equal to the shell circumferential waves: $f_r = \dfrac{1}{2\pi a}\sqrt{\dfrac{E}{\rho}}$. In coupled elasto-acoustic analysis, there is another parameter that is important; the so-called critical frequency or coincidence frequency f_c, which is defined as the frequency at which the acoustic wave-length in the medium is equal to the acoustic wave-length in the shell structure material, or alternatively it is the frequency at which the speed of sound in the fluid equals the bending wave velocity in the structure:

$$f_r = \frac{c_o^2}{2\pi h} \sqrt{\frac{12\rho(1-v^2)}{E}}.$$

The aforementioned two characteristic frequencies determine whether a cylindrical shell is classified as acoustically *thin*, when: $f_r < f_c$, or acoustically *thick*, when: $f_r > f_c$. At this point it is quite interesting to make the distinction that a shell structure can be characterised as thin from a geometric point of view (when $h << \alpha$), but as thick from an acoustic classification (when $f_r > f_c$). The acoustically thin shells are associated with a high modal density. This means that the statistical approach methods for the determination of the acoustic radiation properties may give quite accurate results. However, in many engineering applications acoustically thick shells are used, in which case the modal density is high only in the high frequency band. Consequently, an individual analysis of each mode is required for the extraction of accurate conclusions, with regard to the acoustic radiation efficiency that is also strongly dependent on the geometric properties and boundary conditions of the shell. Generally, it has been observed that the radiation efficiency of acoustically thick shells reaches unity at high frequencies, in which case the elasto-acoustic coupling of the structure and the medium should be especially considered. The acoustic behaviour is also influenced by the nature and type of excitation. For example, road surface excitations on the vehicle and the sudden engagement of the clutch could produce different acoustic responses in the drive shafts.

The natural modes of a cylindrical shell structure (i.e. the hollow vehicle drive-shafts) may be categorised into two main groups, acoustically fast or acoustically slow.

The so-called acoustically fast *supersonic* modes have structural wave numbers that are smaller than the corresponding acoustic wave numbers of the external medium.

The so-called acoustically slow or *subsonic* modes have structural wave numbers that are greater than the corresponding acoustic numbers.

The *subsonic* modes are not as efficient in sound radiation, when compared to the *supersonic* modes. For flat plates the critical frequency identifies the distinction between the *supersonic* and the *subsonic* modes in the frequency domain. For cylindrical shells, there is not such a critical frequency. This means that below the critical frequency there is a mixed frequency area, where *supersonic* and *subsonic* modes will coincide due to the influence of the shell curvature in the characterisation of a mode as being acoustically fast or slow.

As has been mentioned before, structural supersonic modes are usually effective acoustic radiators of noise. Therefore, as a first step, an investigation is required for the separation of the driveshaft modes into the *supersonic* and *subsonic* ranges. This distinction has been achieved by calculating the quantity: $\Delta L(m, n) = k - k_s$, where, k and k_s are the acoustic and structural wave-numbers, respectively.

Figure 2 shows a typical demarcation line between *supersonic* and *subsonic* modes of a driveshaft tube. Note that all the modes of the tube up to a frequency of 5 KHz are included in the analysis, as these fall into the observed and measured clonk behaviour of the drivetrain system for the particular light truck investigated.

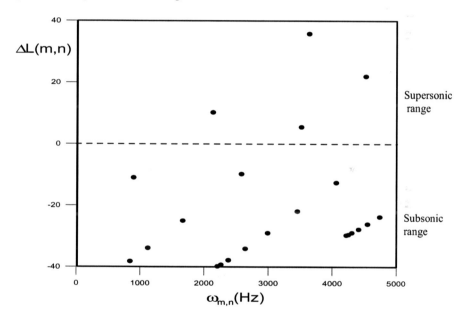

Figure 2. Supersonic and Subsonic Modes: of a driveshaft tube

Note that the value of $\Delta L(m, n) > 0$ correspond to the defined *supersonic* modes. These are the acoustic modes of the structure which should be avoided, as they are efficient resonators.

4.   STRUCTURAL MODAL RESPONSE

It is important to obtain the structural modes of the elastic solids in order to ascertain the possibility of their excitation by the impact energy through the lash zones. If such is to take place, and the modal response of the structure happens to coincide with its *supersonic* acoustic waves existing within very close proximity, it is then clear that fluid-structure interactions would lead to radiated sound. This phenomenon is referred to as elasto-acoustic coupling. In the hierarchical manner described before, the sound propagation is due to molecular action of fluid medium, almost like an impact-strewn environment where the energy is transmitted in a many-body system, or in other words on a nano-scale. Thus, the change in the temperature of the fluidic medium enveloped by the structure (such as the internal cavity of the driveshaft tubes) or surrounding the structure change the manner in which sound propagates (i.e. changes the speed of sound waves, even if by a small amount). The structural modes correspond to vibration amplitudes in the micrometer range. The modal behaviour of the driveshaft pieces is obtained using finite element analysis. For thin tubes shell elements can be used to mesh the structure. It is clear that a large number of such elements are required, at least twice the number of the highest frequency limit set in the analysis.

Now simple modal analysis can be carried out when the correct boundary conditions are applied. For a predetermined frequency area, a number of shape vectors, φ are obtained. The *modal superposition* principle is based on the assumption that the linear deformation of a component's very large number of nodal DOF, u, in a pre-determined frequency area can be captured as a linear combination of a smaller number of shape vectors, φ.

Figure 3. A typical mode shape for a hollow driveshaft tube

Then for the number of mode shapes, m, the physical co-ordinates can be obtained as: $u = \sum_{i=1}^{m} \varphi_i q_i$. This is essentially the transformation from the small set of modal coordinates, q, to the larger set of physical coordinates, u. For a frequency range of interest, indicated by the simple analytical technique in section 3 above, one would need to include the appropriate shape vectors (mode shapes) that are likely to contribute to coincidence of structural waves with the acoustic behaviour. This procedure is referred to as component mode synthesis, leading to the modal matrix ϕ, where: $\phi = \sum\limits_{i=1}^{m} \varphi_i$. Figure 3 shows a typical mode shape.

5.   SYSTEM LEVEL INTEGRATION WITH MULTI-BODY DYNAMICS

Once the elastic response of the nominated flexible members have been determined for the modal co-ordinates, q_j, these can be incorporated within the multi-body analysis represented by the constrained Lagrangian dynamics, where:

$$\frac{d}{dt}\left(\frac{\partial L}{\partial \dot{\xi}_j}\right) - \frac{\partial L}{\partial \xi_j} + \frac{\partial D}{\partial \dot{\xi}_j} + \sum_{k=1}^{n} \lambda_k \frac{\partial C_k}{\partial \xi_j} = 0, \tag{1}$$

where: $\{\xi_j\}_{j=1 \to 6} = \{x, y, z, \psi, \theta, \varphi\}^T$ for the *rigid body* degrees of freedom, $\{\xi_j\}_{j=1 \to 6+m} = \{x, y, z, \psi, \theta, \varphi, q\}^T$ for the *flexible body* structural response (q are the modal coordinates of total number, m), $L = T - V$ is the Lagrangian: the difference between Kinetic and Potential energies and $D = 1/2 \dot{q}^T Z \dot{q}$.

The n constraint functions for the different joints in the driveline model are represented by a combination of holonomic and non-holonomic functions as:

$$\left[\begin{matrix} C_k \\ \dot{\xi}_j \dfrac{\partial C_k}{\partial \xi_j} \end{matrix} \right] = 0, \; (j = 1 \to 6 \quad or \quad 1 \to 6 + m), \; (k = 1 \to n). \tag{2}$$

Note that:

$$\frac{\partial L}{\partial \xi_j} = \frac{\partial T}{\partial \xi_j} - \frac{\partial V}{\partial \xi_j} = F_{\xi_j}.$$

Thus, one needs to include all the body and applied forces. Of particular interest is the impact force (the impulsive action as described in section 2).

The simplest form of impact dynamic analysis that can be employed for impact of gears is the Hertzian impact theory, providing that Hertzian conditions can be assumed. These are that the area of contact/impact is much smaller than the principal radii of curvature of approaching solids, in this case an impacting pair through backlash. Obviously, the profile of the impacting solids should be approximated in the region of impact by ellipsoidal equivalent solids. In the case of contacts of helical involute pairs in vehicle transmission the impacting pair of teeth may be regarded as elastic solids of low conformity under highly loaded condition. The footprint shape has been shown to be elliptical, thus make the use of Hertzian theory valid as a first approximation. When the radii of curvature of such impacting pairs are known at a particular location at a given instance of time, elastostatic conditions may be assumed. Of course, this approach also ignores the lubricant reaction, as well as the frictional nature of the contact.

6.   HERTZIAN IMPACT DYNAMICS

Consider two elastic spheres of masses m_1 and m_2 approaching each other with velocities v_1 and v_2 and impact. Applying Newton's 2nd law of motion to each mass in turn, it follows that:

$$F = m_2 \frac{dv_2}{dt} \text{ and } F = m_1 \frac{dv_1}{dt}. \tag{3}$$

Consider a local elastic Hertzian depression of δ due to the impact, then:
Let $\dot{\delta} = v_1 + v_2$ is the relative velocity just before impact.
Substituting from the above, it follows that:

$$\ddot{\delta} = -F \frac{m_1 + m_2}{m_1 m_2}, \text{ where } F = k\,\delta^{3/2}, \tag{4}$$

$$\therefore \ddot{\delta} = -k\delta^{3/2}.$$

Multiply both sides by $\dot{\delta}$ and integrating, it follows that:

$$\frac{1}{2}(\dot{\delta}^2 - v^2) = -\frac{2}{5}\frac{k}{m}\delta^{5/2}, \tag{5}$$

where v=velocity of approach at start of impact.
If one substitutes: $\delta = 0$ in the above equation, then:

$$\delta = \left(\frac{15m_{eq}v^2}{8E^*\sqrt{R}}\right)^{2/5}.$$ (6)

Thus: $dt = \left\{v^2 - \frac{4k}{5m}\delta^{5/2}\right\}d\delta$ (7)

and $t = \dfrac{2.94328\delta}{v}.$ (8)

This is the contact time for two elastic spheres of same material and dimensions. In this case the contact duration time is proportional to the elastic deflection and inversely proportional to the velocity of approach.

Where $1/m = (1/m_1 + 1/m_2)$, m being the equivalent (i.e. referred to as normalized) mass impacting a rigid flat semi-infinite plane and E^* being the reduced (effective) elastic modulus of impacting solids:

$$\frac{1}{E^*} = (1-\gamma_1^2)/E_1 + (1-\gamma_2^2)/E_2.$$ (9)

The impact force is then obtained as: $F = \dfrac{2}{3}\sqrt{R}E'\left(\delta^*\right)^{3/2}$. This is then added to the multi-body model, together with the modal behaviour of the elastic members. The simulation of this model provides the structural modes of the resonators.

7.   ELASTO-ACOUSTIC COUPLING

Propagating waves can then be obtained using the boundary element approach. These waves can be represented by the von Helmholtz equation in the frequency domain, as [7]:

$$(\nabla^2 + k^2)p = 0$$ (10)

where, $\nabla^2 = \dfrac{\partial^2}{\partial^2 x} + \dfrac{\partial^2}{\partial^2 y} + \dfrac{\partial^2}{\partial^2 z}$ and $k = \dfrac{\omega}{c}$.

The Green's function is given as:

$$G = \frac{e^{-jkr}}{4\pi r},$$
(11)

which is a fundamental solution to

$$\nabla^2 G + k^2 G = -\delta(x, y),$$
(12)

where, $\delta(x, y) = \begin{cases} 0 & when \quad x \neq y \\ 1 & when \quad x = y \end{cases}$.

By applying the indirect method, the pressure and velocity in the boundary surface are given as

$$p = \int_\Gamma \left(G\sigma - p\frac{\partial G}{\partial n}\mu \right) G d\Gamma \text{ and } v = \int_\Gamma \left(\frac{\partial G}{\partial n}\sigma - p\frac{\partial^2 G}{\partial n^2}\mu \right) d\Gamma,$$

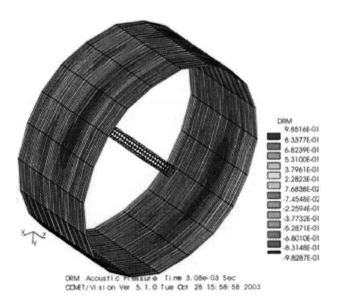

Figure 4. Fringe plot of the acoustic pressure in the environment of the front tube using the Boundary Element Method

respectively, where, $\sigma = \left(\dfrac{\partial p}{\partial n}\right)^{+} - \left(\dfrac{\partial p}{\partial n}\right)^{-}$, $\mu = p^{+} - p^{-}$ and n is the surface normal vector.

The acoustic pressure at a point of interest r_i can be written as:

$$p(r_i) = \int_{\Gamma}\left[\mu(r_{\Gamma})\frac{\partial G(r_i, r_{\Gamma})}{\partial n_{\Gamma}} - G(r_i, r_{\Gamma})\sigma(r_i)\right]dS_{\Gamma}, \tag{13}$$

where, r_{Γ} is a point at the boundary Γ.

The acoustic pressure at any given time for each resonating structure can be obtained as shown in Figure 4.

8.   VALVE TRAIN TRIBO-DYNAMICS: HIERARCHICAL MULTI-PHYSICS APPROACH

The valve train system represents one of the major causes of IC engine mechanical inefficiency, as well as an important source of engine noise. Due to its high speed oscillatory motion, the large range of interacting physical phenomena between the valve-train components cannot be decoupled. Consequently, for a correct understanding of the valve-train behaviour, a multi-physics approach is required.

The present work demonstrates an integrated approach for the tribo-dynamics of a four cylinder in line engine. The general view of the pushrod type mechanism is shown in Figure 5.

The valve train configuration is rather complex, and it includes cam, tappet, valve (the minimum necessary for an overhead valve train system), pushrod and rocker arm.

Figure 5. General view of a line valve trains for a four cylinder engine

Each individual valve-train mechanism can be modelled as a multi-body system, comprising a number of inertial elements connected by restraining elements. Figure 6 shows such a model, in which a two-mass system with three degrees of freedom is considered. One mass concentrates the tappet, pushrod, rocker arm, whilst the second mass concentrates the valve and the proportion of the non-negligible mass of the valve spring. The concentrated masses are connected with elements which represent the system stiffness and damping. The gaps depicted by the horizontal parallel lines in Figure 6 represent the lubricated conjunctions, where the model accounts for the possible contact frictional losses, as well as for the oil film squeeze effect and viscous damping. The impact between the valve and its guide take into account its influence upon the system's tribo-dynamic characteristics.

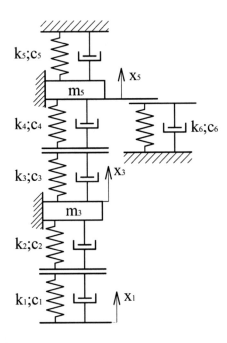

Figure 6. Equivalent model for an individual valve train

Figure 7 shows a schematic view of the camshaft equivalent model. For each section of the camshaft (cam and half of each adjacent journal bearing) the model considers an inertial element, which concentrates the mass (for simulation of bending oscillations in the vertical plane) and the moment of inertia (for the torsional vibrations). The inertial elements are connected with stiffness and damping elements. The bearing deflection on the vertical axis has been considered as well, together with its damping effect. The viscous friction in the journal bearings is accounted for through the resisting torque applied on each bearing.

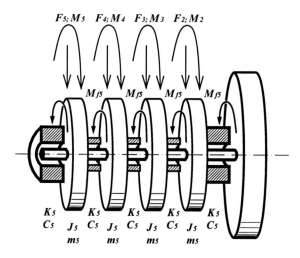

Figure 7. Schematic representation of the camshaft model

Therefore, the model is conceived in a hierarchical multi-physics framework, including inertial dynamics, component compliance and lubricated contact behaviour. It may be termed as a tribo-elasto-dynamic model. The physics of motion across a large spectrum is integrated in such an analysis, from micro-scale (of the order of tenths to several micrometers) in lubricated conjunctions, to a few micrometer to tenths of millimeter structural deformation onto "large" displacement rigid body constrained motions.

For a fundamental understanding of the dynamic coupling between the camshaft and the four valve trains, the model of the entire line of valve-trains is considered in an iterative marching solution scheme. At each step of time, the dynamic response of the camshaft is computed, based upon the vertical force and torque around the centre of the cam predicted by the valve train dynamic modules (described above) at the previous time step. Furthermore, within each step of time, the camshaft dynamic model's predictions are fed back into the individual valve train modules in order to determine the force and the torque applied on each cam-follower pair.

9.   RESULTS OF THE VALVE TRAIN SYSTEM

Figure 8 shows the valve acceleration for each valve-train. Figure 8(a) depicts the model predictions neglecting component elasticity (except for the valve spring), whilst Figure 4 (b) includes the effect of the elastic behaviour of system components.

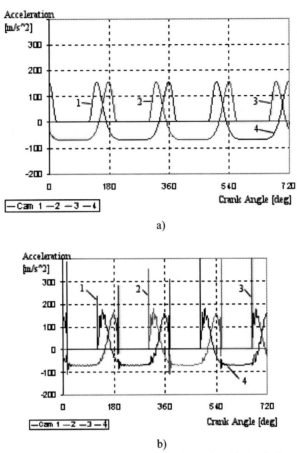

Figure 8. Valve train acceleration a)rigid line of valve trains, b) elastic line of valve trains

The most loaded sections of each cam are generated by the impacts at opening and closing events. At the beginning of the opening event, the cam-tappet clearance is gradually diminished by the ramp-up in the cam profile just prior to its flank. However, due to the high level of vibrations in the system (on the camshaft as well as in the individual valve trains) this is not necessarily the case. At the closing event, the valve hits the valve seat and the impact energy propagates through the whole system. It is noticeable that these predicted impacts in one cam-follower pair propagates through the camshaft to the rest of the valve trains in the line of valves. It is also important to emphasise that the model successfully predicts a larger amplitude for the transmitted impact between neighbouring cams (the 1st and the 2nd) than between distant cams (2nd and 4th) as would be logically expected.

The camshaft dynamic model considers the angular velocity of the cam-gear and determines the torsional and bending vibrations accordingly (see

Figure 9). Consequently, the amplitude of the bending, as well as the torsional vibrations for the first cam are the lowest in magnitude and the corresponding ones for the 4th cam are the largest. The model considers identical bearings between the cams and significantly larger bearings at the ends of the camshaft. The bearing stiffness plays an important role in reducing the influence of the individual valve trains over the global camshaft deformation. Nevertheless, the force applied by one of the cams significantly affects the bearings remotely situated from it. The furthermost bearing from the cam-gear location assumes the role of the most loaded due to a cantilever effect imposed upon the camshaft.

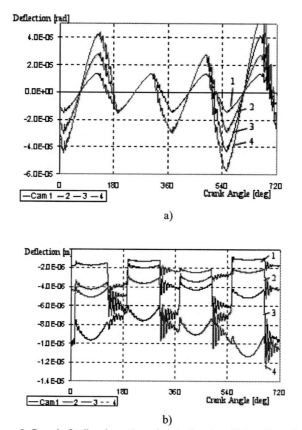

a)

b)

Figure 9. Camshaft vibrations a) torsional vibrations b) bending vibrations

Due to the very high accelerated oscillatory motion of the tappet, the cam-tappet interface accounts for the most loaded contact in the IC engine. Under the usual engine operating conditions this contact alone is responsible for over 85% of the energy lost in the whole valve train system [8]. The significant variations in contact load and lubricant entraining velocity, the regime of lubrication during the cam cycle varies from hydrodynamic at the opening and closing events, all the way to elastohydrodynamics and through to mixed lubrication at the cam nose contact. Consequently, certain regions of the cam tappet contact could experience much heavier wear than the rest, resulting in uneven tappet wear. To overcome this shortcoming, the most commonly used method is to intentionally design the cam eccentric from the tappet vertical axis of symmetry. Therefore, when the cam-tappet contact is on the cam nose, the traction force applied by the cam will have a spinning effect, encouraging lubricant entrainment into the contact, as well as ensuring that the cam-tappet interface will be equally exposed to the very high generated shear forces, and protect the tappet against unequal wear.

Since the eccentricity between the cam and tappet axis is responsible for the tappet dynamics, choosing its optimal value is a very important task.

Figure 10 shows a schematic representation of cam-tappet system geometry. The tappet dynamic behaviour is understood, based upon the instantaneous assumed equilibrium conditions around the vertical axis "z": between the spinning torque (generated by the cam-tappet friction force) and the resisting torque (generated by the tappet-tappet bore friction force as well as by the inertial contribution), together with the tappet equilibrium along x and y axis, which generates the load in the tappet-tappet bore contact and subsequently the triboloby of the tappet-tappet bore [9].

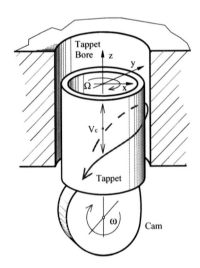

Figure 10. Cam-Tappet-Tappet Bore system

The friction force between the cam and the tappet can be computed, considering its boundary and viscous components, along with the non-Newtonian behaviour of the lubricant trapped in the contact. Figure 11 shows the cam-tappet friction force along the cam width (from –a to a) for the most loaded section of the cam cycle (i.e. between -70° before the cam nose to 70° after the cam nose). The higher peaks around 35° following the cam nose are due to a local decrease in the entrainment velocity, which determines a sharp rise in the local boundary friction (for additional details see [10]).

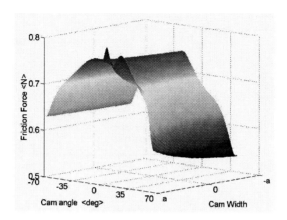

Figure 11. Cam-tappet friction force at 1200 rpm (engine speed)

Since the prevailing fluid film conditions in the concentrated counterformal contact between the cam and the tappet is elastohydrodynamic, there is a marked insensitivity to applied load. Therefore, there is a very small influence from the valve train dynamic behaviour upon the cam tappet friction force.

Figure 12 shows the predictions from the tappet spin model against the experimental measurements. They show similar general trend. However, the measurements show consistent oscillations of the tappet as it spins, which can be due to concurrent slipping and is due to significant instability in the cam-tappet-tappet bore tribo-dynamic behaviour [11]. Continuous oscillations of the tappet along the vertical axis generate fluctuations of the tappet-tappet bore contact force and consequently the tappet-tappet bore regime of lubrication regime. Thus, the resistant torque oscillates as well.

Figure 13 shows the change in tappet spin with engine speed. By increasing the engine speed the tappet spin increases as well. However, by increasing the engine speed, the cam-tappet friction force decreases and consequently the driving torque also follows suit. Since the tappet-tappet bore entrainment velocity is not affected in the same manner, for the example considered here, there is a limiting engine speed (of approximately 1000 rpm) after which the tappet spin tends to decrease. However, the cam-tappet friction force at the beginning and at the end of the cam cycle is purely Newtonian (due to low contact loads). At high engine speed this friction force increases and as a result the tappet spin at high engine speed has a wider range.

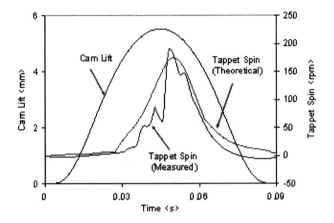

Figure 12. Tappet spin

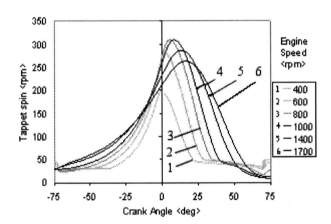

Figure 13. The influence of engine speed on the tappet spin

The accurate prediction for the camshaft vibration amplitude plays an important role in the in-depth understanding of the tribological conditions in the camshaft bearings. For each bearing the forces applied by the adjacent cams have to be considered, in order to determine the total force applied on a bearing. These are shown in Figure 14.

In the journal bearings the prevailing conditions are usually in the range from iso-viscous rigid (hydrodynamic) to iso-viscous elastic, when thin shells of materials of low elastic moduli are used. The contact is conformal, and the combination of these makes for lower pressures. Thus, unlike the case of cam-tappet contact the applied load plays a significant role in the lubricant film thickness and pressure distribution that govern wear and fatigue life of bearings. These predictions can be obtained readily by use of certain simplifying, but justifiable assumptions, such as those embodied in the short bearing theory and determination of the deformation of thin shells, using the column method.

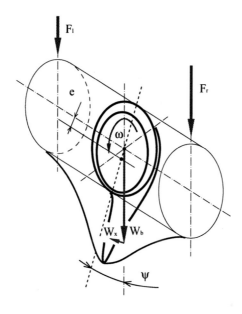

Figure 14. Forces applied to the supporting journal bearings

Thus:

$$p = 3u \frac{\eta_0}{R_j h^3} \frac{\partial h}{\partial \varphi} \left(y^2 - \frac{L_b^2}{4} \right) \tag{14}$$

$$h = c(1 + \varepsilon \cos \varphi) + \delta \tag{15}$$

$$\delta = \frac{(1-2\gamma)(1+\gamma)d}{E^*(1-\gamma)} p \tag{16}$$

Figure 15 shows the three dimensional pressure distribution in a camshaft journal bearing for a given condition. The maximum pressure is around 20 MPa, which indicates stresses in the shell are well within the elastic limit.

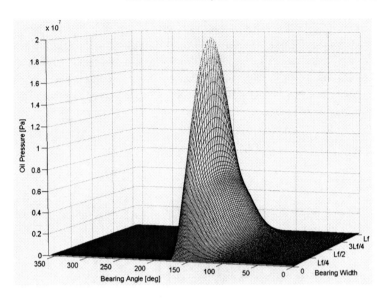

Figure 15. Pressure at 320 deg and 3000 rpm

The corresponding oil film thickness can be observed by *unwrapping* the elastic film shape as shown in Figure 16. To better observe the dip in the oil film shape due to elastic deformation of the bearing shell, the figure is inverted, thus the reason for the negative values along the ordinate of the figure. This shows that the minimum lubricant film thickness observed in this analysis is in the region 2-4 μm. This film thickness determines the gap between the contiguous surfaces (i.e. the journal and the bearing shell). It is then clear that gaps of the order of the root mean square of the roughness of composite surfaces can be bridged by the asperities of contiguous surfaces and promote wear, which gradually deteriorate the functional behaviour of the system. Thus, further physical phenomena can also be brought into the multi-physics approach.

10.   CONCLUSIONS

It is clear that the growing understanding of various phenomena, specifically their interactions, and the enhancing numerical techniques and computational power has brought new opportunities in the holistic, realistic and practical uses of science in the arena of engineering applications. Much of the methodology can be incorporated into single analysis framework, but retention of sufficient detail and dispensing with unnecessary detail. This calls for a fundamental approach, in a logical manner, for which one can take inspiration from natural thinking, which often points to the fundamental principle of parsimony. This approach is referred to here as *multi-physics*.

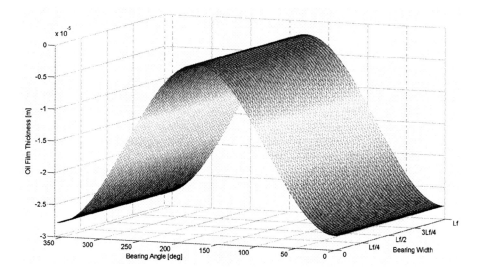

Figure 16. Oil film thickness at 320 deg and 3000 rpm

ACKNOWLEDGEMENT

The authors wish to express their gratitude for financial and technical support extended to their research by Ford motor Company, SKF, UK Department of Trade and Industry, The Engineering and Physical Sciences Research Council and the Vehicle Foresight Directorate.

REFERENCES

1. H. RAHNEJAT, *Multi-body dynamics: historical evolution and application* (The Millennium Paper, Proc. Instn. Mech. Engrs. Part C, Journal of Mechanical Engineering Science, The Special Millennium Issue, 214C1, 2000, pp 149-173).
2. H. RAHNEJAT FOREWORD, *A Tribute to Daniel Bernoulli (1700 – 1782) Mathematical-physical renaissance in mechanics of motion* (Proc. 2nd International Symposium on Multi-body Dynamics: Monitoring & Simulation Techniques, Bradford University, Bradford, UK 2000).
3. A. FARSHIDIANFAR, M. EBRAHIMI, H. RAHNEJAT, M.T. MENDAY, M. MOAVENIAN, *Optimization of the high-frequency torsional vibration of vehicle driveline systems using genetic algorithms* (Proc. Inst. Mech. Engrs, J Multi-body Dynamics, 216 (Part K), 2002, pp 249-262).

4. H. RAHNEJAT, *Multi-body Dynamics: Vehicles, Machines and Mechanisms* (Professional Engineering Publications IMechE, and Society of Automotive Engineers (SAE), USA, July 1998).

5. D. ARRUNDALE, K. HUSSAIN, H. RAHNEJAT, M.T. MENDAY, *Acoustic response of driveline pieces under impacting loads (Clonk)* (Proc. 31st ISATA , Dusseldorf, June 1998, pp 319-331).

6. M. GNANAKUMARR, S. THEODOSSIADES, H. RAHNEJAT, M. MENDAY, *Elasto-Multibody Dynamic Simulation of Impact Induced High Frequency Vehicular Driveline Vibrations* (Proc. 2003 Fall Technical ASME Conference, September 7th – 10th 2003).

7. M. GNANAKUMARR, S. THEODOSSIADES, H. RAHNEJAT, M. MENDAY, *Combined Multi-body dynamics, structural modal analysis, and boundary element method to predict multi-physics interactions of driveline clonk* (Proc. 3rd International Symposium on Multi-body Dynamics: Monitoring & Simulation Techniques, Loughborough University, Loughborough, UK 12th-13th 2004).

8. M. TEODORESCU, D. TARAZA, N.A. HENEIN, W. BRYZIK, *Experimental Analysis of Dynamics and Friction in Valve Train Systems* (SAE Paper 2002-01-0484, 2002).

9. M. TEODORESCU, D. TARAZA, *Numerical prediction and experimental investigation of cam – to flat tappet lubricated contact with tappet spin* (Proc. Instn. Mech. Engrs., Part K: J. Multy-Body Dyn., in review process).

10. D. TARAZA, M. TEODORESCU, N. A. HENEIN, W. BRYZIK, *Two Dimensional Effects in the Lubrication of the Cam-Tappet Contact* (Proc. 2003 Fall Technical ASME Conference, September 7th – 10th 2003).

A COLLABORATIVE SIMULATION ENVIRONMENT FOR MULTIBODY SYSTEM ANALYSIS

M. González and J. Cuadrado
Laboratory of Mechanical Engineering, University of La Coruña, Ferrol, Spain.

Abstract: This article describes a new collaborative simulation environment for multibody system analysis designed to streamline the development, testing and evaluation of new methods for multibody system simulation. The environment, called MbsLab, is made up of 3 components: (1) MbsML, a neutral and extensible XML-based language for describing multibody system models and related information, (2) MbsEngine, an open-source modular simulation tool that uses MbsML as native data format, and (3) MbsBenchmark, a benchmarking suite for multibody system analysis software. The three components can be used together to develop and compare the performance of new simulation methods in a short time. Currently, the environment only supports rigid bodies, but it is expected to grow to support flexible bodies, contact and impact.

Key words: multibody, dynamic simulation, XML, benchmark, software development.

1.   INTRODUCTION

Multibody system simulation is a complex process involving several factors: modeling, dynamic formulation and integration procedures. Cuadrado [6] concluded that the adequate combination of these factors depends heavily on the properties of the system to be analyzed, and therefore there is not a universal solution: different alternatives (modeling techniques, formulations and integration schemes) must be combined, evaluated and compared in order to find the optimal combination for a particular system. Commercial tools do not allow this level of flexibility, since they usually implement just one kind of dynamic formulation and a few integrators. The

D. Talabă and T. Roche (eds.), Product Engineering, 299–310.

capabilities of the solver kernel cannot be modified or replaced with user code. Therefore, custom tools must be developed by the researcher.

The authors have developed a simulation environment designed to streamline the development and testing of new methods for multibody system simulation. The system is made up of three components: (1) MbsML, a neutral XML-based language for describing multibody system models and related information, (2) MbsEngine, an open-source modular simulation tool that uses MbsML as native data format, and (3) MbsBenchmark, a benchmarking suite for multibody system analysis software. The three components are described in the following sections.

2.   MBSML

Multibody system data standardization is still an open research field [18] Available simulation tools, commercial and from academic research groups, differ in model description and data formats. Since there is no common standard language, bidirectional translations and adapting interfaces are required in order to exchange models within a simulation environment made up by different simulation tools. This is a major obstacle for model sharing and cooperation.

Previous approaches to multibody system information modeling fall in two main categories: the data approach and the document approach. In the data approach, information is managed by relational or object-oriented databases. This solution is quite efficient and uses standardized techniques for information storage, search and retrieval [20]. However, databases do not offer god facilities for structured text and the user needs special tools to access to the information in an intuitive way. On the other hand, the document approach has all the advantages that databases lack of: documents can be copied, e-mailed and printed easily, and structured formats, like Dymola [15], can represent very well the structure of the real system. But this approach has also disadvantages: information processing is difficult, since document syntax must be parsed. If the document syntax is very complex, writing compliant parsers will be cumbersome, very few tools will be developed to support that syntax and the language will not became a standard.

Recently, XML [2], has emerged as an increasingly popular format to encode information. A key feature of XML is that it integrates the benefits of the data and document approaches in one single format [13], so we have decided to design an XML-based language for modeling the information used and generated in our simulation environment. The following sections give a brief introduction of XML and MbsML.

2.1   Brief introduction to XML

XML, Extensible Markup Language [2] is a simple, very flexible text format derived from Standard Generalized Markup Language (SGML, ISO 8879) which was developed in the 1970s for the large-scale storage of structured text documents. The XML standard was published in February 1998 by the World Wide Web Consortium, the organization responsible for defining many internet-related standards, most notably HTML. Originally designed to meet the challenges of large-scale electronic publishing, XML is also playing an increasingly important role in the exchange of a wide variety of data on the Web and elsewhere.

2.1.1   Structure

XML documents have a hierarchical structure composed of tag elements. Figure 1 shows an MbsML document that models a pendulum. Elements (like <model> in the example) are the basic building blocks of XML markup and may have attributes (like "id") or nested elements (like <ground> or <body>). XML only defines the syntax of the document; each XML vocabulary defines the valid structure: names for elements and attributes, content type (numeric, text, nested elements, etc.) and cardinality.

2.1.2   Parsing and validation

One of the strong points of XML is the availability of parsers. Parsers read the information contained in a XML file using standardized techniques, thus saving many hours of work to software developers. Currently there are good-quality XML parsers, free or commercial, implemented in Java, C, C++, Visual Basic, Python, etc.

Before processing the information contained in a document, the structure and content must be checked to ensure integrity and data type validity. This is usually a hard task for the programmer, and the resulting code can be very lengthy and obfuscated. XML provides simple and reliable ways to perform this job. The syntax of an XML-based language can be described with a DTD (Document Type Definition), or one of the several schema languages available. A validating parser can use this language definition to validate the content of an XML document.

2.1.3   Available tools and applications

A wide range of technologies are available to work with XML: XSLT [3] specifies transformations of XML into other versions or formats, XPath [3] selects fragments of XML documents, XInclude [11] is a general inclusion mechanism, RDF embeds meta-data inside XML, etc. In addition, all major Relational Database Management Systems have become XML-enabled.

XML has been applied successfully to define languages in several fields: MathML in Mathematics [8], CML in Chemistry [13], CellML in biology [7], GML in Geography [10] etc. In the field of engineering, several organizations are also working with XML: the NIST [14] is working on MatML for material properties modeling; ISTOS [11] is developing femML, for encoding finite element models in XML format; and PDES [17] is developing STEPml, an XML version of some parts of STEP.

2.2   MbsML Goals

MbsML has been designed to be used in a collaborative environment for research in multibody system simulation; therefore the design was driven by the following requirements:
a) Foster data exchange, providing a neutral modeling language independent on the formalism used to perform multibody system simulations. MbsML provides only a physical description of the system. Modeling concepts which are meaningful only to a particular formalism are avoided. Therefore, MbsML models can be used as input data for different simulation tools, and different research groups can exchange models and results easily.
b) Foster data reutilization splitting the description of a simulation into logical sections: model, analysis and method. This organization provides maximum decoupling and allows easy combination of different methods to perform different analysis on different models.
c) Foster extensibility providing a modular language, which users can adapt, configure or extend to fit their particular needs.

2.3   MbsML document structure

Several schema languages are available to define the formal syntax of an XML vocabulary. The most used are DTD (Document Type Definition, a part of the first XML specification) and XML Schema [19], designed to overcome the limitations of DTD. Recently, a new schema language called Relax NG [5] has emerged. Although Relax NG is not widely supported by XML validating and authoring tools, its simplicity and powerful modeling capabilities make it a good choice for rapid development of XML-based languages. The authors use Relax NG to develop MbsML, and final versions are also released in XML Schema format to take advantage of its wide support. The MbsML schema allows five types of documents: model, analysis, method, job and results:

```
<model id="Pendulum">
  <ground id="ground">
    <geometry>
      <point id="Origin">
        <x>0.0</x> <y>0.0</y> <z>0.0</z>
      </point>
    </geometry>
  </ground>
  <body id="rod">
    <geometry>
      <point id="P1">
        <x>0.0</x> <y>0.0</y> <z>0.0</z>
      </point>
      <point id="P2">
        <x>1.0</x> <y>0.0</y> <z>0.0</z>
      </point>
    </geometry>
    <inertia>
      <rod>
        <mass>1.0</mass>
        <fromPoint idref="P1"/>
        <toPoint idref="P2"/>
      </rod>
    </inertia>
  </body>
  <joint.spherical id="theJoint">
    <body idref="ground">
      <centerPoint idref="Origin"/>
    </body>
    <body idref="rod">
      <centerPoint idref="P1"/>
    </body>
  </joint.spherical>
  <load.gravity>
    <direction bodyRef="ground" id="-Z">
      <x>0.0</x> <y>0.0</y> <z>-1.0</z>
    </direction>
    <magnitude>9.81</magnitude>
  </load.gravity>
</model>
```

Figure 1. Example of MbsML document

- A <model> element describes the multibody system: bodies, joints, forces, actuators, constraints, nested sub-models, etc. Angles and distances between components can be defined. Figure 1 shows an example of a model. Every component is identified by a unique "id" attribute value; component can refer to other components using this "id". Currently only rigid bodies described with natural coordinates are supported.
- An <analysis> element describes the analysis to be performed on a

particular multibody system model: type of analysis (kinematic simulation, forward or inverse dynamics, static equilibrium), conditions (for example, initial and final time), etc. The initial conditions of the system and the desired output are also described in this element. This element only describes the kind of analysis, not how to perform it.

- A <method> element describes the method to be used to perform a particular analysis: formulation, integrator type, algorithm parameters (like the integration time step), etc.

- Once the model, the analysis and the method are defined, they are combined in a job, which serves as input file to the simulation tool. The <job> element is made up by one <model> and one or more <task> element; each <task> is made up by an <analysis> and a <method>, and tasks are performed sequentially (the final state of the system after a task is the initial state for the next task). With this decoupling between model, analysis and method, it is easy to build and test combinations: different analysis for the same model, same analysis on different models, different methods for the same analysis, etc.

- The results from a simulation are also encoded in MbsML: time-history of coordinates, reaction forces, final state of the system, etc.

2.4   Other features

Parametric models are essential in optimization, and they are also very useful to build libraries of components. In MbsML, every numeric field (masses, force magnitudes, etc) can be given by a symbolic expression involving parameters. The value of these parameters can be defined in the XML document or by the simulation software.

The XInclude technology [11] can be used to build XML documents from small parts using an <include> element that works like a link: <include href="someXMLfile.xml"/>. This elements inserted in MbsML files are a powerful method to reuse MbsML components (bodies, forces) and combine them in high-level models.

By default all quantities are given in SI units. When other units are used in an element, the user must specify the unit in a "uom" (units of measure) attribute: <mass uom="pound"/> 23 </mass>. An XML dictionary of units is available, which provides conversion factors to the corresponding SI unit. This dictionary can be extended and used by the simulation software to perform unit conversion.

Metadata can be embedded in MbsML documents using RDF. Resource Description Framework is a XML language for representing information about resources in the World Wide Web. It is particularly intended for representing metadata of a document, such as title, author, date of modification, copyright and licensing information. Models encoded in MbsML can carry this information, and when they are shared in a web environment, RDF-enabled search engines can read and use it to classify the resource.

2.5   Modular design

MbsML has been designed with a modular philosophy inspired by XHTML [1] the schema language definition is split into modules; each module defines a particular construction (a body, a joint) or feature (parameters, unit support). These modules are assembled together to build the language. This approach has many advantages: it is very easy to customize the language by removing, substituting or adding modules. For example, if a potential user needs support for flexible bodies (not supported in the current MbsML specification), he can develop a new module and start to use it immediately. Later, the module can be included in the next official version of the language and other researchers can use it. In this way, the language can evolve driven by the needs of the community.

3.   MBSENGINE

MbsEngine is a C++ library for multibody system simulation with a highly modular design that provides different solver components (dynamic formulations, integrators, etc.) which can be easily combined to obtain a particular solver configuration. The main features of the library are the native support for MechML and the collaborative development environment used to implement it.

3.1   Support for MbsML

Reading MbsML files involves using a XML parser and one of the several existing interfaces to access XML data. The main interfaces are DOM [21] and SAX [12]. DOM (Document Object Model) presents an XML document as an in-memory tree structure composed by elements and attributes that replicate the structure of the XML document. The developer can navigate, access or modify this structure easily. SAX has a different approach for reading XML: the parsed reads data sequentially, and signals events to the application with information about the elements and attributes. This method consumes few resources (memory and CPU cycles) than DOM, but it is more complex from the programming point of view.

Both methods (DOM and SAX) present information to the application in the form of element and attributes. When reading MbsML documents, the application must interpret these items and convert them in meaningful objects: bodies, forces, etc. The authors have developed a library that performs this task: each MbsML component has been mapped into a C++ class with methods for reading from XML to memory and writing from memory to XML. Similar to the DOM approach, the library has methods to navigate and modify the resulting object structure. The library is independent from MbsEngine, so it can be used by other researchers to add MbsML support to other simulation tools.

3.2   Collaborative development

MbsEngine is developed using a Free/Open Source Project Hosting Site (FOSPHost). The site provides a centralized source code repository managed by CVS, mailing lists for users and developers, bug tracking system and task management tools. Since the C++ library has a modular design, different research groups can work on different modules and communicate with others easily using the facilities provided by the site. The resulting software is released under the GPL license.

4.   MBS BENCHMARK

Benchmarking is widely used in computer science to measure the performance of hardware and software. This phenomenon is not restricted to vendors: benchmarks are also used in academic research to prove the qualitative advantage of new ideas, and by manufacturers to drive product development. Standard, well-known benchmarks exist for testing hardware (CPU, multi-CPU systems, memory, network, graphics, disk IO ...) and software (primary servers: e-mail, databases, e-Commerce platforms ...). These benchmarks use to be administered by industry associations (like SPEC) or major software vendors (like Oracle or SAP), and results are made public periodically.

In the field of multibody systems, benchmarking is no a standardized practice: academic papers measure performance of new formalisms with different, non-standard problems, making it impossible to compare results. MBS Benchmark (Multi Body Systems Benchmark) is a collaborative project dedicated to develop and maintain a standardized set of problems and procedures which enable easy and objective performance evaluation of multibody systems simulation software.

4.1   Problem collection

The collection is divided in groups of problems; currently there are only 2 groups (Table 1): group A, composed by simple, academic problems, and group B, composed by real-life problems. Both groups involve only rigid bodies. It is expected than new groups will be added in the future to cover flexible bodies, contact, impacts and collisions, etc. Each problem is documented with a detailed description, a model in MbsML format and a reference solution. This reference solution is usually the state of the system at the end of the simulation, and has been obtained using reliable solvers with low integration steps and tolerances (at the cost of a high computational time).

Table 1. Problem collection

Problem		Characteristic
Group A: academic problems with rigid bodies		
A01	Double pendulum	High accelerations
A02	Double four-bar mechanism	Singular positions
A03	Andrew's mechanism	Small time scale
A04	Bricard's mechanism	Redundant equations
A05	Bicycle	Stiff system
Group B: real-life problems with rigid bodies		
B01	Iltis vehicle	Automotive
B02	Dornier antenna	Aerospatial
B03	Human body	Biomechanics
B04	PUMA robot	Robotics (serial)
B05	Stewart platform	Robotics (parallel)

4.2   Performance evaluation

Every problem has a precision constraint: when a problem is solved using a particular software, the relative error between the obtained solution and the reference solution must be below a certain limit (which is given in the problem documentation) in order to consider the obtained solution as valid. Currently the only performance indicator is the CPU time spent in solving the problem, which varies in inverse proportion to the relative error. Therefore, the researcher must tune the software (integration step, tolerances, etc.) to achieve the required precision with a minimum computation time.

Once the problem is solved within the given precision, the results can be submitted to a central database using a web browser. The user must provide information about the environment used to solve the problem: computer (hardware and operational system), software (name, website), build system (compiler, optimization flags and libraries) and method (description of the formalism, formulation and integrator implemented by the software). This information can be submitted to the central database manually (filling HTML forms in a web browser) or automatically (the information can be encoded in XML files and submitted to the server using scripts available from the benchmark website).

4.2.1   Performance comparison

The information stored in the database can be queried and retrieved from the benchmark website using a simple HTML interface. The visitor specifies search criteria and a report is generated. Currently, only two types of reports can be generated: (a) comparison of several methods to solve a particular problem, (b) comparison between two methods in solving a range of problems. In order to compare performance results produced in different computers, a computer performance index computed by a program distributed from the MbsBenchmark website is taken into account. Reports are in HTML format with embedded SVG (XML format) graphs; Figure 2 shows and example. In the future new features will be added: more kinds of reports (relative error versus CPU-time curves for a particular method, obtained with different integration steps or tolerance settings), performance comparisons between hardware configurations and compilers, facilities to export the performance data in XML or plain text format, etc. The web application is implemented using a MySQL database, Java Server Pages (JSP) and XSLT.

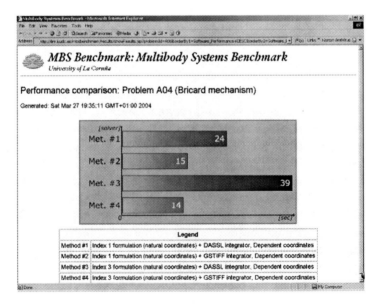

Figure 2. Example of performance report

5.   CONCLUSIONS

The three components (MbsML, MbsEngine and MbsBenchmark) are used together to develop and test new methods for multibody system simulation: different researchers implement new formulations and algorithms within the MbsEngine framework as new library modules; their performance is measured with any of the problems from the benchmark collection, and results are stored in the central database. Then, the alternatives developed by different researchers can be compared easily, and results guide future developments. The evolution of the system is driven by the needs of the community. If researchers want to compare methods to solve a particular kind problem (for example, systems involving contact), MbsML and MbsEngine are extended to model and solve that problem, and new problems are added to the MbsBenchmark collection.

The resulting system streamlines the development, testing and evaluation of new methods for multibody system simulation and facilitates the cooperation between different research groups by providing a neutral modeling language.

Currently, the environment only supports multibody systems made up of rigid bodies, but it is expected to grow to support flexible bodies, contact, impact, etc.

REFERENCES

1. M. ALTHEIM, S. MCCARRON, *Building XHTML Modules* (W3C, 2000, http://www.w3.org/TR/2000/WD-xhtml-building-20000105).
2. T. BRAY, J. PAOLI, C.M. SPERBERG-MCQUEEN, *Extensible Markup Language (XML) 1.0: W3C Recommendation* (W3C, 1998. http://www.w3.org/TR/REC-xml).
3. J. CLARK, *XSL Transformations (XSLT) Version 1.0* (W3C, 1999, http://www.w3.org/TR/xslt).
4. J. CLARK, S. Derose, *XML Path Language (XPath) Version 1.0, W3C Recommendation* (1999, <http://www.w3.org/TR/1999/REC-xpath-19991116>).
5. J. CLARK, M. MAKOTO, *RELAX NG Specification* (OASIS, 2001, http://www.relaxng.org/spec-20011203.html).
6. J. CUADRADO, J. CARDENAL, P. MORER, *Modeling and Solution Methods for Efficient Real-Time Simulation of Multibody Dynamics* (Multibody System Dynamics, Vol. 1, No. 3, pp. 259-280, 1997).
7. W. HEDLEY, M. NELSON, D. BULLIVANT, P. NIELSEN, *A short introduction to CellML* (Philosophical Transactions: Mathematical, Physical & Engineering Sciences, The Royal Society, Vol. 359, No. 1783, pp. 1073-1089, 2001).
8. P. ION, R. MINER, et al, *Mathematical Markup Language (MathML) 1.01 Specification: W3C Recommendation* (W3C, 1999, http://www.w3.org/TR/REC-MathML).
9. ISTOS, *Finite Element Modeling Markup Language (FEMML)* (2001, http://www.istos.org/femML).

10. R. LAKE, *Geography Markup Language (GML) v1.0* (OpenGIS Consortium, Inc (OGC), 2000, http://www.opengis.org/docs/00-029.pdf).
11. J. MARSH, D. ORCHARD, *XML Inclusions (XInclude) Version 1.0* (W3C, W3C Working Draft, 2003, <http://www.w3.org/TR/2003/WD-xinclude-20031110>)
12. D. MEGGINSON, *SAX 1.0 (the Simple API for XML)* (1998, <http://www.saxproject.org>).
13. P. MURRAY-RUST, H.S. RZEPA, *CML (Chemical Markup Language)* (2001, <www.xml-cml.org>).
14. *National Institute of Standards and Technology* (MatML, 2001, <http://www.matml.org>).
15. M. OTTER, H. ELMQVIST, F.E. CELLIER, *Modeling of multibody systems with the object-oriented modeling language Dymola* (Nonlinear Dynamics, Vol. 9, No. 1-2, pp. 91-112, 1996).
16. M. OTTER, M. HOCKE, A. DABERKOW, G. LEISTER, *An Object-Oriented Data Model for Multibody Systems* (Advanced Multibody System Dynamics - Simulation and Software Tools, Ed. Schiehlen, W., Kluwer Academic Publisher, 1993).
17. PDES Inc. (STEPml, 2001, <http://www.stepml.org>).
18. W. SHIEHLEN, *Multibody System Dynamics: Roots and Perspectives* (Multibody System Dynamics, Vol. 1, No. 2, pp. 149-188, 1997).
19. H.S. THOMPSON, D. BEECH, M. MALONEY, N. MENDELSOHN, *XML Schema Part 1: Structures* (W3C, 2001, <http://www.w3.org/TR/2001/PR-xmlschema-1-20010330>).
20. C. TISELL, K. ORSBORN, *A system for multibody analysis based on object-relational database technology* (Advances in Engineering Software, Vol. 31, No. 12, pp. 971-984, 2000).
21. L. WOOD, *Document Object Model (DOM) Level 1 Specification* (W3C Recommendation, 1998, <http://www.w3.org/TR/1998/REC-DOM-Level-1-19981001>).

DESIGN EVALUATION OF MULTIBODY SYSTEMS IN VIRTUAL ENVIRONMENT

Cs. Antonya and D. Talabă
Transilvania University of Brasov, Romania

Abstract: Visualization in virtual environment is the use of computer graphics to create visual images that aid in the understanding of various simulation processes. This may be the output of real-time numerical simulation or can be based on pre-recorded data or scenarios. This paper presents two virtual reality interfaces and applications which enables the user to make modifications on various multibody systems, by changing the position and orientation of elements and joints. The modification can be easily followed and evaluated in the virtual environment. In comparison to traditional systems for design reviews, the applications presented shows that user-interfaces can be made more user-friendly with new visualization techniques.

Key words: multibody system, virtual reality, visual evaluation, virtual environment.

1.   INTRODUCTION

Virtual environments (VE) can be used as an effective tool for training, education and in the design process in various fields. Visualization in VE is the use of computer graphics to create visual images that aid in the understanding of various simulation processes. This may be the output of real-time numerical simulation or can be based on pre-recorded data or scenario. Non-immersive interactive visualization systems implemented for the conventional desktop and mouse are effective for moderately complex problems, but virtual reality displays provides a rich set of spatial and depth cues.

Interaction plays a very important role in virtual environments. Virtual reality devices, such as haptic devices, provide users with powerful new interfaces where they can interact with digital models in 3D. Interactive systems, in software terms, are traditionally associated with the relationship

311

D. Talabă and T. Roche (eds.), Product Engineering, 311–320.
© 2004 *Springer. Printed in the Netherlands.*

between the user and the software product through the system's interface [5]. Virtual reality interfaces allow the rapid and intuitive exploration of the simulations, enabling the phenomena at various places in the volume to be explored, as well as provide control of the visualization environment through interfaces integrated into the environment. In user interaction with virtual worlds, realistic behaviour of objects is very important. Realistic behaviour can help to achieve consistency and predictability, specially when dealing with designing mechanical systems. Linking simulation and virtual reality will enhance the design process, offering powerful visual evaluation tools.

Various software (like ADAMS) offers parameterization possibilities, design evaluation, but they usually compare the results using graphics and tables of data. The applications presented are for visual evaluation of modification of mechanical systems, using all the facilities offered by the virtual environment (stereoscopic visualization, walk-through, taking measurements). The interfaces presented combine various computer programs, for computation of the kinematics and dynamics of the mechanical system, for introducing design modification by altering positions of elements and joints and for exporting simulation data into the virtual environment in which visual evaluation can be easily done.

2.   THE COMPUTATION MODULE

The computational module is a kinematic and dynamic analysis software based on the multibody system theory. The mechanical system is considered a collection of rigid bodies, connected by mass-less links. The applications are processing a text file (the input file) which contains all the data needed about the mechanism in the simulation process. The input file contains data about the analysis type (kinematic or dynamic), time interval of the simulation (including the computation step), initial position and orientation of each element, coordinates of point which describes various element (joints, forces, springs, dampers), joints (revolute, translational and composite joints), driving motions, mass properties of the elements, external forces, springs and dampers [2, 6]. For example, the data about the bodies that compose the system are stored in matrixes, each line containing the number of the element and the numeric values of the x and y position of the origin of the references frame attached to the body and the angle between the global and the local reference frame (orientation angle of the body). In the case of points, they are stored also in matrixes which has on each line the identification number of the point, the coordinates measured in local reference of each body and the number of the body to which they belong.

The developed programs are using a Cartesian coordinate approach [3].

The constraints on the motions of the bodies due to the joints are formulated in terms of Cartesian generalized coordinates (position and orientation of a mobile reference frame assigned to each element).

For the kinematic analysis, in order to control the motion of the system, kinematic drivers are introduced. The system of equation (1), composed by n_{ce} geometric constraint equations and n_{de} driving constraint equations (which is equal to the mobility of the system) is solved using the Newton-Raphson method for the vector of generalized coordinates.

$$\Phi(q,t) = \begin{bmatrix} \Phi^{n_{ce}}(q) \\ \Phi^{n_{de}}(q,t) \end{bmatrix}_{n_c \times 1} = 0 \qquad (1)$$

The behaviour of the system due to the applied forces is computed according the Newton-Euler laws. Introducing the Lagrange multipliers (λ), the differential – algebraic system of equation for the mechanism ca be written as (2):

$$\begin{bmatrix} M & J^T \\ J & 0 \end{bmatrix} \cdot \begin{bmatrix} \ddot{q} \\ \lambda \end{bmatrix} = \begin{bmatrix} F^A \\ \gamma \end{bmatrix}, \qquad (2)$$

in which the geometric constraints corresponding to the joints are represented by the algebraic equations of the generalized accelerations and the internal forces are included in the generalized forces vector $[F^A]$. In the computational modules (2) is solved with Adams-Bashforth-Moulton predictor - corrector method, and Baumgarte stabilization for reducing the constraints violation.

The computed data of the displacement and rotation of every element of the mechanical system can be stored in data sheets and imported into the VE with special command buttons on the interface or they can be used for animation, modifying the position of elements in the VE right in the computation loop.

3. DESIGN MODIFICATION

In case of the planar linkages design modification, without altering the number of bodies which defines the mechanical system and the type of connections between them, can be done by modifying the position of a joint or modifying the position of a body.

Making a design modification on the mechanical system with the developed interface means that the input data file has to be changed accordingly. Because no change has been considered in the structure of the system, the only modifications in the input data file will be at the initial position of the elements and at the coordinates of point which defines the joints.

Modifying the position of a revolute joint implies the modification of the local coordinates of the two points which define the revolute joint, other data about the mechanism remains unaffected (in the equations of revolute joints it is usually considered that two point of two bodies are in the same place during the simulation). The new coordinates of the two points (P_i, P_j) in their local frame (fig. 1) can be computed knowing their original coordinates (x_{Pi}^i, y_{Pi}^i, x_{Pj}^j, x_{Pj}^j,), the orientation angle of the two bodies (φ_i, φ_j) and the revolute joint modification values (Δx_r, Δy_r):

$$\begin{bmatrix} x_{Pi}^i \\ y_{Pi}^i \end{bmatrix}_{new} = \begin{bmatrix} x_{Pi}^i \\ y_{Pi}^i \end{bmatrix} + \begin{bmatrix} \cos\varphi_i & -\sin\varphi_i \\ \sin\varphi_i & \cos\varphi_i \end{bmatrix} \cdot \begin{bmatrix} \Delta x_r \\ \Delta y_r \end{bmatrix}, \tag{3}$$

$$\begin{bmatrix} x_{Pj}^j \\ y_{Pj}^j \end{bmatrix}_{new.} = \begin{bmatrix} x_{Pj}^j \\ y_{Pj}^j \end{bmatrix} + \begin{bmatrix} \cos\varphi_j & -\sin\varphi_j \\ \sin\varphi_j & \cos\varphi_j \end{bmatrix} \cdot \begin{bmatrix} \Delta x_r \\ \Delta y_r \end{bmatrix}. \tag{4}$$

The translational joint is usually defined by 4 point (P_i, Q_i, P_j, Q_j,), two an each body (Figure 2), which lies on the same line. Modifying the position of the joint in a parallel position has no effect on the kinematics and little on the dynamics. More radical consequence on the behaviour can be achieved by rotating the joint's line (Figure 2).

Presuming that the point P_i is not changing its position (is in the centre of the rotation), than, in the input data file, the local coordinates of the other 3 points has to be changed. If the joint's rotation angle is Δa_t (the original angle is α), then new coordinates will be:

$$\begin{bmatrix} x_{Qi}^i \\ y_{Qi}^i \end{bmatrix}_{new} = \begin{bmatrix} x_{Pi}^i \\ y_{Pi}^i \end{bmatrix} + \begin{bmatrix} d_{PiQi} \cdot \cos(\alpha + \Delta a_t) \\ d_{PiQi} \cdot \sin(\alpha + \Delta a_t) \end{bmatrix} =$$

$$= \begin{bmatrix} \cos\Delta a_t & -\sin\Delta a_t \\ \sin\Delta a_t & \cos\Delta a_t \end{bmatrix} \cdot \begin{bmatrix} x_{Qi}^i - x_{Pi}^i \\ y_{Qi}^i - y_{Pi}^i \end{bmatrix}, \tag{5}$$

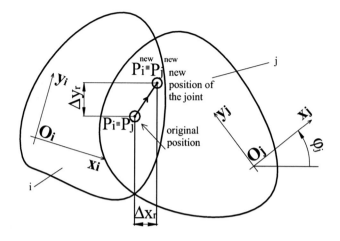

Figure 1. Modification of a revolute joint

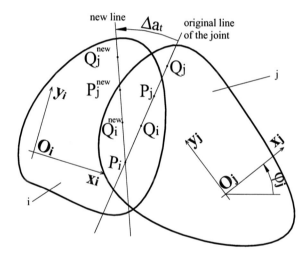

Figure 2. Modification of a translational joint

$$
\begin{bmatrix} x_{Pj}{}^{j} \\ y_{Pj}{}^{j} \end{bmatrix}_{new} = \begin{bmatrix} \cos\varphi_j & \sin\varphi_j \\ -\sin\varphi_j & \cos\varphi_j \end{bmatrix} \cdot
$$

$$
\cdot \left(\begin{bmatrix} x_{Pi} \\ y_{Pi} \end{bmatrix} + \begin{bmatrix} \cos\Delta a_t & -\sin\Delta a_t \\ \sin\Delta a_t & \cos\Delta a_t \end{bmatrix} \cdot \begin{bmatrix} x_{Pj} - x_{Pi} \\ y_{Pj} - y_{Pi} \end{bmatrix} - \begin{bmatrix} x_{Oj} \\ y_{Oj} \end{bmatrix} \right), \tag{6}
$$

$$\begin{bmatrix} x_{Qj}{}^j \\ y_{Qj}{}^j \end{bmatrix}_{new} = \begin{bmatrix} \cos\varphi_j & \sin\varphi_j \\ -\sin\varphi_j & \cos\varphi_j \end{bmatrix} \cdot \left(\begin{bmatrix} x_{Oi} - x_{Oj} \\ y_{Oi} - y_{Oj} \end{bmatrix} + \begin{bmatrix} \cos\varphi_i & \sin\varphi_i \\ -\sin\varphi_i & \cos\varphi_i \end{bmatrix} \begin{bmatrix} x_{Pi}{}^i \\ y_{Pi}{}^i \end{bmatrix} \right.$$

$$\left. \cdot \begin{bmatrix} \cos\Delta a_t & -\sin\Delta a_t \\ \sin\Delta a_t & \cos\Delta a_t \end{bmatrix} \begin{bmatrix} x_{Qj} - x_{Pi} \\ y_{Qj} - y_{Pi} \end{bmatrix} \right). \tag{7}$$

The consequence of the modification of a body's position or orientation angle will be that other elements, which are linked to this, will have their geometry altered. Because joint are described using points, in the input data file only the coordinates of those points will be modified which describes the modified body's joints. Position of external loads, springs and damper are also described by point. If these elements position is influenced by the position of the body on which they act, than the coordinates of those points will also change accordingly.

If body i position is modified by Δx_e, Δy_e (Figure 3) and it is linked with a revolute joint to body j, than only the P_j's coordinates will change, P_i's coordinates, in the local reference frame, remaining the same. The new coordinate of P_j can be written as:

$$\begin{bmatrix} x_{Pj}{}^j \\ y_{Pj}{}^j \end{bmatrix}_{new} = \begin{bmatrix} x_{Pj}{}^j \\ y_{Pj}{}^j \end{bmatrix} - \begin{bmatrix} \Delta x_e{}^j \\ \Delta y_e{}^j \end{bmatrix} = \begin{bmatrix} x_{Pj}{}^j \\ y_{Pj}{}^j \end{bmatrix} - \begin{bmatrix} \cos\varphi_j & -\sin\varphi_j \\ \sin\varphi_j & \cos\varphi_j \end{bmatrix} \cdot \begin{bmatrix} \Delta x_e \\ \Delta y_e \end{bmatrix}. \tag{8}$$

In case of the modification of the orientation angle (φ_i) of the body i with Δu_a and this element is linked by a translational joint to body j, than the coordinates of P_j and Q_j will be modified in the input data file. For P_j the coordinates are (for Q_j is similar):

$$\begin{bmatrix} x_{Pj}{}^j \\ y_{Pj}{}^j \end{bmatrix}_{new} = \begin{bmatrix} x_{Pj}{}^j \\ y_{Pj}{}^j \end{bmatrix} - \begin{bmatrix} \cos(\varphi_j - \varphi_i) & -\sin(\varphi_j - \varphi_i) \\ \sin(\varphi_j - \varphi_i) & \cos(\varphi_j - \varphi_i) \end{bmatrix} \left(\begin{bmatrix} \cos\Delta u_a & -\sin\Delta u_a \\ \sin\Delta u_a & \cos\Delta u_a \end{bmatrix} \right.$$

$$\left. \cdot \begin{bmatrix} x_{Pi}{}^i \\ y_{Pi}{}^i \end{bmatrix} - \begin{bmatrix} x_{Pi}{}^i \\ y_{Pi}{}^i \end{bmatrix} \right) \tag{9}$$

Other modifications in the position of joints can be written in the same manner.

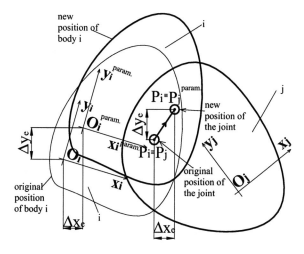

Figure 3. Modification of a body's position

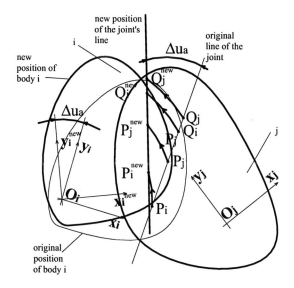

Figure 4. Modification of a body's orientation

4. VIRTUAL REALITY INTERFACE FOR DESIGN EVALUATION

The virtual reality interface (Figure 5) contains command buttons, scroll bars, check boxes and field. It is developed in Excel, using Visual Basic for Applications to assign various commands and can be easily used on a PC with two displays (one for the interface, another for the VR model). The interface enables the user to compute the kinematic or dynamic behaviour of the mechanical system (Compute button) described in the input text file (stored separately) by accessing a stand-alone multibody analysis software written a C and described in Chapter 2 of this paper. The computed simulation data, stored in Excel sheets, is linked with the virtual environment with the aid of the Delfoi Integrator™, a PC based message broker. The geometry of the elements of the mechanism, obtained from a CAD model, is stored in a VRLM file. Using VRML viewers like Bitmanagement, Blaxxun or Cortona the stereoscopic simulation of the mechanism can be achieved by linking the simulation data with the geometrical information (by accessing the translational and rotational nodes in the VRML file). The interface enables the user to load separately the simulation of the mechanism's elements, adjust the speed, freeze the animation and hide elements.

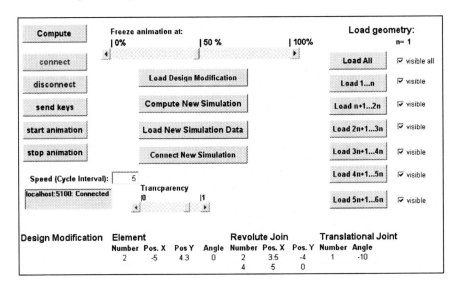

Figure 5. The VR interface

The interface allows the user to evaluate modifications of various type. The different alteration of the position and orientation angle of elements and joint can be introduced via the interface. It allows the combination of multiple changes: modification of the position and orientation of elements, position of revolute joints and angle of translational joints. The design modification values first change the input text file (the Load Design Modification button), after that the simulation can be run again (Compute New Simulation). The newly computed simulation is ready to be loaded into the virtual environment, using the existing geometrical model, and it can be superimposed with the original (unaffected) one. Running in the same environment the two graphical simulation, the design modification can be easily evaluated. The transparency of the two systems and the simulation speed can be modified also, and the simulation can be stopped to take measurement in the virtual environment.

The interface has been tested with several mechanical systems. The input text file can be obtained from the CAD model of the elements and this also can be used for obtaining the VRML model of the system. For a quick-return mechanism, a simulation frame is presented in Figure 6.

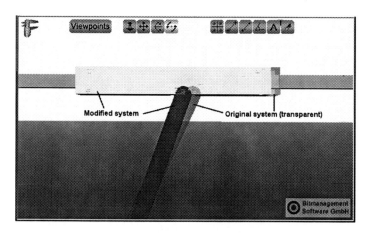

Figure 6. VR evaluation of a simulation

5. INTERACTIVE DESIGN EVALUATION

For interactive design evaluation in virtual environment, an other computational program has been developed based on Matlab programming language. The input data file is similar to the previous example, and also the kinematic and dynamic computational modules. In this case design modifications for multibody systems are made interactively, with the mouse, using touch sensors implemented in the VRML file. The user can modify position of bodies and joints with the mouse, during the computational procesess. The communication with the VE is made by the Matlab's Virtual Reality Toolbox. This resulted in the integration of a near real-time dynamics and kinematics computational engine with the interactive VRML world. In Figure 7 such an interactive simulation frame is presented. For an automotive suspension mechanism the revolute joint's position of the lower arm can be interactively changed with the mouse in the touch sensor aria during the computation process.

6.   CONCLUSIONS

With some minor modification on the multibody system by moving one or more bodies, or changing the position of one or more joints the entire behaviour (kinematic or dynamic) can be changed. The modification can be easily followed and evaluated in the virtual environment due to the facilitates offered by the interfaces: the two animation can be superimposed, various elements can be hidden, the transparency of the models can be modified for better visual evaluation of the consequences of the changes or the changes can be made interactively with the mouse. The visualisation software also offers measuring tools for distances and angles. As a result the virtual environments can be developed in a way that all work can be done there without having to leave the environment.

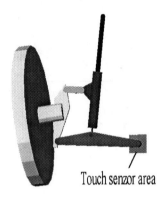

Touch senzor area

Figure 7. Interactive simulation

REFERENCES

1. CS. ANTONYA, D. TALABĂ, *Dynamic simulation of civil engineering structures in virtual reality environment* (MBD-MST 2004, 3rd International Symposium on Multi-body Dynamics: Monitoring & Simulation Techniques, Loughborough 2004 in press).
2. CS. ANTONYA, *Graphical simulation of rigid body systems* (in Romanian) (Transilvania Univ. Press, 2004).
3. H. GRANT, C. K. LAI, *Simulation modeling with artificial reality technology (SMART): an integration of virtual reality and simulation modelling* (Proceedings of the 1998 Winter Simulation Conference, pp. 437–441, 1998).
4. E. J. HAUGH, *Computer Aided Kinematics and Dynamics of Mechanical Systems* (Vol. I: Basic Methods, Allyn and Bacon, 1989).
5. M. I. SANCHEZ-SEGURA, J.J CUADRADO, A. M. MORENO, A. DE AMESCUA, A. DE ANTONIO, O. MARBAN, *Virtual reality systems estimation vs. traditional systems estimation* (The Journal of Systems and Software 72, pp. 187–194, 2004).
6. I. VIŞA, D. TALABĂ, CS. ANTONYA, *Software for kinematic and dynamic analysis of mechanisms* (in Romanian) (National Symposium PRASIC, Vol. I, pp. 223–234, Braşov 1998).

SPIRAL ELEVATOR MODELLING AND ANALYSIS USING ADAMS SOFTWARE

I. Batog
Transilvania University of Brasov, Romania

Abstract: An ADAMS multi-body dynamic model of the spiral elevator has been developed including elastic mounted working member and two unbalanced masses. In order to perform the simulation the unbalanced masses have been an applied rotational motion. The time history plots for vertical and angular displacement of the system, obtained as a result of simulation are very similar to the plots obtained by experimental means. The values obtained for the amplitude of steady-state oscillations of the system are very close to those obtained by analytical and by experimental means. The magnitude of system oscillations during the starting period has been determined, being of major interest as they may endanger the supporting springs.

Key words: spiral elevators, dynamic analysis.

1. INTRODUCTION

The vibrating spiral elevators are designed for upward conveyance of bulk materials along a helical path also used for process applications such as cooling, heating and drying. They consist of an upright tube around which winds a helical conveying-path. Vibrations of working member are usually excited by two unbalanced vibrators which are mounted in diametrical positions and at a certain angle either at the bottom or at the top of the tube. Due to the vibration induced to the working member solid particles are transported upward along the helical conveying trough.

The motion of the conveying trough of the spiral elevator is the response to the excitation consisting of an alternating driving force and an alternating driving moment applied to the working member of the conveyor.

D. Talabă and T. Roche (eds.), Product Engineering, 321–332.

1.1   Equations of motion

The conveyor can be simply modelled as a spring-mass-damper system, subjected to an alternating driving force and an alternating driving moment generated by unbalanced motors (Figure 1).

A set of initial assumptions have to be mentioned, for further analysis:
a) the supporting elastic elements are considered identical and massless;
b) the nonlinear character of the elastic elements is ignored;
c) the dissipative forces have a viscous character;
d) the external damping is ignored.
e) the resultants of the elastic, dissipative and driving forces (moments) coincide with the vertical symmetry axis of the conveyor;

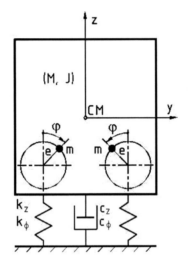

Figure 1. Simplified model

One may show [2] that for identical supporting elements located symmetrically about the vertical z-axis, the translation along z-axis and the rotation about z-axis are each independent of all other modes. The vectors of the driving force and the driving moment oriented on the z-axis will excite only these two independent modes of vibration.

Using a more refined model considering the position of eccentric masses, the generalized coordinates considered are: the vertical displacement Z, the angular displacement about vertical axis, Φ and the angular displacement of eccentric masses, φ. The equations of motion for the system are

$$(M+m_o)\ddot{Z}+c_z\dot{Z}+k_zZ = m_oe\sin\psi\left(\ddot{\varphi}\sin\varphi+\dot{\varphi}^2\cos\varphi\right), \tag{1}$$

$$(J+m_o a^2)\ddot{\Phi}+c_\phi\dot{\Phi}+k_\phi\Phi=m_o ea\cos\psi(\ddot{\varphi}\sin\varphi+\dot{\varphi}^2\cos\varphi)-m_o e^2\ddot{\varphi}\cos\psi,\quad (2)$$

where:
- M is the mass of the working member of the machine;
- J is the mass moment of inertia of the working member;
- m_o is the mass of unbalanced weights;
- e is the eccentricity of unbalanced weights;
- φ is the angular displacement, $\dot{\varphi}$ is the angular velocity and $\ddot{\varphi}$ is the angular acceleration of masses m;
- Z is the displacement of the working member along z-axis; $\dot{Z}$ is the velocity and $\ddot{Z}$ is the acceleration of the working member, along z-axis;
- c_z is the translational damping coefficient;
- k_z is the total translational spring constant of supporting springs;
- Φ is angular displacement $\dot{\Phi}$ is the angular velocity and $\ddot{\Phi}$ is the angular acceleration of the working member about z-axis;
- c_ϕ is the torsional damping coefficient;
- k_ϕ is the total torsional spring constant of supporting springs.

These equations can describe both the steady-state and the transient state. The Eq. (2) has a simplified form but ensure a very good approximation [1, 3].

The steady-state oscillation of the system is characterized by the constant angular velocity (ω) of the eccentric masses of the vibrator. For the generalized coordinate φ this means

$$\begin{cases} \varphi=\omega t \\ \dot{\varphi}=\omega \\ \ddot{\varphi}=0 \end{cases} \qquad (3)$$

Substituting the Eq. (3) in Eq. (1) and Eq. (2) the differential equations of motion become

$$(M+m_o)\ddot{Z}+c_z\dot{Z}+k_z Z=m_o e\omega^2\sin\psi\cos\omega t, \qquad (4)$$

$$(J+m_o a^2)\ddot{\Phi}+c_\phi\dot{\Phi}+k_\phi\Phi=m_o ea\omega^2\cos\psi\cos\omega t. \qquad (5)$$

The particular solution of Eq. (4) representing the translational steady state oscillation of the system and is given by

$$Z=Z_o\cos(\omega t-\gamma_z), \qquad (6)$$

where Z_o is the amplitude and γ_z is the phase angle of translational oscillation.

$$Z_o = \frac{m_o e\omega^2 \sin\psi}{\sqrt{\left[k_z - (M + m_o)\omega^2\right]^2 + (c_z\omega)^2}},\tag{7}$$

$$\tan\gamma_z = \frac{c_z\omega}{k_z - (M + m_o)\omega^2}.\tag{8}$$

The particular solution of Eq. (5) representing the rotational steady state oscillation of the system is

$$\Phi = \Phi_o \cos(\omega t - \gamma_\phi),\tag{9}$$

where Φ_o is the amplitude and γ_ϕ is the phase angle of rotational oscillation.

$$\Phi_o = \frac{m_o e a\omega^2 \cos\psi}{\sqrt{\left[(k_\phi - (J + m_o a^2)\omega^2\right]^2 + (c_\phi\omega)^2}},\tag{10}$$

$$\tan\gamma_\phi = \frac{c_\phi\omega}{k_\phi - (J + m_o a^2)\omega^2}.\tag{11}$$

At start, the angular velocity $\dot{\varphi}$ of eccentric masses grows from zero to the steady-state value ω. The angular acceleration $\ddot{\varphi}$ remains constant during 80% of the starting period, as justified by experimental verifications [4]. On this basis the angular acceleration can be expressed by

$$\ddot{\varphi} = \varepsilon = \frac{\omega}{t_o},\tag{12}$$

where t_o is the time needed for angular speed to reach the value ω. The equations (1) and (2) expressed in terms of ε and t become

$$(M + m_o)\ddot{Z} + c_z\dot{Z} + k_z Z = m_o e\sin\psi\left(\varepsilon\sin\frac{\varepsilon t^2}{2} + (\varepsilon t)^2 \cos\frac{\varepsilon t^2}{2}\right),\tag{13}$$

$$\left(J+m_{o}a^{2}\right)\ddot{\Phi}+c_{\phi}\dot{\Phi}+k_{\varphi}\Phi=m_{o}ea\cos\psi\left(\varepsilon\sin\frac{\varepsilon t^{2}}{2}+(\varepsilon t)^{2}\cos\frac{\varepsilon t^{2}}{2}\right)-$$
$$-m_{o}e^{2}\varepsilon\cos\psi \tag{14}$$

The equations (13) and (14) describe the system movement during the starting period but they cannot be solved by analytical means.

2. COMPUTER MODELLING AND SIMULATION

The value of computer modelling and simulation is in the rapid process of the evaluation of the effect of parameters and a reduction in the expense of extensive experimental investigations. Computer models are simplified representations of real systems. Therefore, virtual prototyping depends on the particular application and on the appropriate and accurate use of developed models.

A model consists of parts geometry and mass properties, reference frame definitions, compliance description, constraints and externally applied forces. First step in modelling is to define reference frames and coordinate systems. The global reference frame (GRF) fixed in a grounded body is a sufficient approximation to an inertial system. Then, physical components of the mechanical system are modelled. Bodies are elements which possess geometric, mass and inertia properties. A body reference frame (BRF) is defined for each part and is fixed in that body.

The next step is to describe the possible relative motions between the bodies by specifying the types of mechanical joints. The joints are perfectly rigid constraints allowing only a relative motion of the two bodies in accordance to the joint type. Joints can be defined between two bodies, referred as body I and body J. The number of degrees of freedom (DOF) of a mechanical system represents the number of independent motions allowed by the system. If the DOF of the model is larger than the number of independent motions the system is underconstrained and an analysis gives no useful results. If the DOF of the model is smaller than the number of independent motions the system is overconstrained and that may not be solved for a unique solution. For eliminating the overconstraints, some joints differ from the ones used in the real mechanism. The effect of the model joints is to render the same articulation degrees of freedom.

Having defined the bodies and the constraints, the next step is to define internal compliance elements (spring, dampers, etc) and the external forces influencing the behaviour of the system.

Markers are orthogonal triades of unit vectors which mark points of interest on a part. The markers are necessary in order to define joints, forces, centre of mass and inertia, etc.

2.1 The conveyor model

The model developed consists of four rigid bodies. Part 1 is ground and part 2 represents the working member. Parts 3 and 4 are modelling the eccentric masses. Simple shapes were used when creating the geometry of model parts. The mass and inertia properties of model parts are then redefined using user input option. The centre-of-mass marker is redefined for each part according to the centre-of-mass real position. The parts of the model are described in Table 1.

Table 1. Model parts

No.	Part name	Center of mass marker	Mass [kg]	DOF added
1	ground	–	–	–
2	wk_member	wm_cm	4300	6
3	ecc_mass_lh	eccm_lh_cm	63.6	6
4	ecc_mass_rh	eccm_rh_cm	63.6	6

The inertia properties of model parts computed at the centre of mass and aligned with the output coordinate system are described in Table 2.

Table 2. Inertia properties of model parts

No.	Part name	Inertia marker	Inertia [kg-m^2]					
			Ixx	Iyy	Izz	Ixy	Iyz	Izx
1	wk_member	wm_cm	2100	2100	2000	–	–	–
2	ecc_mass_lh	eccm_lh_cm	0.59	0.547	0484	0.043	0.099	0.099
3	ecc_mass_rh	eccm_rh_cm	0.59	0.547	0484	0.043	0.099	0.099

Constraints define how parts are attached to one another and how they are allowed to move relative to each other.

The joints used in the model allow the same motions as in the actual machine. Two revolute joints connect the eccentric masses to the working member. A cylindrical joint connects the working member to ground allowing relative translation and rotation between the two bodies about z-axis. The joints are described in Table 3.

Table 3. Model joints

No.	Constraint name	Constraint type	Body I	Body J	DOF removed
1	rev_joint_lh	revolute	ecc_mass_lh	wk_member	5
2	rev_joint_rh	revolute	ecc_mass_rh	wk_member	5
3	cyl_joint	cylindrical	wkg_member	ground	4

The motion of the two eccentric masses is modelled using motion generators described in Table 4.

Table 4. Motion generators

No.	Motion name	Motion type	Aplied to	Motion magnitude definition	DOF removed
1	rot_ecc_lh	joint motion	rev_joint_lh	function expression	1
2	rot_ecc_lh	joint motion	rev_joint_rh	function expression	1

The working member is connected to ground through elastic elements which are modelled using two flexible connectors: a translational spring-damper element and a torsion spring, having the equivalent stiffness and damping coefficient as the supporting springs. Flexible connectors do not absolutely prohibit any part movement and, therefore, do not remove any degrees of freedom from the model. Flexible connectors do typically resist movements between parts by applying spring and damper forces to the connected bodies. The flexible connectors are described in Table 5.

Table 5. Flexible connectors

No.	Connector name	Connector type	Body I	Body J	Stiffness	Damping coef.
1	t_spring_damp	translational	wk_member	ground	13187136 N/m	10500 N-sec/m
2	r_spring_damp	torsional	wk_member	ground	2042000 N-m/radian	1626 N-m-sec/radian

The model presented in Figure 2 has two DOF: the translation of working member along z-axis and the rotation of working member about z-axis.

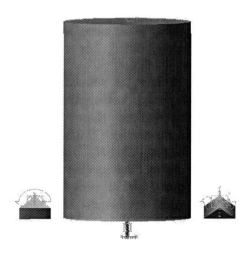

Figure 2. ADAMS model

2.2   Simulation results

A first simulation has carried out in order to investigate the oscillations of the system. In order to perform this simulation the motion magnitude of the two eccentric masses was defined in terms of rotational velocity, expressing it as a function of angular acceleration. The function expression used to define the rotational velocity was (IF (time −2 : 62.8*time, 125.6, 125.6)). This function describes the evolution of the rotational velocity during the starting period and further, when the velocity becomes stable. Figure 3 shows the time history for angular velocity and angle of the motion.

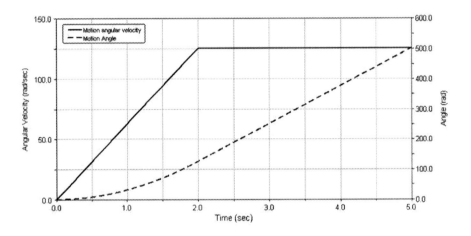

Figure 3. Angular speed of unbalanced masses

The simulation was run using 2000 steps for a simulation time of 5 seconds, with a value of 2 seconds for t_o. The time required by simulation was 33 seconds.

Figure 4 shows the time history for vertical displacement of working member during the simulation period. The response time history for vertical displacement indicates a duplex frequency response as shows the fast Fourier transform of the displacement time history (Figure 5). This is the result of superimposing the free damped translational vibration of the system and the forced translational vibration having the frequency of the driving force.

The plot presented in Figure 4 shows an increase in the magnitude of system oscillation under transient conditions. The oscillation has a larger magnitude at frequencies close to the natural frequency of the system. The plot shows a peak at 1.1 seconds. After 4 seconds from the start the natural damped oscillation of the system almost disappears and the amplitude of the system oscillation is close to the steady state amplitude.

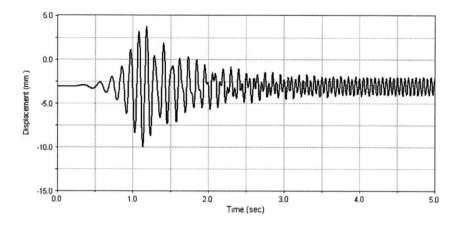

Figure 4. Vertical displacement of working member

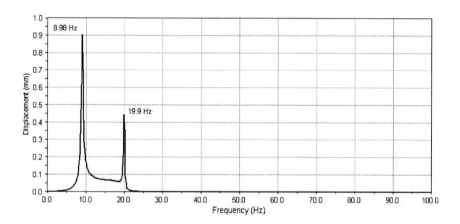

Figure 5. FFT of vertical displacement of working member

Figure 6 shows the time history for angular displacement of working member during the simulation period. The response time history for angular displacement indicates also a duplex frequency response as shows the fast Fourier transform of the angular displacement time history (Figure 7). This is the result of superimposing the free damped rotational vibration of the system and the forced rotational vibration having the frequency of the driving force.

The plot in Figure 6 shows an increase in the magnitude of system oscillation under transient conditions. The oscillation has a larger magnitude at frequencies close to the natural frequency of the system. The plot indicates a peak at 0.78 seconds. Due to the lighter torsional damping the natural oscillation of the system is still active after 5 seconds from the start.

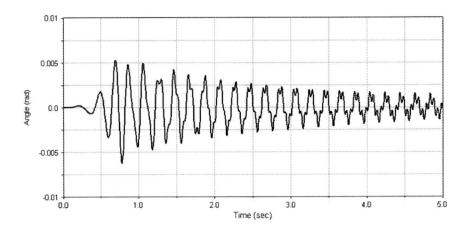

Figure 6. Angular deformation of torsion spring

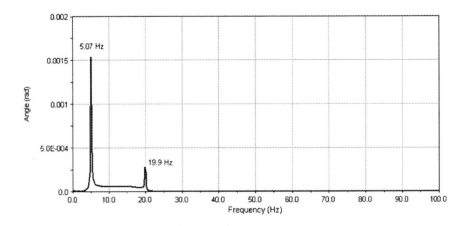

Figure 7. FFT of angular deformation of torsion spring

A second simulation has been carried out in order to investigate the steady-state oscillations of the system. The simulation was run using 5000 steps for a simulation time of 15 seconds. The values obtained for the amplitude of vertical steady-state oscillation and for angular steady-state oscillation of the system are very close to those obtained by analytical and by experimental means. Table 6 displays the values of amplitude and acceleration for the two steady-state oscillations, obtained by simulation, computed and measured on a real conveyor, for a forcing frequency of 125.6 rad/sec (20Hz).

Table 6. Amplitude and acceleration of oscillations

Parameter	Value		
	simulated	computed	measured
Translational amplitude [mm]	0.92	0.92	0.9
Translational acceleration [m/sec^2]	14.5	14.51	14.4
Rotational amplitude [radian]	0.00108	0.001	0.001
Rotational acceleration [radian/sec^2]	16.8	15.77	15.24

3.   CONCLUSIONS

An ADAMS multi-body dynamic model of the spiral elevator has been developed, including elastic mounted working member and two unbalanced masses. A rotational motion has been applied to the unbalanced masses in order to perform the simulation. This model was found more accurate than a mass-spring-damper model subjected to a harmonic force and a harmonic torque, generated by the unbalanced vibrators.

Using the model for analysis, the oscillations of the system have been studied. The plots obtained for displacement along vertical axis and for angular displacement about the same axis show an increase in the magnitude of system oscillations at a frequency close to the corresponding natural frequency of the system. Though the aspect of plots is as expected, the peak values of amplitude are of major interest for the design of conveyor as they may endanger the supporting springs. The values obtained for the amplitude of the system steady-state oscillations are very close to those obtained by analytical and by experimental means.

The future development of the model may focus on the following points:
a) using a more refined function or curve definition to describe the angular velocity of the eccentric masses during the starting period;
b) modeling each supporting spring instead of using equivalent characteristics of elastic mounting;
c) development of a customized interface allowing rapid re-modelling and modification of the conveyor parameters.

REFERENCES

1. I. BATOG, *Elevatoare vibratoare elicoidale* (Editura Universitatii Transilvania, Brasov, 2003).
2. P. BRATU, *Sisteme elastice de rezemare pentru masini si utilaje* (Bucuresti, Editura Tehnica, 1990).

3. C. YQUING, L. CUIYING, *Helical Vibratory Conveyors for Bulk Materials* (Bulk Solids Handling, volume 1, number 1, 1987, pp 103-112).

4. M. MUNTEANU, *Introducere in dinamica masinilor vibratoare* (Editura Academiei, Bucuresti, 1986).

5. R.R. RYAN, *ADAMS multibody system analysis software* (MDI Technical Publications, Vol. 2, 1989).

6. D. TALABĂ, I. BATOG, *A method to Represent the Rigid Bodies Orientation for Multibody Systems Analysis* (International Computer Science Conference "microCAD '97", Miscolc, 1997, Section L, pp. 9-13).

7. E. VOLTERRA, E. C. ZACHMANOGLOU, *Dynamics of Vibrations* (Charles E. Merrill Books, Inc., Columbus, Ohio, 1965).

A NEW MODEL TO ESTIMATE FRICTION TORQUE IN A BALL SCREW SYSTEM

D. Olaru[1], G. C. Puiu[2], L. C. Balan[1] and V. Puiu[2]
[1]Technical University "Gh. Asachi" of Iasi, Departament of Machine Design and Mechatronics, Romania; [2]University of Bacău, Departament of Machine Design, Romania

Abstract: A new model to estimate all friction losses between the balls and the circular - arch grooves races (two contact ball screw) has been developed. Based on the equilibrium of the forces and moments acting on the ball, the sliding forces between ball and both screw and nut races are determined without integrating the shear stress on contact ellipses. The sliding forces include the rolling forces on the two contacts, the friction moments due to the races curvature, the elastic resistance moments in ball-races contacts, the contact force between the balls and the friction moment between the balls. The program was adapted for a ball – screw system and the influences of the speed and load on total friction torque were investigated. The efficiency of the system was computed and good agreement with the research literature was obtained.

Key words: ball screw system, rolling friction, friction torque, friction coefficient.

1.   INTRODUCTION

In a ball screw system the total friction consist of rolling and sliding friction in rolling contacts, friction in the return zones and recirculation channels, friction generated by lubricant and friction generated by sealing systems. The total friction torque generated in a ball screw system can be approximated by some empirical formulas depending on load and preload, rotational speed, lubricant, lubrication regimes, temperatures and sealing systems. In many applications it is necessary to accurately estimate the friction torque acting on the nut or on the screw, for various operating conditions. Houpert [1, 2, 3] developed simple and powerful relationships to evaluate the friction moments for radial and angular contact ball bearings.

333

D. Talabă and T. Roche (eds.), Product Engineering, 333–346.

Also, Olaru et al. developed analytical models to estimate friction forces in linear rolling guidance systems [4, 5].

Based on the friction models for rolling bearings and for linear rolling guidance systems, the authors developed a new analytical model to evaluate the friction torque in a ball screw system considering only the friction losses between the balls and races and between the balls. The numerical results were compared with experimental results obtained in literature [6, 7, 8] and good agreement was obtained. Also, the authors analytically evaluated the efficiency of a ball screw system.

2.   FORCES AND MOMENTS ACTING ON BALL IN A TWO POINTS CONTACTS BALL SCREW SYSTEM

In Figure 1 the position of a ball and the position of the normal load Q in a ball screw system, when the screw is rotating with an angular speed ω and the nut is moving with an axial speed v is presented.

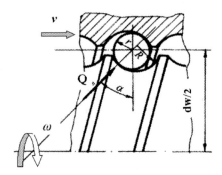

Figure 1. The contacts between ball and the two races in a ball screw system

2.1   Tangential forces on ball-race contact ellipses

The tangential forces developed between the balls and the races of the screw and of the nut are presented in Figure 2. So, between the ball and nut act the following forces: hydrodynamic rolling forces FRn, pressure forces FPn and $FPnb$, friction forces FSn. Also, between the ball and screw act the following forces: hydrodynamic rolling forces FRs, pressure forces FPs and $FPsb$, friction forces FSs. Between the balls acts the force FB. The directions of all above-mentioned forces are presented in Figure 2.

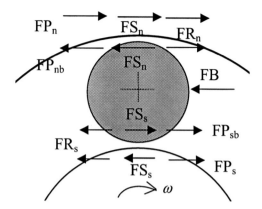

Figure 2. The forces acting on the ball and on the two races in a ball-screw system

FR is hydrodynamic rolling force due to Poiseuille flow of the lubricant contact. In the EHD lubrication condition this force can be computed by relationship (1), (2), (4):

$$F_R = 2.86 \cdot E \cdot Rx^2 \cdot k^{0.348} \cdot G^{0.022} \cdot U^{0.66} \cdot W^{0.47}, \tag{1}$$

where *E* is the equivalent Young's modulus of the materials in contact, R_X is the equivalent radius in the rolling direction, *k* is the radii ratio, *G, U, W* are the dimensionless material, speed and load parameters.

For ball – nut race contact:

$$\frac{1}{Rxn} = \frac{2}{d_w} - \frac{2 \cdot \cos \alpha}{d_m + d_w \cdot \cos \alpha}. \tag{2}$$

For ball - screw race contact:

$$\frac{1}{Rxs} = \frac{2}{d_w} + \frac{2 \cdot \cos \alpha}{d_m - d_w \cdot \cos \alpha}. \tag{3}$$

The radii ratio *k* is given by relation:

$$k = \frac{Ry}{Rx}, \tag{4}$$

where the transversal equivalent radius *Ry* is given by relations:

$$Rys = \frac{fs \cdot d_w}{2 \cdot fs - 1} \text{ and } Ryn = \frac{fn \cdot d_w}{2 \cdot fn - 1}, \tag{5}$$

where fs and fn are the curvature parameters for the screw and nut races, respectively and d_w is the ball diameter. Usually $f = 0.515$ to 0.54.

G is the dimensionless material parameter,

$$G = E \cdot \alpha_p, \tag{6}$$

where α_p is the piesoviscozity coefficient.

U is the dimensionless speed parameter,

$$U = \frac{\eta_0 \cdot v}{E \cdot Rx}, \tag{7}$$

where η_0 is the viscosity of the lubricant at the atmospheric pressure and at the contact temperature, v is the tangential speed in the ball-races, in the rolling direction.

The tangential speed in ball- nut race and ball- screw race contacts is given by relation:

$$v = \pi \cdot \frac{n}{30} \cdot \left[1 - \left[\frac{d_w \cdot \cos \alpha}{d_m} \right]^2 \right] \cdot \frac{d_m}{4}. \tag{8}$$

W is the dimensionless load parameter definite by the relation:

$$W = \frac{Q}{E \cdot Rx^2}, \tag{9}$$

where Q is the normal load in the ball-race contact.

FP is pressure forces due to the horizontal component of the lubricant pressure in the rolling direction.

For a ball – ring contact, the pressure force acting on the center of the ball can be expressed as a function of hydrodynamic rolling force F_R [1, 2, 4]

$$FPb = 2 \cdot FR \cdot \frac{R_R}{R_R + R_b}, \tag{10}$$

where R_b is the ball radius and R_R is the ring radius.

Relation (10) applied to ball-races contacts gives:
a) for ball-nut contact

$$FPnb = 2 \cdot FRn \cdot \left[1 + \frac{d_w \cdot \cos \alpha}{d_m} \right]. \tag{11}$$

a) for ball-screw contact

$$FPsb = 2 \cdot FRs \cdot \left[1 - \frac{d_w \cdot \cos \alpha}{d_m} \right].$$

(12)

Fb is the contact force between two balls. In this analysis *Fb* is imposed.

The friction forces *FSn* and *FSs* on the two contacts are the sliding traction forces due to local micro-slip occurring in the contact, and can be calculated explicitly as the integral of the shear stress τ over the contact area:

$$FS = \int \tau \cdot dA .$$

(13)

The variation of the shear stresses τ on contact ellipse is a complex problem depending of film thickness, rheological parameters of lubricant, sliding and micro-sliding speed and local contact pressure.

In this analysis the sliding traction forces are computed by the equilibrium of forces and moments acting on the ball.

2.2   The moments developed in the ball-race contact ellipses

In the contact ellipses following two moments are developed: the moments due to elastic resistance generated in ball races contacts *MER* and the moments generated by the curvature friction *MC*. Also, the resistance moment acting on the ball due to frictions between balls *Mb*. Figure 3 presents the moments which act on the ball and are generated on the contact ellipses: *MCs*, *MERs* are developed in the ball – screw race contact ellipse and *MCn*, *MERn* are developed in the ball – nut race contact ellipse.

MC is the curvature friction moment due to the local shear stresses τ developed in the slip region. The curvature friction moment is definite by relation [2, 4]:

$$MC = \int \tau \cdot z \cdot dA = \int \mu \cdot p \cdot z \cdot dA ,$$

(14)

where z is the distance between the position of local shear stress and the line of pure rolling, as is presented in the Figure 4, µ is the local friction coefficient and p is local contact pressure.

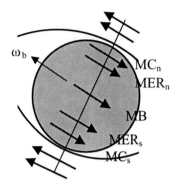

Figure 3. The Moments acting on the ball and on the two races in a ball-screw system;

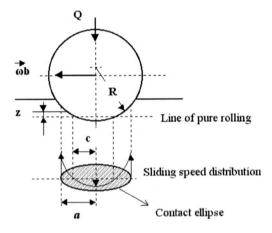

Figure 4. Sliding speed distribution in a ball race contact ellipse

Using a constant coefficient of friction μ_m and a Hertzian distribution of the contact pressure, equation (9) can be integrated and an analytical relation for *MC* results [1, 2, 4]:

$$MC = 0.1 \cdot \mu_m \cdot \frac{Q \cdot a^2}{Rd} \cdot (1 - 5 \cdot Y^3 + 3 \cdot Y^5),\qquad(15)$$

where *Rd* is the deformed radius in ball-race contact. For the ball-screw and ball-nut contacts the following relations can be used:

$$Rd_{s,n} = \frac{2 \cdot d_w \cdot f_{s,n}}{2 \cdot f_{s,n} + 1}.\qquad(16)$$

In equation (15) Y is a dimensionless distance between the point of pure rolling and the center of contact ellipse.

$$Y = \pm \frac{c}{a}, \tag{17}$$

where c is the distance from center of ellipse and points of pure rolling and can be in the positive or negative direction and a is the semi major axis of the contact ellipse (See Figure 4).

The semi-major and semi-minor axis of contact ellipse are determined for steel-steel contacts by relations [2, 3]:

$$a \approx 1.1552 \cdot Rx \cdot k^{0.4676} \cdot \left(\frac{Q}{E \cdot Rx^2} \right)^{\frac{1}{3}}, \tag{18}$$

$$b \approx 1.1502 \cdot Rx \cdot k^{-0.1876} \cdot \left(\frac{Q}{E \cdot Rx^2} \right)^{\frac{1}{3}}. \tag{19}$$

MER is the "elastic" resistance of pure rolling. One can use the relation of Snare [2, 3, 4]:

$$MER = 7.48 \cdot 10^{-7} \cdot \left(\frac{d_w}{2} \right)^{0.33} \cdot Q^{1.33} \cdot \left[1 - 3.519 \cdot 10^{-3} \cdot (k-1)^{0.806} \right]. \tag{20}$$

MB is a braking moment applied about the center of the rolling element and tt can be computed as:

$$MB = \mu_b \cdot \frac{d_w}{2} FB, \tag{21}$$

where μ_b is the friction coefficient between the contacted balls and *Fb* is the contact force between the balls.

3.   DETERMINATION OF SLIDING FORCES

It is possible to express sliding forces *FSs* and *FSn* analytically to satisfy the equilibrium of forces and moments around the center of the ball

$$\sum F = 0 , \tag{22}$$

$$\sum M = 0 . \tag{23}$$

According to the forces and moments acting on the ball and presented in Figure 2 and Figure 3, the above two equations lead to the following expressions for the forces FSc and FSg:

$$FSs = \frac{MCn + MCs + MERn + MERs + MB}{d_w} + FRn + \frac{(FRs + FRn) \cdot d_w \cdot \cos\alpha}{d_m} + \frac{FB}{2} , \tag{24}$$

$$FSn = \frac{MCn + MCs + MERn + MERs + MB}{d_w} + FRs - \frac{(FRs + FRn) \cdot d_w \cdot \cos\alpha}{d_m} - \frac{FB}{2} . \tag{25}$$

4.   FILM THICKNESS IN BALL - RACES CONTACTS

Minimum film thickness between ball and screw or nut race can be computed by Hamrock – Dowson relation [9]:

$$h = 3.63 \cdot R_X \cdot U^{0.66} \cdot G^{0.49} \cdot W^{-0..073} \cdot (1 - exp(-0.68 \cdot \frac{a}{b})) . \tag{26}$$

In order to evaluate the lubrication regime, the lubricant parameter λ must be computed with following relation [10]:

$$\lambda = \frac{h}{\sqrt{\left(\sigma_b^2 + \sigma_R^2\right)}} , \tag{27}$$

where σ_b is the ball roughness and σ_R are the screw and nut race surface roughness.

5.   FRICTION COEFFICIENT IN CONTACT ELLIPSES

In the ball – screw systems, according to the rotational speed of the screw, the lubrication regime in the ball- races contacts can be limited, mixed or full EHD.

In these conditions the friction coefficient has a large field of values. It is realistic to assume, in mixed lubrication, that the total friction is generated both by the contact roughness and by the lubricant film [10, 11].

The friction coefficient due by the roughness in the balls-races contacts can be estimated by the relation proposed by Zhou [11]:

$$\mu_a = \mu_0 \cdot exp(-B \cdot \lambda^C),$$ (28)

where μ_0 is the friction coefficient due only to the contact of the rugosity (when the lubricant parameter $\lambda \to 0$), λ is the lubricant parameter and B, C are constants depending of the roughness of the surfaces in contacts. For the ball and race roughness used in ball-screw systems the following values can be used: $B = 1.42$ and $C = 0.8$. For μ_0 is adopted the value of 0.2 [11].

The friction generated by the lubricant film, for low speed correspond with small values for friction coefficient (smaller that 0.01 - 0.02) and can be neglected in relation to the friction coefficient due to the roughness.

When lubricant parameter λ increases over 2.5 - 3, the friction coefficient μ_a becomes smaller than 0.02. In these conditions the predominant friction is due only to shearing the lubricant and it is realistic to assume a mean value for friction coefficient.

In this analysis the total friction coefficient on ball-races contact μ_m was computed as:

1. If $\mu_a \ge \mu_0$ (for very small speed), $\mu_m = \mu_0$ (29)
2. If $\mu_0 \ge \mu_a \ge 0.02$ (for normal speed), $\mu_m = \mu_a$ (30)
3. If $\mu_a \le 0.02$ (for high speed), $\mu_m = 0.02$. (31)

6. FRICTION TORQUE DEVELOPED IN BALL - SCREW AND BALL - NUT CONTACTS

According to Figure 2 the total tangential force between a ball and the screw **Fs** is the algebraic sum of the tangential contact forces in the rolling direction as $Fs = FRs + FSs$. According to Eq. (24) results:

$$Fs = \frac{MCn + MCs + MERn + MERs + MB}{d_w} + FRn + FRs +$$
$$+ \frac{(FRs + FRn) \cdot d_w \cdot \cos \alpha}{d_m} + \frac{FB}{2}.$$ (32)

The friction torque generated by a ball-screw contact is:

$$Ms = Fs \cdot Rs, \tag{33}$$

where **Rs** is the radius of the ball- screw contact given by relation:

$$Rs = \frac{d_m}{2} - \frac{d_w \cdot \cos\alpha}{2}. \tag{34}$$

The total tangential force between a ball and the nut Fn is the algebraic sum of the tangential contact forces in the rolling direction as $Fn = FRn + FSn$. According to equation (25) results:

$$Fn = \frac{MCn + MCs + MERn + MERs + MB}{d_w} + FRn + FRs - $$
$$- \frac{(FRs + FRn) \cdot d_w \cdot \cos\alpha}{d_m} - \frac{FB}{2}. \tag{35}$$

The friction torque generated by a ball- nut contact is:

$$Mn = Fn \cdot Rn, \tag{36}$$

where **Rn** is the radius of the ball- nut contact given by relation:

$$Rn = \frac{d_m}{2} + \frac{d_w \cdot \cos\alpha}{2}. \tag{37}$$

For a number of z loaded balls, the total friction torque acting on the screw is obtained by summing all friction torques in the ball-screw contacts:

$$MS = \sum_{1}^{z} Ms_i. \tag{38}$$

Also, the total friction torque acting on the nut is obtained by summing of all friction torques in the ball- nut contacts:

$$MN = \sum_{1}^{z} Mn_i. \tag{39}$$

7.   ESTIMATION OF THE EFFICIENCY IN A BALL SCREW SYSTEM

In a ball- screw system, neglecting the friction, to transmit an axial load F must act on the spindle with an active torque Msa given by relation:

$$MSa = F \cdot Rs \cdot \frac{p}{\pi \cdot d_m}, \qquad (40)$$

where p is the axial pitch of the screw.

Considering only the friction losses in ball – races contacts, the efficiency of the ball-screw system can be expressed by the relation:

$$\eta = \frac{MSa}{MSa + MS}. \qquad (41)$$

8.   NUMERICAL RESULTS

A numerical program was developed in order to analyze the variation of the friction torque according to the lubricant viscosity, rotational speed of the screw, axial load and races conformities.

The following geometrical parameters were considered:
a) the average diameter of the screw $d_m = 40$ mm;
b) the ball diameter dw $= 3.175$ mm;
c) the pitch of the screw p $= 5$ mm
d) the number of active balls $z = 56$;
e) contact angle $\alpha = \pi/4$
f) ball roughness is $\sigma b = 0.08 \mu m$;
g) screw and nut races roughness $\sigma_s = \sigma_n = 0.3 \ \mu m$
h) friction coefficient betweens balls $\mu_b = 0.1$

The viscosity of the lubricant is considered as $\eta_0 = 0.05 \ldots 0.5 \ Pa.s$ and the friction coefficient between balls is assumed at a constant value $\mu_b = 0.1$. Also, the contact load between balls is assumed to be $FB = 1/z$ Newtons.

The dimensionless parameter Y is assumed to be 0.34, both for ball – screw and ball- nut contacts. The conformity of the screw and nut races was considered to have values of 0.515 . The rotational speed n was considered between zero and 1000 rpm. and axial load F was between 100 to 4 000 Newtons.

According to relations (38) and (39) the friction torque acting on screw and on nut was calculated. For a value of FB= 1/z Newtons differences between the two above mentioned torques are no relevant. Increasing of the ball - ball contact forces (FB = 2/z Newtons) lead to an increase the differences between the two torques, as is presented in Figure 5 for an axial load F = 4000 N. Based on this observation the screw torque *MS* was considered in the efficiency relation (41).

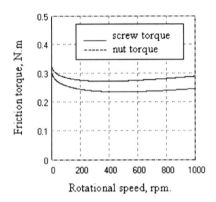

Figure 5. Total friction torque acting on the screw and on the nut

The variation of the friction torque MS function on the rotational speed of the screw *n* is presented in Figure 6 for an axial load F = 4000 N and a lubricant viscosity of 0.1 Pa.S. Also, for the same conditions, in Figure 7 is presented the variation of the ball- screw efficiency with rotational speed of the screw.

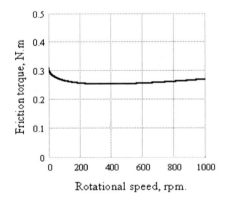 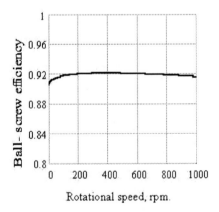

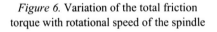

Figure 6. Variation of the total friction torque with rotational speed of the spindle

Figure 7. Variation of the ball screw efficiency with rotational speed of the spindle

In Figure 6 shows that by increasing of the rotational speed from zero to 200 – 300 rpm an decrease of the friction torque appear caused by the reduction of the friction coefficient in ball-race contact with the conversion from a limit lubrication regime to mixed lubrication regime (See Eq. 15).

Increasing the rotational speed over 200-300 rpm lead to increasing of the hydrodynamic forces FR and an increasing of the friction torque. In Figure 7 it can be observed that the efficiency of the ball- screw system has values between 0.9 to 0.92, according to the values indicated in literature.

The influence of the contact load Q on friction torque and on the efficiency are presented in the Figures 8 and 9 respectively, for a rotational speed of n = 600 rpm. In Figure 8 it can be observed that increasing the axial load (increasing of the contact load Q) lead to an increase of the friction torque. The variation of the efficiency with the contact load has two zone: first zone correspond to low values of normal load (Q = 0 to 20 N) when the efficiency increases from zero to 0.9 and the second zone with Q higher than 20 N with a constant value for the efficiency (0.9 – 0.92).

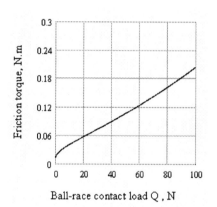

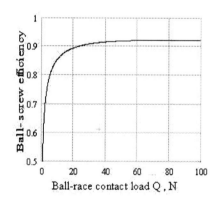

Figure 8. Variation of the total friction torque with ball–race contact load

Figure 9. Variation of the ball screw efficiency with ball–race contact load

9.   CONCLUSIONS

Based on the equilibrium of the forces and moments acting on the ball, the sliding forces between ball and both screw and nut in a ball - screw system was determined without an integrating of the shear stress on contact ellipses. The sliding forces include the rolling forces on the two contacts, the friction moments due to the curvature, the elastic resistance moments in ball-races contacts, the contact force between the balls and the friction moment between the balls.

Summing the sliding and rolling forces on the contact ellipses was obtained the total resistance forces acting on the screw and nut as effect of one loaded ball. Based on the total resistance forces acting on contact ellipses the friction torque for screw and nut was determined. Summing the friction torques for all loaded balls, the total friction torque acting on the screw and nut has been obtained. Also, the analytical expression for ball- screw efficiency has been developed.

The friction model includes a lot of parameters as: ball- races contact geometry, lubricant viscosity, normal loads, rotational speed, contact forces between balls, friction coefficient between balls and friction coefficient in contact ellipses. The developed model lead to realistic values both for the friction torque and for the efficiency of a ball – screw system.

REFERENCES

1. L. HOUPERT, P. LEENDERS, *A theoretical and experimental investigation into Rolling Bearing Friction* (Proc. of Eurotrib Conference, Lyon, 1985).
2. L. HOUPERT, *Numerical and Analytical Calculations in Ball Bearings* (Proc. of Congres Roulements, Toulouse, 5-7 Mai 1999, pp. 1-15).
3. L. HOUPERT, *Ball Bearing and Tapered Roller Bearing Torque: Analytical, Numerical and Experimental Results* (Proc of STLE Annual Meeting, Houston, May 19 – 23, 2002).
4. D. N. OLARU, P. LORENZ, D. RUDY, SP. CRETU, GH. PRISACARU, *Tribology Improving the Quality in the Linear Rolling Guidance Systems* (Part 1- "Friction in Two Contact Points Systems", Proc. of the 13th International Colloquium on Tribology, Esslingen (Germany), 15-17 January, 2002).
5. D. N. OLARU, P. LORENZ, D. RUDY, *Friction in the Circular – Arc Grooves Linear Rolling Guidance System* (Tribologie und Schmierungstechnik, 2, 2004, pp. 9 –14).
6. V. PUIU, *Ball Helical Transmission* (Buletinul Institutului Politehnic Iaşi, vol. XLV (XLIX), fasc. 1-2, secţia V (Construcţii de maşini), 1999, p. 35-40).
7. V. PUIU, D. N. OLARU, *Film Thickness in Ball Screw* (Buletinul Institutului Politehnic Iaşi, vol.XLV (XLIX), fasc. 3-4, secţia V (Construcţii de maşini), 1999, pp. 17-20).
8. V. PURICE, *Influenţa tipului de lubrifiant şi a parametrilor funcţionali asupra momentului de frecare din şuruburile cu bile* (TRIBOTEHNICA'84, Vol. II, Iaşi, 1984, pp.105 – 110).
9. B. J. HAMROCK, D. DOWSON, *Isothermal EHD Lubrication of Point Contacts* (Part III, Trans of. ASME, Journal of Lubrication Technology, 99, 1977, pp. 15-23).
10. T. HARRIS, *Rolling Bearing Analysis* (John Wiley & Soons, Ed., 1991).
11. R. S. ZHOU, M. R. HOEPRICH, *Torque of Tapered Roller Bearings* (Trans. of ASME, Journal of Tribology, 113, 1991, pp. 590-597).

Part 4

ROBOTICS AND MANUFACTURING

1.   INVITED LECTURES

2.   CONTRIBUTIONS

TRANSLATIONAL PARALLEL ROBOTS WITH UNCOUPLED MOTIONS
A structural synthesis approach

G. Gogu

Laboratory of Mechanics and Engineering, University Blaise Pascal and French Institute of Advanced Mechanics, France

Abstract: The paper presents a structural synthesis approach of translational parallel manipulators (TPMs) with uncoupled motions (UM) and revolute actuators (RA) situated on the fixed base (TPMs-UM-RA). Parallel robotic manipulators (PMs) show, in general, desirable characteristics like a large payload to robot weight ratio, considerable stiffness, low inertia and high dynamic performances. With respect to serial manipulators, disadvantages consist in a smaller workspace, complex command and a lower dexterity due to a high motion coupling and multiplicity of singularities inside their workspace. A TPM is a 3-DOM (degree of mobility) parallel mechanism whose output link, called platform, can achieve three independent orthogonal translational motions with respect to the fixed base. The manipulators presented in this paper have three legs connecting the moving platform and the base (fixed platform). Only one pair per leg is actuated and the motors are situated on the fixed base. Linear and revolute actuators can be used. Due to space limitations, in this paper we consider only rotational actuators (RA). A one-to-one correspondence exists between the actuated joint space and the operational space of the moving platform. The Jacobian matrix of TPM-UM is a 3×3 diagonal matrix throughout the entire workspace. A method is proposed for structural synthesis based on the theory of linear transformations and an evolutionary morphology approach. The method allows us to obtain all structural solutions of TPMs-UM-RA in a systematic manner. Overconstrained/isostatic solutions with elementary/complex and identical/different legs are obtained. TPM-UM have the advantage of simple command and important energy-saving due to the fact that, for a unidirectional translation, only one motor works as in a serial translational manipulator. Only a small number of the solutions of TPMs-UM-RA presented in this paper are actually known in the literature.

Key words: translational parallel robots, structural synthesis, uncoupled motions, isotropy.

D. Talabă and T. Roche (eds.), Product Engineering, 349–378.

1. INTRODUCTION

The industrial and economic relevance of parallel robotic manipulators (PMs) is continuously growing. This technology can be considered a well-established option for many different applications of manipulation, machining, guiding, testing, control, etc. Parallel mechanisms have been the subject of study of much robotic researche during the last two decades. Some important studies are now available to support the design of parallel manipulators, such as the pioneering works of Hunt [1-3] that showed various kinematic architectures of parallel robots, the comprehensive enumerations presented by Merlet [4] and Tsai [5]. Rigidity, accuracy, high speed, and high load-to-weight ratio are their main merits. With respect to serial manipulators, disadvantages include a smaller workspace, complex commands and a lower dexterity due to a high motion coupling and multiple singularities inside their workspace. PMs with uncoupled motions (PMs-UM) overcome these disadvantages. They are singularity free throughout the entire workspace having a very simple command and achieving important energy-saving due to the fact that for a unidirectional motion only one motor works as in serial manipulators.

We know that the Jacobian matrix of a robotic manipulator is the matrix mapping (i) the actuated joint velocity space and the end-effector velocity space, and (ii) the static load on the end-effector and the actuated joint forces or torques. Isotropicity of a robotic manipulator is related to condition number of its Jacobian matrix, which can be calculated as the ratio of the largest and the smallest singular values. A robotic manipulator is fully-isotropic if its Jacobian matrix is isotropic throughout the entire workspace, i.e., the condition number of the Jacobian matrix is one. Thus, the condition number of the Jacobian matrix is an interesting performance index characterizing the distortion of a unit ball under this linear mapping. The condition number of the Jacobian matrix was first used by Salisbury and Craig [6] to design mechanical fingers and developed by Angeles [7] as a kinetostatic performance index of robotic mechanical systems. The isotropic design improves kinematic and dynamic performance [8].

Parallel manipulators (PMs) are composed of an end effector (mobile platform) connected to the base (fixed platform) by at least two kinematic chains (legs). Translational parallel manipulators (TPMs) enable three independent translations of the mobile platform along the x, y and z axes. TPMs are used in applications that require positioning of a body in space with a constant orientation.

Various architectures are known in the literature to obtain three-legged TPMs [9-64]. The following types of pairs are used: revolute (R), prismatic (P), helical (H), cylindrical (C), spherical (S), planar contact (E), universal joint (*U*) as well as the parallelogram loop (*Pa*) which can be considered as a complex pair of circular translation [53, 57, 60,]. As a matter of fact, some architectures of TPMs are quite popular already, for instance the DELTA robots (*3-RRPaR*, *3-RUU*, *3-PRPaR*) proposed by Clavel [9-11, 31] and realized in many different configurations by Hervé (the *3-RHPaR* Y-star [13, 14], the *3-RPPR* Prism-Robot [16], the *2-RHPaR* + *1PRPaR* H-robot [14]), Tsai and Stamper (the *3-RRPaR* University of Maryland manipulator [19]), Company and Pierrot (the *3-PUU* Urane Sx [36]), Chablat and Wenger (the *3-PRPaR* Orthoglide [30, 52]). Industrial implementations of these solutions are already reached by Demareux, ABB, Hitachi, Mikron, Renault-Automation Comau, etc. Various architectures have been proposed to achieve three pure translational motions of the platform, such as the *3-UPU* by Tsai and Joshi [23], the *3-URC* by Di Gregorio [27], the *3-RRC* by Zhao and Huang [25], the *3-RER* by Zlatanov and Gosselin [61] and others with different theoretical approaches and practical results. Hervé and Sparacino [12] introduced a new class of TPMs with 4-DoF legs constraining two rotations of the platform. This class is addressed as Hervé family. The entire class of TPMs with 5-DoF legs was exhaustively presented by Tsai [5], Frisoli, Checcacci, Salsedo and Bergamasco [22] and Carricato and Parenti-Castelli [26]. This class is addressed as Tsai/Frisoli/Carricato family. Kong and Gosselin [62] presented the entire class of TPMs with elementary legs including one or two inactive joints per leg. We address this clas as Gosselin/Kong family.

Caricato and Parenti-Castelli [32] presented for the first time a topological synthesis and classification of fully-isotropic symmetric parallel mechanisms with three translational DoFs of the mobile platform. They founded their approach on constraint and direct singularity investigation. In [32] the isotropy is defined by taking into account only the parallel Jacobian. Kong and Gosselin [46] presented a structural synthesis approach of linear TPMs, whose forward displacement analysis can be performed by solving a set of linear equations. They presented 5 solutions of TPMs-UM-RA for which three translations along three orthogonal directions can be controlled independently by three actuators [47]. The full-isotropic conditions necessary in assembling the three legs to the fixed and mobile platforms are not analysed.

Kim and Tsai [44] presented a fully-isotropic Cartesian Parallel Manipulator (*3-PRRR*), developed in the Department of Mechanical Engineering at University of California. Gosselin and Kong [40, 45, 47, 64] presented fully-isotropic *3-CRR* and *3-PRR* (call Orthogonal Tripteron) translational parallel manipulators developed in the Department of Mechanical Engineering at University of Laval.

The general methods used previously for structural synthesis of parallel manipulators can be divided into two approaches: methods based on displacement group theory [12-14, 16, 57] and methods based on screw algebra [22, 29, 53, 58, 62].

The method used in this paper for structural synthesis of TPMs-UM is founded on the theory of linear transformations and an evolutionary morphology approach proposed by the author of this paper [65]. The method allows us to obtain all structural solutions of TPMs-UM with linear or revolute actuators in a systematic manner by taking into account parallel and serial Jacobians. Overconstrained/isostatic solutions with elementary/ complex and identical/different legs are obtained. All solutions with linear actuators are fully-isotropic and singularity free. In these cases, the Jacobian matrix is a 3×3 identity matrix throughout the entire workspace. Due to space limitations, we reduced our presentation to TPMs-UM-RA. The method proposed in this paper gathers the families of Hervé, Tsai/Frisoli/Carricato, Gosselin/Kong and Carricato/Parenti-Castelli into a general and organic campus of singularity free TPMs-UM-RA including some new families presented for the first time in the literature. We will show that some families of TPMs considered in [32] as fully-isotropic are in fact fully-isotropic only in relation with the parallel Jacobian and non fully-isotropic in relation with the serial Jacobian. In our approach we use the global Jacobian that includes the parallel and the serial Jacobians.

2.   ISOTROPY AND MOTION DECOUPLING

For parallel manipulators, the velocities of the moving platform are related to the velocities of the actuated joints $[\dot{q}]$ by the general equation:

$$[A]^{p}\begin{bmatrix} v \\ \omega \end{bmatrix}_{H} = [B][\dot{q}], \qquad (1)$$

where: $[v]=[v_x\ v_y\ v_z]^T$ is the velocity of a point H belonging to the moving platform, $[\omega]=[\omega_x\ \omega_y\ \omega_z]^T$ - angular velocity of the moving platform, $[A]$ - parallel (direct) Jacobian, $[B]$ - serial (inverse) Jacobian and p is the coordinate system in which the velocities of the moving platform are expressed.

Eq. (1) can also be written as a linear mapping between joint velocities and the operational velocities (the velocities of the moving platform)

$$^{p}\begin{bmatrix} v \\ \omega \end{bmatrix}_{H} = [J][\dot{q}], \qquad (2)$$

where $[J]=[A]^{-1}[B]$ (3)

is the (global) Jacobian including the parallel and the serial Jacobians. For many architectures of parallel manipulators it is more convenient to study the conditioning of the Jacobian matrix that is related to the inverse transformation, i.e., $[J]^{-1}=[B]^{-1}[A]$.

The isotropic conditions should apply to either both Jacobian matrix A and B or only to Jacobian matrix J or J^{-1}. A Jacobian matrix $(A, B, J$ or $J^{-1})$ is said to be isotropic when its condition number attains its minimum value of one [7], that is when its singular values are all identical. If we have a matrix M (which can be A, B, J or J^{-1}) whose entries all have the same units, we can define its condition number $\kappa(M)$ as a ratio of the largest singular value σ_1 of M and the smallest one σ_2. We note that $\kappa(M)$ can attain values from 1 to infinity. The condition number attains its minimum value of unity for matrices with identical singular values. Such matrices map the unit ball into another ball, although of different size, and are, thus, called isotropic.

We know that the singular values of M are given by the square roots of the eigenvalues of $[M]^T[M]$. A matrix M is isotropic if $[M]^T[M]$ is proportional to an identity matrix I. So, the Jacobian matrices A, B, J or J^{-1} are isotropic if

$$[A]^T[A] = \tau_1^2[I],$$ (4)

$$[B]^T[B] = \tau_2^2[I],$$ (5)

$$[J]^T[J] = \tau_3^2[I],$$ (6)

$$[J]^{-1}[J]^T = \tau_4^2[I],$$ (7)

where τ_i (i=1,.., 4) is a scalar.

By extension, the manipulators whose Jacobian matrices can attain isotropic values are called isotropic as well. The manipulators whose Jacobian matrices have condition number with unity value throughout the entire workspace are called fully-isotropic. We distinguish the following types of fully-isotropic PM:

a) fully-isotropic PM with respect to its parallel (direct) Jacobian, this PM respects Eq. (4) throughout the entire workspace,

b) fully-isotropic PM with respect to its serial (indirect) Jacobian, this PM respects Eq. (5) throughout its entire workspace,

c) fully-isotropic PM, this PM is fully-isotropic with respect to its parallel (direct), serial (indirect) and the global Jacobians and it respects all Eqs. (4)-(7).

The condition number of a Jacobian matrix J indicates the distortion in the operational velocity space of a unit ball defined in the joint velocity space. A higher condition number indicates a larger distortion. The worst-conditioned Jacobians are those that are singular. In this case, one of the semiaxes of the ellipsoid vanishes and the ellipsoid degenerates into what would amount to an elliptical disk in the 3-dimensional space.

If the entries of a Jacobian matrix have different units, Angeles [7] proposed to define a characteristic length, by which we divide the Jacobian entries that have units of length, thereby producing a new Jacobian that is dimensionally homogeneous. Tsai and Huang [67] proposed a dimensionally homogeneous Jacobian whose entries are defined by the work done between two screws.

Eq. (2) can be considered as the design equation of parallel robotic manipulators and the Jacobian matrix J as the design matrix in terms of axiomatic design [67]. We define the following types of parallel manipulators (PMs) by taking into account the form of the Jacobian matrix J:

– fully-isotropic parallel manipulator (PM-IM), if the Jacobian matrix is a diagonal matrix with identical diagonal elements,
– PM with uncoupled motions (PM-UM), if the Jacobian matrix is a diagonal matrix,
– PM with decoupled motions (PM-DM), if the Jacobian matrix is a triangular matrix,
– PM with coupled motions (PM-CM), if the Jacobian matrix is neither diagonal nor triangular.

If the Jacobian J is a diagonal matrix the singular values are equal to the diagonal elements. Consequently, a PM with uncoupled motions is fully isotropic if all diagonal elements of its Jacobian matrix are identical.

For a TPM, the linear mapping between the actuated joint velocity space $[\dot{q}]=[\dot{q}_1,\dot{q}_2,\dot{q}_3]^T$ and the end-effector velocity space $^0[v]_H=^0[v_x,v_y,v_z]^T_H$ becomes:

$$^0[v]_H = [J][\dot{q}].$$
(8)

We denoted by $^0[v]_H$ the velocity of a point H belonging to the moving platform with respect to a reference coordinate system $(O_0x_0y_0z_0)$ situated to the fixed platform (0). The TPM constrains the rotations of the moving platform and angular velocity is zero.

A TPM with uncoupled motions is fully isotropic if

$$v_x = \lambda\dot{q}_1,$$
(9)

$$v_y = \lambda\dot{q}_2,$$
(10)

$$v_z = \lambda \dot{q}_3, \tag{11}$$

where λ is the value of the diagonal elements. A mechanism Q respecting the conditions (9)-(11) is fully-isotropic and implicitly it has uncoupled motions. The mechanism Q achieves a homothetic transformation of coefficient λ between the velocity of actuated joints and the velocity of the moving platform. When $\lambda = 1$ the Jacobian matrix becomes the identity 3×3 matrix ($I_{3\times 3}$). In this case, the condition number and the determinant of the Jacobian matrix being equal to one, the manipulator performs very well with regard to force and motion transmission.

We illustrate motion decoupling by using *3-PRPaR* Orthoglide [30, 52], belonging to Clavel family when the actuated prismatic joints are orthogonal. The underlined symbol indicates the actuated joint in each leg. Orthoglide (Figure 1) is a TPM-CM with three coupled motions. Eq. (7) gives just one isotropic position when $x=y=z=0$, when the parallelogram mechanisms integrated in the 3 legs are situated in three orthogonal planes.

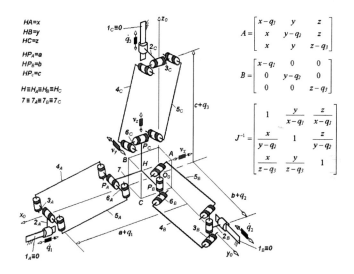

$$
J^{-1}J^{-T} =
\begin{bmatrix}
1+\dfrac{y^2+z^2}{(x-q_1)^2} & \dfrac{x}{y-q_2}+\dfrac{y}{x-q_1}+\dfrac{z^2}{(x-q_1)(y-q_2)} & \dfrac{x}{z-q_3}+\dfrac{y^2}{(x-q_1)(z-q_3)}+\dfrac{z}{x-q_1} \\[3ex]
\dfrac{x}{y-q_2}+\dfrac{y}{x-q_1}+\dfrac{z^2}{(x-q_1)(y-q_2)} & 1+\dfrac{x^2+z^2}{(y-q_2)^2} & \dfrac{x^2}{(y-q_2)(z-q_3)}+\dfrac{y}{z-q_3}+\dfrac{z}{y-q_2} \\[3ex]
\dfrac{x}{z-q_3}+\dfrac{y^2}{(x-q_1)(z-q_3)}+\dfrac{z}{x-q_1} & \dfrac{x^2}{(y-q_2)(z-q_3)}+\dfrac{y}{z-q_3}+\dfrac{z}{y-q_2} & 1+\dfrac{x^2+y^2}{(z-q_3)^2}
\end{bmatrix}
$$

Figure 1. Orthoglide - a TPM-CM (*3-PRPaR*) with three coupled motions

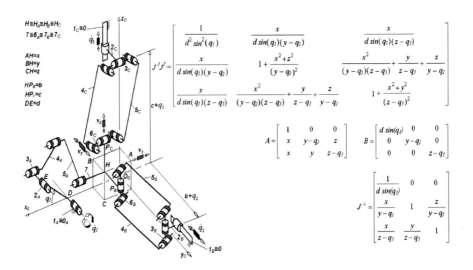

Figure 2. TPM-CM of type *1-RRRRR* + *2-PRPaR* with one uncoupled motion

We can see that for any translational motion of the moving platform (including a pure unidirectional motion on x, y or z) the three linear actuators must act in parallel. This is one of the main disadvantages of the TPMs-CM

Figure 2 presents a TPM-CM (*1-RRRRR* +*2-PRPaR*) with one uncoupled motion. This TPM-CM could be obtained from Orthoglide by replacing one. *PRPaR* leg with a *RRRRR*-leg. The axes of the last three revolute joints are perpendicular to the axes of the first two revolute joints in the *RRRRR*-leg. The unidirectional translation of the moving platform on x direction is uncoupled. This motion is achieved by only one actuator that is the revolute actuator (*R*) in the *RRRRR*-leg. Eq. (7) gives the single isotropic position when $x=y=z=0$, $d=1$ and $q_1=\pi/2$.

Figure 3 presents a TPM-DM (*2-RRRRR+1-PRPaR*) with two uncoupled motions. This TPM-DM could be obtained from Orthoglide by replacing 2-*PRPaR* legs with 2-*RRRRR*-leg. Each unidirectional translation of the moving platform on x or y is achieved independently by only one actuator. Eq. (7) gives a single isotropic position when $x=y=0$, $d=f=1$ and $q_1=q_2=\pi/2$.

Figure 4 presents a TPM-UM-RA (*3-RRRRR*) with 3 uncoupled motions. This TPM-UM could be obtained from Orthoglide by replacing the *3-PRPaR* legs with *3-RRRRR*-legs. Each unidirectional translation of the moving platform on x or y or z is achieved independently by only one actuator. This TPM-UM-RA is fully isotropic with respect to only matrix A. For this reason this solution was considered fully-isotropic in [32, 54]. In fact, Eqs. (5)-(7) indicate that *3-RRRRR* (Figure 4) has just a single isotropic position when $d=f=g=1$ and $q_1=q_2=q_3=\pi/2$.

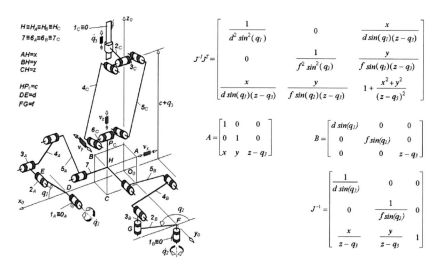

Figure 3. TPM-DM of type 2-$\underline{R}RRRR$ + 1-$\underline{P}RPaR$ with two uncoupled motions

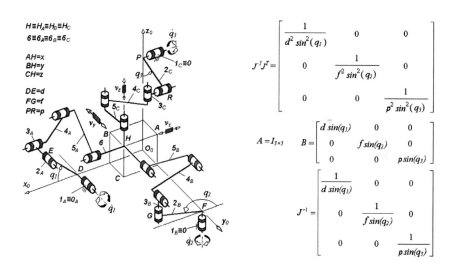

Figure 4. TPM-UM-RA of type 3-$\underline{R}RRRR$ with three uncoupled motions

3. STRUCTURAL SYNTHESIS

The basic kinematic structure of a TPM-UM is obtained by concatenating three legs $A_1(1\equiv0-...-n_{A1}\equiv n)$, $A_2(1\equiv0-...-n_{A2}\equiv n)$, and $A_3(1\equiv0-...-n_{A3}\equiv n)$. The first link of each leg is the fixed platform (0) and the final link is the moving platform ($n_{Ai}\equiv n$). A leg can be represented by an elementary open kinematic chain (elementary leg - EL) or by a complex open kinematic chain (complex leg - CL). An EL is obtained by serial concatenation of joints. A CL is obtained by serial and parallel concatenation of joints integrating one or more closed loops. The three translational motions are uncoupled, that is each unidirectional translation of the moving platform is achieved by only one actuator. We consider that the first joint of each leg A_i $(i=1,...,3)$ is actuated. The actuator of the leg A_1 achieves the translation of the moving platform on direction x, the actuators of the legs A_2 and A_3 on the directions y and z respectively.

The general conditions for structural synthesis of TPMs-UM proposed in this paper are founded on the theory of linear transformations [69]. The type synthesis of various legs obeying to these conditions is founded on an evolutionary morphology approach formalized by a 6-tuple of final objectives, primary elements, morphological operators, evolution criteria, morphologies and a termination criterion [65].

3.1 General conditions for structural synthesis via linear transformations

Let us consider a mechanism L that can be elementary or complex, open or closed. It achieves a linear transformation F from the finite dimensional vector space U (called joint velocity space) into the finite dimensional vector space W (called external velocity space). The linear transformation F, that could be defined by Eq. (2), assigns to each vector $\dot{q}$, in the joint vector space U, a uniquely determined vector $F(\dot{q})$ in the vector space W. The vector $F(\dot{q})$ in W is the image of the vector $\dot{q}$ in U under the transformation F. The image of U in W represents the range R_F of F. In this paper, the range R_F of F is called operational space. We can write

$$\dim(R_F)=\text{rank}(F)=\text{rank}[J]. \tag{12}$$

We define the relative spatiality ($S_{a/b}^L$) between two elements (a and b) in the mechanism L as the number of independent relative velocities allowed by the mechanism between the two elements. It is given by the dimension of the vector space $R_{a/b}^L$ of the relative velocities between elements a and b in the mechanism L

$$S_{a/b}^L = dim(R_{a/b}^L). \tag{13}$$

We note that the range $R_{a/b}^L$ is a sub-space of R_L and implicitly a sub-space of W. Relative spatiality characterizes the connectivity of the two elements (a and b) in the mechanism L having the degree of mobility M_L.

The mechanism associated to a TPM-UM is denoted by Q. The existence of this mechanism involves the following conditions for the degree of mobility, the relative spatiality between the moving and the fixed platform and for the base of the vector space of relative velocities of the mobile platform:

1. general conditions for any position of the mechanism when $\dot{q}_1 \neq 0$, $\dot{q}_2 \neq 0$, and $\dot{q}_3 \neq 0$

$$3 \leq M_{Ai} = S_{nAi/1}^{Ai} \leq 5, \text{ (i=1,2,3)}, \tag{14}$$

$$M_Q = 3 \tag{15}$$

$$S_{n/1}^Q = 3 \tag{16}$$

$$(R_{n/1}^Q) = (v_x, v_y, v_z) \tag{17}$$

2. particular condition when one actuator is locked $\dot{q}_i = 0$ (i=1,2,3)

$$S_{n/1}^Q = 2 \tag{18}$$

$$(R_{n/1}^Q) = (v_y, v_z) \quad if \quad \dot{q}_1 = 0 \tag{19}$$

$$(R_{n/1}^Q) = (v_x, v_z) \quad if \quad \dot{q}_2 = 0 \tag{20}$$

$$(R_{n/1}^Q = (v_x, v_y) \quad if \quad \dot{q}_3 = 0 \tag{21}$$

The base $(R_{n/1}^Q)$ must be the unique base of the vector space of the relative velocities between mobile and fixed platform. It is given by:

$$(R_{n/1}^Q) = (R_{nA1/1}^{A1} \cap R_{nA2/1}^{A2} \cap R_{nA3/1}^{A3}). \tag{22}$$

We recall that by M_L we denoted the degree of mobility of the kinematic chain L, by $S_{a/b}^L$ we denoted the relative spatiality between the elements a and b in the kinematic chain L and by $(R_{a/b}^L)$ we denoted the base of the vector space $R_{a/b}^L$ of the relative velocities between elements a and b in the kinematic chain L. The elements a and b represents the extreme links of L; a is the last element (n_L) and b is the fixed base ($1 \equiv 0$). In the previous notations the kinematic chain L could be A_1, A_2, A_3, Q and the last element n_L could be n_{A1}, n_{A2}, n_{A3}, n.

3.2   Legs for TPMs-UM-RA

In this section we present various architectures of elementary and complex legs with rotational actuators without/with idle mobilities obeying to conditions (14)-(21).

An idle mobility is a joint mobility that is inactive supposing a perfect mechanism (a mechanism without geometrical flaws and perfect assembly with respect to parallelism and orthogonality of joint axes/directions).

Planar closed loops (simply denoted by planar loops) are integrated in the complex legs. These loops can have one, two or three DOM and are named monomobile (1-DOM - parallelogram or rhombus loops), bimobile (2-DOM) or trimobile (3-DOM) loop. Usually these planar loops are overconstrained and they contain only revolute and prismatic pairs. All axes of revolute pairs are parallel and all dirctions of prismatic pairs are perpendicular to the revolute axes. An isostatic (not overconstrained) planar loop is obtained by introducing 3 idle mobilities. We can do it in various ways, for example, by replacing two revolute pairs by a cylindrical pair and a spherical pair. The planar loops are denoted by *Pa* (parallelogram loop), *Rb* (rhombus loop), *Pn2* (bimobile loop) and *Pn3* (trimobile loop). The same notations are used for overconstrained and isostatic loops too. When any special mention is made, we refer implicitly to an overconstrained planar loop.

These architectures can be grouped into:

I.   legs without idle mobilities:

   I.1) elementary leg with $M_A = S_{nA//I}^A = 4$ - Figure 5,

   I.2) elementary leg with $M_A = S_{nA//I}^A = 5$ - Figure 6,

   I.3) complex leg with $M_A = S_{nA//I}^A = 3$ and one parallelogram loop - Figure 7 (a-b),

   I.4) complex leg with $M_A = S_{nA//I}^A = 3$ and two parallelogram loops - Figure 7 (c-f),

   I.5) complex leg with $M_A = S_{nA//I}^A = 3$ and three parallelogram loops - Figure 7 (d),

   I.6) complex leg with $M_A = S_{nA//I}^A = 4$ and one parallelogram loop - Figure 8 (a-f),

   I.7) complex leg with $M_A = S_{nA//I}^A = 4$ and two parallelogram loops - Figure 8 (g-h),

   I.8) complex leg with $M_A = S_{nA//I}^A = 5$ and one parallelogram loop - Figure 9,

   I.9) complex leg with $M_A = S_{nA//I}^A = 4$ containing a parallelogram loop and n ($n=1,2,...$) rhombus loops - Fig. 10, $n=1$ (a) and $n=2$ (b),

I.10) complex leg with $M_A = S_{nA/1}^A = 5$ containing n $(n=1,2,...)$ rhombus loops - Figure 11, $n=1$ (a) and $n=2$ (b),

I.11) complex leg with $M_A = S_{nA/1}^A = 4$ containing one parallelogram loop and one bimobile planar loop - Figure 12,

I.12) complex leg with $M_A = S_{nA/1}^A = 5$ and a bimobile planar loop - Figure 13,

I.13) complex leg with $M_A = S_{nA/1}^A = 4$ containing one parallelogram loop and one trimobile planar loop - Figure 14,

I.14) complex leg with $M_A = S_{nA/1}^A = 5$ and a trimobile planar loop - Figure 15,

II. legs with idle mobilities:

II.1) elementary leg with $M_A = S_{nA/1}^A = 5$ and one idle mobility,

II.2) complex leg with $M_A = S_{nA/1}^A = 4$ having one extra loop idle mobility and p $(p=1,2,3)$ parallelogram loops from which $m \leq p$ $(m=0,1,2,3)$ are isostatic,

II.3) complex leg with $M_A = S_{nA/1}^A = 5$ having two extra loop idle mobilities and p $(p=1,2,3)$ parallelogram loops from which $m \leq p$ $(m=0,1,2,3)$ are isostatic,

II.4) complex leg with $M_A = S_{nA/1}^A = 5$ having one extra loop idle mobility and p $(p=1,2,3)$ parallelogram loops from which $m \leq p$ $(m=0,1,2,3)$ are isostatic,

II.5) complex legs with $M_A = S_{nA/1}^A = 5$ having one extra loop idle mobility and $(1+n)$ loops (one parallelogram and n rhombus loops) from which m loops $(m=0,1,2,...,n+1)$ are isostatic,

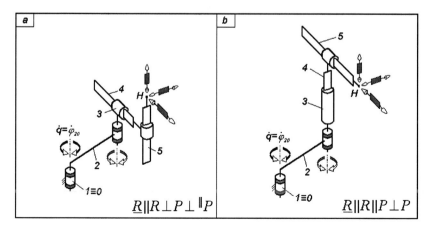

Figure 5. Elementary legs without idle mobilities $(M_A = S_{nA/1}^A = 4)$

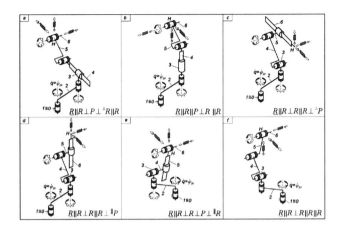

Figure 6. Elementary legs without idle mobilities $(M_A = S^A_{nA/1} = 5)$

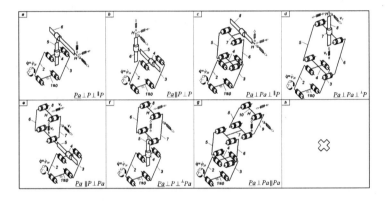

Figure 7. Complex legs with parallelogram loops without idle mobilities $(M_A = S^A_{nA/1} = 3)$

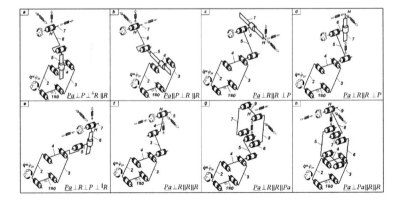

Figure 8. Complex legs with parallelogram loops without idle mobilities $(M_A = S^A_{nA/1} = 4)$

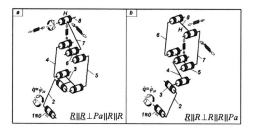

Figure 9. Complex legs with parallelogram loops without idle mobilities $(M_A = S_{nA/1}^A = 5)$

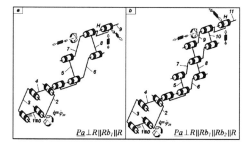

Figure 10. Complex legs with rhombus and parallelogram loops without idle mobilities $(M_A = S_{nA/1}^A = 4)$

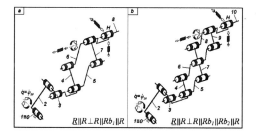

Figure 11. Complex legs with rhombus loops without idle mobilities $(M_A = S_{nA/1}^A = 5)$

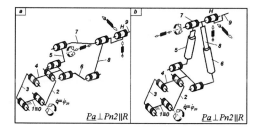

Figure 12. Complex legs with bimobile planar loops without idle mobilities $(M_A = S_{nA/1}^A = 4)$

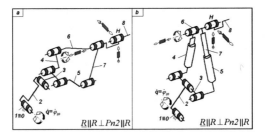

Figure 13. Complex legs with bimobile planar loops without idle mobilities $(M_A = S_{nA/1}^A = 5)$

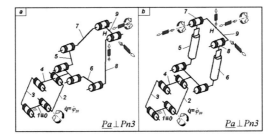

Figure 14. Complex legs with trimobile planar loops without idle mobilities $(M_A = S_{nA/1}^A = 4)$

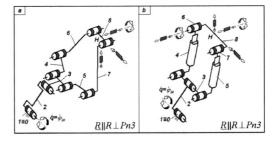

Figure 15. Complex legs with trimobile planar loops without idle mobilities $(M_A = S_{nA/1}^A = 5)$

II.6) complex leg with $M_A = S_{nA/1}^A = 5$ having one extra loop idle mobility and two different loops (one parallelogram and one bimobile planar loop) from which *m* loops (*m=0,1,2*) are isostatic,

II.7) complex leg with $M_A = S_{nA/1}^A = 5$ having one extra loop idle mobility and two different loops (one parallelogram and one trimobile planar loop) from which *m* loops (*m=0,1,2*) are isostatic.

As we can see, mobility and spatiality of a leg are invariant with respect to overconstrained or isostatic loops integrated in leg structure. Only the idle mobilities introduced in the joints situated outside of closed loops (called extra loop idle mobilities) influence leg mobility and spatiality.

3.3   Architectures of TPMs-UM-RA

To obtain a TPM-UM-RA, three elementary or complex legs, presented in the previous section, are assembled by respecting conditions (15)-(21).

Tables 1-5 present the joint arrangement in each leg A_i ($i=1,2,3$) of a TPM-UM-RA without idle mobilities. The indice associated to the joint symbol in Tables 1-5 denotes the direction of the joint axis. The index associated with the planar loop symbol (*Pa, Rb, Pn2, Pn3*) denotes the direction of the revolute joint axes in the planar loop. By various associations of the the three legs we could obtain $(21+2n)^3$ structural types of TPM-UM-RA without idle mobilities from which $21+2n$ types have structural identical leg type. By n we denoted the number of rhombus loops in a leg ($n=0,1,2,...$). By taking tnto consideration the various solutions associated to each leg type we could obtain $(150+4n)^3$ structural solutions of TPM-UM-RA without idle mobilities from which $150+4n$ have identical structural leg solutions.

Two conditions are necessary to obtrain an isostatic (not overconstrained) TPM-UM:

$$\sum_{i=1}^{3} S_{nAi/1}^{Ai} = 15 \tag{23}$$

and each planar loop integrated in leg structure must be isostatic (if planar loops of type *Pa, Rb, Pn2* or *Pn3* are used).

If the TPM-UM has identical type of leg, condition (23) becomes:

$$S_{nAi/1}^{Ai} = 5 \, , \text{i}=1,2,3. \tag{24}$$

We can see that condition (24) could be respected by only $8+n$ types of legs with $66+2n$ distinct solutions of legs. In these solutions all the planar loops integrated in the leg structure must be isostatic.

To enlarge the number of isostatic solutions of TPM-UM-RA various types of elementary and complex legs with one or two extra loop idle mobilities can also be used. These solutions (see Table 6) are obtained from the solutions with $M_A = S_{nA/1}^{A} = 3$ and $M_A = S_{nA/1}^{A} = 4$ presented in Tables 1-5 by introducing one or two idle mobilities outside the closed loops integrated in the kinematic chain associated with the same leg. These extra loop idle mobilities are introduced in the revolute pairs denoted by R^*. For example the TPM-UM-RA in Figure 16 (b) of type 3-$\underline{RRR}^*PP$ is obtained from the solution 3-$\underline{RRPP}$-type in Figure16 (a) by introducing an idle mobility in the third joint of each leg.

Table 1. Elementary legs for TPMs-UM-RA without idle mobilities

No	Type	$M_A = S_{nA/1}^A$	Joint arrangement	Base $(R_{n/1}^Q)$	Special conditions	No. sol./leg
1	$\underline{RRPP}$	4	$\underline{R}\|\|R \perp P \perp^\| P$		$c \neq d$	2
			Leg A_1: $\underline{R}_c$-R_c-P_d-P_c	(v_x,v_y,v_z,ω_c)	$c=y,z;\ d=y,z$	
			Leg A_2: $\underline{R}_c$-R_c-P_d-P_c	(v_x,v_y,v_z,ω_c)	$c=x,z;\ d=x,z$	
			Leg A_3: $\underline{R}_c$-R_c-P_d-P_c	(v_x,v_y,v_z,ω_c)	$c=x,y;\ d=x,y$	
			$\underline{R}\|\|R \|\|P \perp P$		$c \neq d$	2
			Leg A_1: $\underline{R}_c$-R_c-P_c-P_d	(v_x,v_y,v_z,ω_c)	$c=y,z;\ d=y,z$	
			Leg A_2: $\underline{R}_c$-R_c-P_c-P_d	(v_x,v_y,v_z,ω_c)	$c=x,z;\ d=x,z$	
			Leg A_3: $\underline{R}_c$-R_c-P_c-P_d	(v_x,v_y,v_z,ω_c)	$c=x,y;\ d=x,y$	
2	$\underline{RRPRR}$	5	$\underline{R}\|\|R \perp P \perp^\perp R \|\|R$		$c \neq d$	2
			Leg A_1: $\underline{R}_c$-R_c-P_d-R_x-R_x	$(v_x,v_y,v_z,\omega_x,\omega_c)$	$c=y,z;\ d=y,z$	
			Leg A_2: $\underline{R}_c$-R_c-P_d-R_y-R_y	$(v_x,v_y,v_z,\omega_y,\omega_c)$	$c=x,z;\ d=x,z,$	
			Leg A_3: $\underline{R}_c$-R_c-P_d-R_z-R_z	$(v_x,v_y,v_z,\omega_z,\omega_c)$	$c=x,y;\ d=x,y$	
			$\underline{R}\|\|R\|\|P \perp R \|\|R$			2
			Leg A_1: $\underline{R}_c$-R_c-P_c-R_x-R_x	$(v_x,v_y,v_z,\omega_x,\omega_c)$	$c=y,z$	
			Leg A_2: $\underline{R}_c$-R_c-P_c-R_y-R_y	$(v_x,v_y,v_z,\omega_y,\omega_c)$	$c=x,z$	
			Leg A_3: $\underline{R}_c$-R_c-P_c-R_z-R_z	$(v_x,v_y,v_z,\omega_z,\omega_c)$	$c=x,y$	
3	$\underline{RRRRP}$	5	$\underline{R}\|\|R \perp R\|\|R \perp^\perp P$		$c \neq d$	2
			Leg A_1: $\underline{R}_c$-R_c-R_x-R_x-P_d	$(v_x,v_y,v_z,\omega_x,\omega_c)$	$c=y,z;\ d=y,z$	
			Leg A_2: $\underline{R}_c$-R_c-R_y-R_y-P_d	$(v_x,v_y,v_z,\omega_y,\omega_c)$	$c=x,z;\ d=x,z$	
			Leg A_3: $\underline{R}_c$-R_c-R_z-R_z-P_d	$(v_x,v_y,v_z,\omega_z,\omega_c)$	$c=x,y;\ d=x,y$	
			$\underline{R}\|\|R \perp R\|\|R \perp^\| P$			2
			Leg A_1: $\underline{R}_c$-R_c-R_x-R_x-P_c	$(v_x,v_y,v_z,\omega_x,\omega_c)$	$c=y,z$	
			Leg A_2: $\underline{R}_c$-R_c-R_y-R_y-P_c	$(v_x,v_y,v_z,\omega_y,\omega_c)$	$c=x,z$	
			Leg A_3: $\underline{R}_c$-R_c-R_z-R_z-P_c	$(v_x,v_y,v_z,\omega_z,\omega_c)$	$c=x,y$	
4	$\underline{RRRPR}$	5	$\underline{R}\|\|R \perp R \perp P \perp^\| R$			2
			Leg A_1: $\underline{R}_c$-R_c-R_x-P_{yz}-R_x	$(v_x,v_y,v_z,\omega_x,\omega_c)$	$c=y,z$	
			Leg A_2: $\underline{R}_c$-R_c-R_y-P_{xz}-R_y	$(v_x,v_y,v_z,\omega_y,\omega_c)$	$c=x,z$	
			Leg A_3: $\underline{R}_c$-R_c-R_z-P_{xy}-R_z	$(v_x,v_y,v_z,\omega_z,\omega_c)$	$c=x,y$	
5	$\underline{RRRRR}$	5	$\underline{R}\|\|R \perp R\|\|R\|\|R$			2
			Leg A_1: $\underline{R}_c$-R_c-R_x-R_x-R_x	$(v_x,v_y,v_z,\omega_x,\omega_c)$	$c=y,z$	
			Leg A_2: $\underline{R}_c$-R_c-R_y-R_y-R_y	$(v_x,v_y,v_z,\omega_y,\omega_c)$	$c=x,z$	
			Leg A_3: $\underline{R}_c$-R_c-R_z-R_z-R_z	$(v_x,v_y,v_z,\omega_z,\omega_c)$	$c=x,y$	

Table 2. Complex legs with parallelogram loops for TPMs-UM-UR without idle mobilities

No	Type	$M_A = S_{nA/1}^A$	Joint arrangement	Base ($R_{n/1}^Q$)	Special conditions	No. sol/ leg
1	*PaPP*	3	$\underline{Pa} \perp P \perp {}^{\parallel}P$		$c \neq d$	2
			Leg A_1: $\underline{Pa_c}$-P_d-P_c	(v_x,v_y,v_z)	$c=y,z;\ d=y,z$	
			Leg A_2: $\underline{Pa_c}$-P_d-P_c	(v_x,v_y,v_z)	$c=x,z;\ d=x,z$	
			Leg A_3: $\underline{Pa_c}$-P_d-P_c	(v_x,v_y,v_z)	$c=x,y;\ d=x,y$	
			$\underline{Pa} \| P \perp P$		$c \neq d$	2
			Leg A_1: $\underline{Pa_c}$-P_c-P_d	(v_x,v_y,v_z)	$c=y,z;\ d=y,z$	
			Leg A_2: $\underline{Pa_c}$-P_c-P_d	(v_x,v_y,v_z)	$c=x,z;\ d=x,z$	
			Leg A_3: $\underline{Pa_c}$-P_c-P_d	(v_x,v_y,v_z)	$c=x,y;\ d=x,y$	
2	*PaPaP*	3	$\underline{Pa} \perp Pa \perp {}^{\parallel}P$			2
			Leg A_1: $\underline{Pa_c}$-Pa_x-P_c	(v_x,v_y,v_z)	$c=y,z$	
			Leg A_2: $\underline{Pa_c}$-Pa_y-P_c	(v_x,v_y,v_z)	$c=x,z$	
			Leg A_3: $\underline{Pa_c}$-Pa_z-P_c	(v_x,v_y,v_z)	$c=x,y$	
			$\underline{Pa} \perp Pa \perp {}^{\perp}P$		$c \neq d$	2
			Leg A_1: $\underline{Pa_c}$-Pa_x-P_d	(v_x,v_y,v_z)	$c=y,z;\ d=y,z$	
			Leg A_2: $\underline{Pa_c}$-Pa_y-P_d	(v_x,v_y,v_z)	$c=x,z;\ d=x,z$	
			Leg A_3: $\underline{Pa_c}$-Pa_z-P_d	(v_x,v_y,v_z)	$c=x,y;\ d=x,y$	
3	*PaPPa*	3	$\underline{Pa} \| P \perp Pa$			2
			Leg A_1: $\underline{Pa_c}$-P_c-Pa_x	(v_x,v_y,v_z)	$c=y,z$	
			Leg A_2: $\underline{Pa_c}$-P_c-Pa_y	(v_x,v_y,v_z)	$c=x,z$	
			Leg A_3: $\underline{Pa_c}$-P_c-Pa_z	(v_x,v_y,v_z)	$c=x,y$	
			$\underline{Pa} \perp P \perp {}^{\perp}Pa$		$c \neq d$	2
			Leg A_1: $\underline{Pa_c}$-P_d-Pa_x	(v_x,v_y,v_z)	$c=y,z;\ d=y,z$	
			Leg A_2: $\underline{Pa_c}$-P_d-Pa_y	(v_x,v_y,v_z)	$c=x,z;\ d=x,z$	
			Leg A_3: $\underline{Pa_c}$-P_d-Pa_z	(v_x,v_y,v_z)	$c=x,y;\ d=x,y$	
4	*PaPaPa*	3	$\underline{Pa} \perp Pa \| Pa$			2
			Leg A_1: $\underline{Pa_c}$-Pa_x-Pa_x	(v_x,v_y,v_z)	$c=y,z$	
			Leg A_2: $\underline{Pa_c}$-Pa_y-Pa_y	(v_x,v_y,v_z)	$c=x,z$	
			Leg A_3: $\underline{Pa_c}$-Pa_z-Pa_z	(v_x,v_y,v_z)	$c=x,y$	
5	*PaPRR*	4	$\underline{Pa} \perp P \perp {}^{\perp}R \| R$		$c \neq d$	2
			Leg A_1: $\underline{Pa_c}$-P_d-R_x-R_x	(v_x,v_y,v_z,ω_x)	$c=y,z;\ d=y,z$	
			Leg A_2: $\underline{Pa_c}$-P_d-R_y-R_y	(v_x,v_y,v_z,ω_y)	$c=x,z;\ d=x,z$	
			Leg A_3: $\underline{Pa_c}$-P_d-R_z-R_z	(v_x,v_y,v_z,ω_z)	$c=x,y;\ d=x,y$	

continued

No	Type	M_A $= S_{nA/1}^A$	Joint arrangement	Base ($R_{n/1}^Q$)	Special conditions	No. sol/ leg
			$\underline{Pa}\|\|P \perp R\|\|R$			2
			Leg A_1: $\underline{Pa_c}$-P_c-R_x-R_x	(v_x,v_y,v_z,ω_x)	$c=y,z$	
			Leg A_2: $\underline{Pa_c}$-P_c-R_y-R_y	(v_x,v_y,v_z,ω_y)	$c=x,z$	
			Leg A_3: $\underline{Pa_c}$-P_c-R_z-R_z	(v_x,v_y,v_z,ω_z)	$c=x,y$	
6	$\underline{Pa}RPR$	4	$\underline{Pa} \perp R \perp P \perp^{\|}R$			2
			Leg A_1: $\underline{Pa_c}$-R_x-P_{yz}-R_x	(v_x,v_y,v_z,ω_x)	$c=y,z$	
			Leg A_2: $\underline{Pa_c}$-R_y-P_{xz}-R_y	(v_x,v_y,v_z,ω_y)	$c=x,z$	
			Leg A_3: $\underline{Pa_c}$-R_z-P_{xy}-R_z	(v_x,v_y,v_z,ω_z)	$c=x,y$	
7	$\underline{Pa}RRP$	4	$\underline{Pa} \perp R\|\|R \perp P$			4
			Leg A_1: $\underline{Pa_c}$-R_x-R_x-P_d	(v_x,v_y,v_z,ω_x)	$c=y,z; d=y,z$	
			Leg A_2: $\underline{Pa_c}$-R_y-R_y-P_d	(v_x,v_y,v_z,ω_y)	$c=x,z; d=x,z$	
			Leg A_3: $\underline{Pa_c}$-R_z-R_z-P_d	(v_x,v_y,v_z,ω_z)	$c=x,y; d=x,y$	
8	$\underline{Pa}RRR$	4	$\underline{Pa} \perp R\|\|R\|\|R$			2
			Leg A_1: $\underline{Pa_c}$-R_x-R_x-R_x	(v_x,v_y,v_z,ω_x)	$c=y,z$	
			Leg A_2: $\underline{Pa_c}$-R_y-R_y-R_y	(v_x,v_y,v_z,ω_y)	$c=x,z$	
			Leg A_3: $\underline{Pa_c}$-R_z-R_z-R_z	(v_x,v_y,v_z,ω_z)	$c=x,y$	
9	$\underline{Pa}PaRR$	4	$\underline{Pa} \perp Pa\|\|R\|\|R$			2
			Leg A_1: $\underline{Pa_c}$-Pa_x-R_x-R_x	(v_x,v_y,v_z,ω_x)	$c=y,z$	
			Leg A_2: $\underline{Pa_c}$-Pa_y-R_y-R_y	(v_x,v_y,v_z,ω_y)	$c=x,z$	
			Leg A_3: $\underline{Pa_c}$-Pa_z-R_z-R_z	(v_x,v_y,v_z,ω_z)	$c=x,y$	
10	$\underline{Pa}RRPa$	4	$\underline{Pa} \perp R\|\|R\|\|Pa$			2
			Leg A_1: $\underline{Pa_c}$-R_x-R_x-Pa_x	(v_x,v_y,v_z,ω_x)	$c=y,z$	
			Leg A_2: $\underline{Pa_c}$-R_y-R_y-Pa_y	(v_x,v_y,v_z,ω_y)	$c=x,z$	
			Leg A_3: $\underline{Pa_c}$-R_z-R_z-Pa_z	(v_x,v_y,v_z,ω_z)	$c=x,y$	
11	$\underline{RR}PaRR$	5	$\underline{R}\|\|\underline{R} \perp Pa\|\|R\|\|R$			2
			Leg A_1: $\underline{R_c}$-R_c-Pa_x-R_x-R_x	$(v_x,v_y,v_z,\omega_x,\omega_c)$	$c=y,z$	
			Leg A_2: $\underline{R_c}$-R_c-Pa_y-R_y-R_y	$(v_x,v_y,v_z,\omega_y,\omega_c)$	$c=x,z$	
			Leg A_3: $\underline{R_c}$-R_c-Pa_z-R_z-R_z	$(v_x,v_y,v_z,\omega_z,\omega_c)$	$c=x,y$	
12	$\underline{RR}RRPa$	5	$\underline{R}\|\|R \perp R\|\|R\|\|Pa$			2
			Leg A_1: $\underline{R_c}$-R_c-R_x-R_x-Pa_x	$(v_x,v_y,v_z,\omega_x,\omega_c)$	$c=y,z$	
			Leg A_2: $\underline{R_c}$-R_c-R_y-R_y-Pa_y	$(v_x,v_y,v_z,\omega_y,\omega_c)$	$c=x,z$	
			Leg A_3: $\underline{R_c}$-R_c-R_z-R_z-Pa_z	$(v_x,v_y,v_z,\omega_z,\omega_c)$	$c=x,y$	

Table 3. Complex legs with rhombus loops for TPMs-UM-RA without idle mobilities

No	Type	M_A $=$ $S_{nA/1}^A$	Joint arrangement	Base $(R_{n/1}^Q)$	Sp. cond.	No sol/ leg
1	$\underline{R}RRRb...RbR$	5	$\underline{R}||R \perp R||Rb_1||...||Rb_n||R$			$2n$
			Leg A_1:			
			$\underline{R}_c\text{-}R_c\text{-}R_x\text{-}Rb_{1x}\text{-}...\text{-}Rb_{nx}\text{-}R_x$	$(v_x,v_y,v_z,\omega_x,\omega_c)$	$c=y,z$	
			Leg A_2:		$c=x,z$	
			$\underline{R}_c\text{-}R_c\text{-}R_y\text{-}Rb_{1y}\text{-}...\text{-}Rb_{ny}\text{-}R_y$	$(v_x,v_y,v_z,\omega_y,\omega_c)$	$c=x,y$	
			Leg A_3:			
			$\underline{R}_c\text{-}R_c\text{-}R_z\text{-}Rb_{1z}\text{-}...\text{-}Rb_{nz}\text{-}R_z$	$(v_x,v_y,v_z,\omega_z,\omega_c)$		
2	$\underline{Pa}RRb...RbR$	4	$\underline{Pa} \perp R ||Rb_1||...||Rb_n||R$			$2n$
			Leg A_1:			
			$\underline{Pa}_c\text{-}R_x\text{-}Rb_{1x}\text{-}...\text{-}Rb_{nx}\text{-}R_x$	(v_x,v_y,v_z,ω_x)	$c=y,z$	
			Leg A_2:			
			$\underline{Pa}_c\text{-}R_y\text{-}Rb_{1y}\text{-}...\text{-}Rb_{ny}\text{-}R_y$	(v_x,v_y,v_z,ω_y)	$c=x,z$	
			Leg A_3:			
			$\underline{Pa}_c\text{-}R_z\text{-}Rb_{1z}\text{-}...\text{-}Rb_{nz}\text{-}R_z$	(v_x,v_y,v_z,ω_z)	$c=x,y$	

n is the number of rombhus loops (n=1,2,...)

Table 4. Complex legs with bimobile planar loops for TPMs-UM-RA without idle mobilities

No	Type	M_A $= S_{nA/1}^A$	Joint arrangement	Base $(R_{n/1}^Q)$	Special conditions	No. sol./ leg
1	$\underline{R}RPn2R$	5	$\underline{R}||R \perp Pn2||R$		–	22
			Leg A_1: $\underline{R}_c\text{-}R_c\text{-}Pn2_x\text{-}R_x$	$(v_x,v_y,v_z,\omega_x,\omega_c)$	$c=y,z$	
			Leg A_2: $\underline{R}_c\text{-}R_c\text{-}Pn2_y\text{-}R_y$	$(v_x,v_y,v_z,\omega_y,\omega_c)$	$c=x,z$	
			Leg A_3: $\underline{R}_c\text{-}R_c\text{-}Pn2_z\text{-}R_z$	$(v_x,v_y,v_z,\omega_z,\omega_c)$	$c=x,y$	
2	$\underline{Pa}Pn2R$	4	$\underline{Pa} \perp Pn2||R$			22
			Leg A_1: $\underline{Pa}_c\text{-}Pn2_x\text{-}R_x$	(v_x,v_y,v_z,ω_x)	$c=y,z$	
			Leg A_2: $\underline{Pa}_c\text{-}Pn2_y\text{-}R_y$	(v_x,v_y,v_z,ω_y)	$c=x,z$	
			Leg A_3: $\underline{Pa}_c\text{-}Pn2_z\text{-}R_z$	(v_x,v_y,v_z,ω_z)	$c=x,y$	

Pn2:

$(R_c||R_c||R_c||R_c||R_c)$, $(P \perp R_c||R_c||R_c||R_c)$, $(R_c \perp P \perp {}^{\parallel}R_c||R_c||R_c)$,

$(R_c||R_c \perp P \perp {}^{\parallel}R_c||R_c)$,

$(P \perp P \perp {}^{\perp}R_c||R_c||R_c)$, $(R_c \perp P \perp {}^{\perp}P \perp {}^{\perp}R_c||R_c)$, $(P \perp R_c \perp {}^{\perp}P \perp {}^{\parallel}R_c||R_c)$,

$(R_c \perp P \perp {}^{\parallel}R_c \perp P \perp {}^{\parallel}R_c)$, $(P \perp R_c||R_c \perp P \perp {}^{\parallel}R_c)$, $(P \perp R_c||R_c||R_c \perp P)$,

$(P \perp P \perp {}^{\perp}R_c||R_c \perp P)$.

c=x (leg A_1), c=y (leg A_2), c=z (leg A_3).

Table 5. Complex legs with trimobile planar loops for TPMs-UM-RA without idle mobilities

No	Type	$M_A = S_{nA/1}^A$	Joint arangement	Base ($R_{n/1}^Q$)	Special conditions	No. sol./ leg
1	*RRPn3*	5	$R \| \| R \perp Pn3$			28
			Leg A_1: $\underline{R_c}$-R_c-$Pn3_x$	$(v_x, v_y, v_z, \omega_x, \omega_c)$	$c=y,z$	
			Leg A_2: $\underline{R_c}$-R_c-$Pn3_y$	$(v_x, v_y, v_z, \omega_y, \omega_c)$	$c=x,z$	
			Leg A_3: $\underline{R_c}$-R_c-$Pn3_z$	$(v_x, v_y, v_z, \omega_z, \omega_c)$	$c=x,y$	
2	*PaPn3*	4	$Pa \perp Pn3$			28
			Leg A_1: $\underline{Pa_c}$-$Pn3_x$	$(v_x, v_y, v_z, \omega_x)$	$c=y,z$	
			Leg A_2: $\underline{Pa_c}$-$Pn3_y$	$(v_x, v_y, v_z, \omega_y)$	$c=x,z$	
			Leg A_3: $\underline{Pa_c}$-$Pn3_z$	$(v_x, v_y, v_z, \omega_z)$	$c=x,y$	

Pn3:

$(R_c \| \| R_c \| \| R_c \| \| R_c \| \| R_c \| \| R_c)$, $(P \perp R_c \| \| R_c \| \| R_c \| \| R_c \| \| R_c)$,
$(R_c \perp P \perp {}^\| R_c \| \| R_c \| \| R_c \| \| R_c)$, $(R_c \| \| R_c \perp P \perp {}^\| R_c \| \| R_c \| \| R_c)$,
$(P \perp P \perp {}^+ R_c \| \| R_c \| \| R_c \| \| R_c)$, $(R_c \perp P \perp {}^+ P \perp {}^+ R_c \| \| R_c \| \| R_c)$,
$(P \perp R_c \perp {}^+ P \perp {}^\| R_c \| \| R_c \| \| R_c)$, $(R_c \perp P \perp {}^\| R_c \perp P \perp {}^\| R_c \| \| R_c)$,
$(P \perp R_c \| \| R_c \perp P \perp {}^\| R_c \| \| R_c)$, $(P \perp R_c \| \| R_c \| \| R_c \perp P \perp {}^\| R_c)$,
$(R_c \| \| R_c \perp P \perp {}^+ P \perp {}^+ R_c \| \| R_c)$, $(P \perp R_c \| \| R_c \| \| R_c \| \| R_c \perp P)$,
$(R_c \perp P \perp {}^\| R_c \| \| R_c \perp P \perp {}^\| R_c)$,
$(P \perp P \perp {}^+ R_c \| \| R_c \| \| R_c \perp P)$,

$$c=x \text{ (leg } A_1), \quad c=y \text{ (leg } A_2), \quad c=z \text{ (leg } A_3).$$

Table 6. Complex legs with idle mobilities for TPMs-UM-RA

No	Type	$M_A = S_{nA/1}^A$	No. sol./leg	No	Type	$M_A = S_{nA/1}^A$	No. sol./leg
1	*PaR*PP*	4	12	20	*PaRRR*P*	5	8
2	*PaPR*P*	4	12	21	*PaRRPR**	5	8
3	*PaPPR**	4	12	22	*PaR*RPR*	5	4
4	*PaR*PaP*	4	12	23	*PaRPRR**	5	4
5	*PaPaR*P*	4	12	24	*PaR*RRR*	5	4
6	*PaPaPR**	4	12	25	*PaRRRR**	5	8
7	*PaR*PPa*	4	12	26	*PaR*RRPa*	5	8
8	*PaPR*Pa*	4	12	27	*PaRRR*Pa*	5	8
9	*PaPPaR**	4	12	28	*PaRRPaR**	5	8
10	*PaR*PaPa*	4	6	29	*PaR*PaRR*	5	8
11	*PaPaR*Pa*	4	6	30	*PaPaR*RR*	5	8
12	*PaPaPaR**	4	6	31	*PaPaRRR**	5	8
13	*RRR*PP*	5	8	32	*PaR*RRb...RbR*	5	4n
14	*RRPR*P*	5	8	33	*PaRRb...RbRR**	5	4n
15	*RRPPR**	5	8	34	*PaR*Pn2R*	5	88
16	*PaR*PRR*	5	8	35	*PaPn2RR**	5	88
17	*PaPR*RR*	5	8	36	*PaR*Pn3*	5	112
18	*PaPRRR**	5	8	37	*PaPn3R**	5	112
19	*PaR*RRP*	5	8				

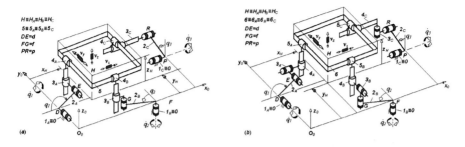

Figure 16. TPM-UM-RA with elementary legs: 3-RRPP-type without idle mobilities (a) and 3-RRR*PP-type with one idle mobility in each leg (b)

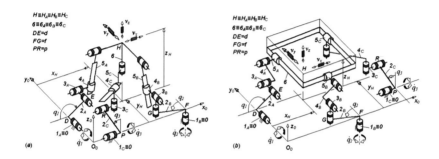

Figure 17. Isostatic TPM-UM-RA with elementary legs and without idle mobilities: 3-RRRPR-type (a) and 3- RRRRP-type (b)

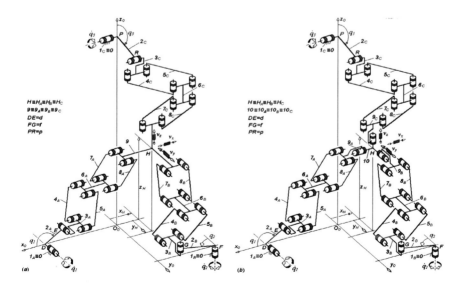

Figure 18. TPM-UM-RA with parallelogram loops: 3-RRPaPa-type without idle mobilities (a) and 3-RRPaPaR*-type with an idle mobility in each leg (b)

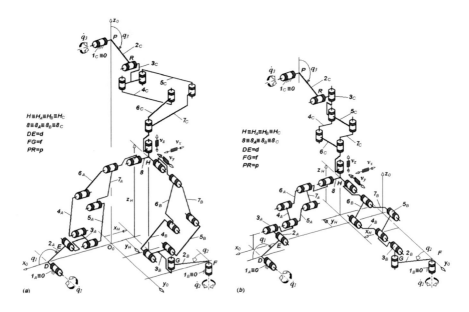

Figure 19. TPM-UM-RA with complex legs and without idle mobilities: 3-R̲RPn2R -type (a)
and 3-R̲RRRbR-type (b)

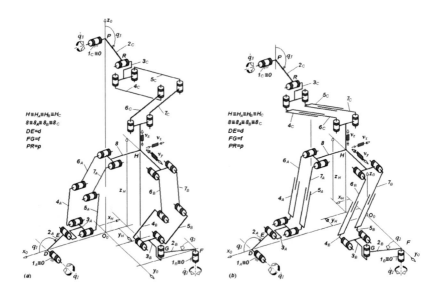

Figure 20. TPM-UM-RA with complex legs 3-R̲RPn3-type with trimobile planar loops of type
RRRRRR (a) and RPRRPR (b)

The idle mobility increases leg mobility and spatiality but they do not
affect mobility spatiality and singularities of TPM-UM-RA if the following
special geometric conditions are respected:

a) the axes/directions of all pairs (including the revolute pairs with idle mobilities) adjacent to the same link must be parallel or perpendicular,
b) three axes of revolute pairs with idle mobilities from different legs must not be parallel,
c) the orientation of the axes of the revolute pairs (including the revolute pairs with idle mobilities) supposing a perfect mechanism must be invariant during robot motion in the entire workspace.

All solutions of legs presented in Tables 1-6 give fully-isotropic TPM-UM-RA with respect to parallel (direct) Jacobian and obey to Eq. (4). By various associations of the the the three legs with/without idle mobilities we get $(76+4n)^3$ structural types of TPM-UM-RA with $(742+12n)^3$ different solutions.

The TPM-UM-RA presented in Figure 4 is an example of an isostatic TPM-UM-RA with elementary legs without idle mobilities (see type number 5 from Table 1). Other examples of isostatic TPM-UM-RA with elementary legs without idle mobilities are presented in Figure 17. In Figures 18-20 we present examples of TPM-UM-RA with complex legs containing parallelogram (Figure 18), rhombus (Figure 19, b) bimobile (Figure 18, a) and trimobile planar loops (Figure 20). The solutions presented in Figures 18 (b), 19 (a and b) and 20 (a and b) contain overconstrained planar loops. For this reason, they are not isostatic even though they have $M_A = S_{nA/1}^A = 5$. Equivalent isostatic solutions can be obtained by introducing 3 idle mobilities in the planar closed loops integrated in each leg, for example, by replacing two revolute pairs by a cylindrical and a spherical pair.

4. CONCLUSIONS

A general approach has been proposed based on the theory of linear transformations for structural synthesis of TPMs-UM-RA. The relation between motion coupling and isotropy has been discussed and the general conditions necessary to obtain a fully-isotropic TPM have been established.

The complete set of structural solutions of overconstrained and isostatic TPMs-UM-RA with elementary or complex legs integrating planar loops (parallelogram, rhombus, bimobile and trimobile planar loops) have been obtained by our approach. All these solutions are fully-isotropic with respect to the parallel Jacobian. The solutions of TPMs-UM-RA with complex legs containing rhombus, bimobile and trimobile planar loops are presented for the first time in the literature.

Special geometric conditions that allow idle mobilities to increase leg mobility and spatiality without affecting mobility spatiality and singularities of TPM-UM-RA have been defined for the first time. In this way, idle mobilities contribute to obtaining isostatic solutions without diminishing singularity-free workspace. The conditions necessary to obtain isostatic TPM-UM-RA have also been established.

The proposed approach can also be applied to structural synthesis of other types of parallel manipulators with uncoupled motions. The results presented in this paper are usual to design of new translational parallel manipulators with uncoupled motions and revotational actuators.

ACKNOWLEDGEMENTS

This work was supported by CNRS (French National Council of Scientific Research) in the framework of the projects ROBEA-MAX (2002-2003) and ROBEA-MP2 (2003-2004) and by EU GROWTH Programme in the framework of the project ADEPT (2003-2004).

REFERENCES

1. K. H. HUNT, *Constant-velocity shaft couplings: a general theory* (Trans. of the ASME, J. of Eng. For Industry, Vol 95B, pp. 455-464, 1973).
2. K. H. HUNT, *Structural kinematics of in-parallel-actuated robot arms* (Trans. of the ASME, J. of Mech. Design, Vol 105, pp. 705-712, 1983).
3. K. H. HUNT, *Kinematic Geometry of Mechanisms* (Oxford University Press, 1978).
4. J.-P. MERLET, *Les robots parallèles* (Paris: Hermès, 1st ed.1990, 2nd ed. 1997).
5. L.-W. TSAI, *Mechanism Design: Enumeration of kinematic structures according to function*, (CRC Press, 2000).
6. J. K. SALSBURY, J. J. CRAIG, *Articulated hands: force and kinematic issues* (Int. J. of Robotics Research, vol. 1/1, pp. 1-17, 1982).
7. J. ANGELES, *Fundamentals of Robotic Mechanical Systems: Theory, Methods, and Algorithms* (New York: Springer, 1987).
8. A. FATTAH, A. M. HASAN GHASEMI, *Isotropic design of spatial parallel manipulators* (Int. J. of Robotics Research, vol. 21/9, pp. 811-824, 2002).
9. R. CLAVEL, *Delta a fast robot with parallel geometry* (Proc. of the 18th Int. Symp. on Industrial Robots, Lausanne, Switzerland, pp 91-100, 1988).
10. R. CLAVEL, *Device for the movement and positioning of an element in space* (US Patent No. 4,976,582, 1990).
11. F. PIERROT, C. REYNAUD, A. FOURNIER, *DELTA: a simple and efficient parallel robot* (Robotica, Vol. 8, pp. 105-109, 1990).
12. J. M. HERVÉ, F. SPARACINO, *Structural synthesis of parallel robots generating spatial translation* (Proc. of the 5th IEEE Int. Conf. on Advanced Robotics, Pisa, Italy, pp 808-813, 1991).
13. J. M. HERVÉ, F. SPARACINO, *Star, a new concept in robotics* (Proc. of the 3rd Int. Workshop on Advances in Robot Kinematics, Ferrara, Italy, pp 176-183, 1992).
14. J. M. HERVÉ, F. SPARACINO, *Synthesis of parallel manipulators using Lie-groups: Y-STAR and H-ROBOT* (Proc. IEEE Intl. Workshop on Advanced Robotics, Tsukuba, Japon,

pp. 75-80, 1993).
15. W. T. APPLEBERRY, *Anti-rotation positioning mechanism* (US Patent No. 5,156,062, 1992).
16. J. M. HERVÉ, *Design of parallel manipulators via the displacement group* (Proc. of the 9th World Congress on the Theory of Machines and Mechanisms, Milan, Italy, pp 2079-2082, 1995).
17. Z. HUANG, W. S. TAO, Y. F. FANG, *Study on the kinematic characteristics of 3 dof in-parallel actuated platform mechanisms* (Mech. Mach. Theory, Vol. 31 (8), pp. 999-1007, 1996).
18. L. W. TSAI, *Kinematics of a three-dof platform with three extensible legs* (Recent Advances in Robot Kinematics, J. Lenarčič and V. Parenti-Castelli, eds., Kluwer Academic Publishers, pp 401-410, 1996).
19. L. W. TSAI, R. E. STAMPER, *A parallel manipulator with only translational degrees of freedom* (Proc. of the 1996 ASME Design Engineering Technical Conf., Irvine, CA, MECH-1152, 1996).
20. R. DI GREGORIO, V. PARENTI-CASTELLI, *A translational 3-dof parallel manipulator* (Advances in Robot Kinematics: Analysis and Control, J. Lenarčič and M. L. Husty, eds., Kluwer Academic Publishers, 49-58, 1998).
21. P.C. SHELDON, *Three-axis machine structure that prevents rotational movement* (US patent n° 5 865 063, Feb.2, 1999).
22. A. FRISOLI, D. CHECCACCI, F. SALSEDO, M. BERGAMASCO, *Synthesis by screw algebra of translating in-parallel actuated mechanisms* (Advances in Robot Kinematics, J. Lenarčič and M. M. Stanišić, eds., Kluwer Academic Publishers, pp 433-440, 2000).
23. L.-W. TSAI, S. JOSHI, *Kinematics and optimization of a spatial 3-UPU parallel manipulator* (Trans. of the ASME, J. of Mech. Design, Vol 122, pp. 439-446, 2000).
24. P.WENGER, D.CHABLAT, *Kinematic analysis of a new parallel machine tool: the Orthoglide, Advances in Robot Kinematics* (J. Lenarčič and M. M. Stanišić eds., Kluwer Academic Publishers, pp 305-314, 2000).
25. T. S. ZHAO, Z. HUANG, *A novel three-dof translational platform mechanism and its kinematics* (Proc. of the 2000 ASME Design Engineering Technical Conf., Baltimore, MD, MECH-14101, 2000).
26. M. CARRICATO, V. PARENTI-CASTELLI, *A family of 3-dof translational parallel manipulators* (Proc. of the 2001 ASME Design Engineering Technical Conf., Pittsburgh, PA, DAC- 21035, 2001).
27. R. DI GREGORIO, *Kinematics of the translational 3-URC mechanism* (Proc. of the 2001 IEEE/ASME Int. Conf. on Advanced Intelligent Mechatronics, Como, Italy, 147-152, 2001).
28. Q. JIN, T. L. YANG, *Position analysis for a class of novel 3-dof translational parallel robot mechanisms* (Proc. of the 2001 ASME Design Engineering Technical Conf., Pittsburgh, PA, DAC-21151, 2001).
29. X. KONG, C. M. GOSSELIN, *Generation of parallel manipulators with three translational degrees of freedom based on screw theory* (Proceedings of CCToMM Symposium on Mechanisms, Machines and Mechatronics, Montreal, Canada, 2001).
30. P. WENGER, D. CHABLAT, C. M. GOSSELIN, *A comparative study of parallel kinematic architectures for machining applications* (Proc. 2nd Workshop on Computational Kinematics, Seoul, Korea, pp. 249-258, 2001).
31. L. BARON, X. WANG, G. CLOUTIER, *The isotropic conditions of parallel manipulators of Delata topology* (in Advances in robot kinematics, J. Lenarčič and F. Thomas, eds, Kluwer Academic Publishers, pp. 357-366, 2002).
32. M. CARRICATO, V. PARENTI-CASTELLI, *Singularity-free fully-isotropic translational parallel mechanisms* (Int. J. of Robotics Research, 21/2: 161-174, 2002).
33. M. CARRICATO, V. PARENTI-CASTELLI, *Comparative position, workspace and singularity analyses of two isotropic translational parallel manipulators with three 4-dof*

legs (Proc.of MuSMe 2002, Int. Symp. on Multibody Systems and Mechatronics, Mexico City, Mexico, Paper No. M22, 2002).

34. M. CARRICATO, V. PARENTI-CASTELLI, *Singularity free fully-isotropic translational parallel manipulators* (2002 ASME Design Engineering Technical Conferences, Montreal, Canada, Sept. 29-Oct. 2, 2002, DETC2001/MECH-34301).

35. D. CHABLAT, PH. WENGER, J. MERLET, *Workspace analysis of the Orthoglide using interval analysis* (in Advances in robot kinematics, J. Lenarčič and F. Thomas, eds, Kluwer Academic Publishers, 2002, pp. 397-406).

36. O. COMPANY, F. PIERROT, *Modelling and design issues of a 3-axis parallel machine-tool* (Mechanism and Machine Theory, Vol. 37, pp. 1325–1345, 2002).

37. R. DI GREGORIO, V. PARENTI-CASTELLI, *Mobility analysis of the 3-UPU parallel mechanism assembled for a pure translational motion* (ASME Journal of Mechanical Design, 124(2): pp 259-264, 2002).

38. R. DI GREGORIO, *Kinematics of a new translational parallel manipulator* (Proc. of the 11th Int. Workshop on Robotics in Alpe-Adria-Danube Region, Balatonfured, Hungary, pp 249-254, 2002).

39. R. DI GREGORIO, V. PARENTI-CASTELLI, *Mobility analysis of the 3-UPU parallel mechanism assembled for a pure translation motion* (ASME J. of Mechanical Design, 124(2): pp 259-264, 2002).

40. C. M. GOSSELIN, X. KONG, *Cartesian parallel manipulators* (International patent WO 02/096605 A1, 2002).

41. S. GUÉGAN, W. KHALIL, *Dynamic Modeling of the Orthoglide* (in Advances in robot kinematics, J. Lenarčič and F. Thomas, eds, Kluwer Academic Publishers, pp. 387-396, 2002).

42. Z. HUANG, Q.C. LI, *Some novel lower-mobility parallel mechanisms* (Proceedings of ASME Design Engineering Technical Conference, Montreal, Canada, 2002).

43. H. S. KIM, L.-W. TSAI, *Design optimization of a Cartesian parallel manipulator* (Proceedings of ASME Design Engineering Technical Conferences, Montreal, Canada, 2002).

44. H. S. KIM, L-W. TSAI, *Evaluation of a Cartesian parallel manipulator* (in Advances in robot kinematics, J. Lenarčič and F. Thomas, eds, Kluwer Academic Publishers, pp. 21-28, 2002).

45. X. KONG, C. M. GOSSELIN, *Kinematics and singularity analysis of a novel type of 3-CRR 3-dof translational parallel manipulator* (Int. J. of Robotics Research, 21/9: pp 791-798, 2002).

46. X. KONG, C. M. GOSSELIN, *A class of 3-DOF translational parallel manipulators with linear input-output equations* (Proceedings of the Workshop on Fundamental Issues and Future Research Directions for Parallel Mechanisms and Manipulators, Québec, October 3–4, pp. 25–32, 2002).

47. X. KONG, GOSSELIN, C. M., *Type synthesis of linear translational parallel manipulators* (in Advances in robot kinematics, J. Lenarčič and F. Thomas, eds, Kluwer Academic Publishers, pp. 453-462, 2002).

48. L. W. TSAI, S. JOSHI, *Kinematic analysis of 3-dof position mechanisms for use in hybrid kinematic machines* (Trans of the ASME J.of Mechanical Design, 124(2): pp 245-253, 2002).

49. L. W. TSAI, S. JOSHI, *Jacobian analysis of limited-dof parallel manipulator* (Trans of the ASME J.of Mechanical Design, 124(2): pp 254-258, 2002).

50. M. CALLEGARI, M. TARANTINI, *Kinematic analysis of a novel translational platform* (Transactions of ASME, Journal of Mechanical Design, Vol. 125, pp. 308-315, 2003).

51.M. CALLEGARI, P. MARZETTI, *Kinematics of a family of parallel translating mechanism* (Proceedings of RAAD'03, 12th International Workshop on Robotics in Alpe-Adria-Danube Region, Cassino, May 7-10, 2003).

52.D. CHABLAT, Ph. WENGER, *Architecture Optimization of a 3-DOF Parallel Mechanism for Machining Applications, the Orthoglide* (IEEE Trans. Robotics and Automation, Vol. 19(3), pp. 403-410, 2003).

53.Z. HUANG, Q.C. LI, *Type synthesis of symmetrical lower-mobility parallel mechanisms using the constraint-synthesis method*, Int. J. of Robotics Research, vol. 22(1), pp. 59-79, 2003.

54.M. CARRICATO, V. PARENTI-CASTELLI, *On the topological and geometrical synthesis and classification of translational parallel mechanisms* (Proceedings of the 11th World Congress in Mechanism and Machine Science, Vol. 4, pp. 1624-1628, China Machine Press, 2004).

55.D. CHABLAT, P. WENGER, J.-P. MERLET, *A comparative study between two three-dof parallel kinematic machines using kinetostatic criteria and interval analysis* (Proceedings of the 11th World Congress in Mechanism and Machine Science, Vol. 3, pp. 1209-1213, China Machine Press, 2004).

56.R. DI GREGORIO, V. PARENTI-CASTELLI, *Comparison of 3-dof parallel manipulators based on new dynamic performance indices* (Proceedings of the 11th World Congress in Mechanism and Machine Science, Vol. 4, pp. 1684-1688, China Machine Press, 2004).

57.J. M. HERVÉ, *New translational parallel manipulators with extensible parallelogram* (Proceedings of the 11th World Congress in Mechanism and Machine Science, Vol. 4, pp. 1599-1603, China Machine Press, 2004).

58.Z. HUANG, *The kinematics and type synthesis of lower-mobility parallel robot manipulators* (Proceedings of the 11th World Congress in Mechanism and Machine Science, Vol. 1, pp. 65-76, China Machine Press, 2004).

59.X. KONG, C. M. GOSSELIN, *Type synthesis of analytic translational parallel manipulators* (Proceedings of the 11th World Congress in Mechanism and Machine Science, Vol. 4, pp. 1642-1646, China Machine Press, 2004).

60.W. LIU, X. TANG, J. WANG, *A kind of three translational-dof parallel cube-manipulator* (Proceedings of the 11th World Congress in Mechanism and Machine Science, Vol. 4, pp. 1582-1587, China Machine Press, 2004).

61.D. ZLATANOV, C. M. GOSSELIN, *On the kinematic geometry of the 3-RER parallel mechanisms* (Proceedings of the 11th World Congress in Mechanism and Machine Science, Vol. 1, pp. 226-230, China Machine Press, 2004).

62.X. KONG, C. M. GOSSELIN, *Type synthesis of 3-dof translational parallel manipulators based on screw theory* (Journal of Mechanical Design, Vol. 126, pp. 83-92, 2004).

63.C.M. GOSSELIN, X. KONG, S. FOUCAULT, I. A. BONEV, *A fully-decoupled 3-dof translational parallel mechanism* (Parallel Kinematic Machines in Research and Practice, The 4th Chemnitz Parallel Kinematics Seminar, pp.595-610, 2004).

64.X.-J. LIU, J. WANG, *Some new parallel mechanisms containing the planar four-bar parallelogram* (Int. J. of Robotics Research, 22/9: pp 717-732, 2003).

65.G. GOGU, *Evolutionary morphology: a structured approach to inventive engineering design* (Invited paper, Proceedings of the 5th International Conference on Integrated Design and Manufacturing in mechanical Engineering, Bath, 5-7 April 2004).

66.K. Y. TSAI, K. D. HUANG, *The design of isotropic 6-DOF parallel manipulators using isotropy generators* (Mechanism and Machine Theory, Vol. 38, pp. 1199-1214, 2003).

67.N. P. SUH, *The Principles of Design* (Oxford University Press, New York, 1990).

68.F. DUDITA, D. V. DIACONESCU, G. GOGU *Mecanisme articulate : inventica si cinematica in abordare filogenetica* (Ed. Tehnica, Bucuresti, ISBN 973-31-0119-2, 1989).

69.G. GOGU, P. COIFFET, A. BARRACO, *Mathématiques pour la robotique : Représentation des déplacements des robots* (Editions Hermès, Paris, 1997).

DESIGN OF NEW HIGH SPEED MACHINING MACHINES

P. Ray
Laboratory of Mechanics and Engineering, French Institute of Advanced Mechanics, France

Abstract: In the manufacturing industry, machine tools play a very important role. Many companies are searching for these machines in order to increase their competitiveness. They should be flexible, highly accurate, productive, ecological and cost efficient. At present, the traditional machine tool cannot fulfil such demands as they achieve either high flexibility or high accuracy. However, since the 90's, Parallel Kinematics Machines (PKM) appear in real production. These machines have many advantages over the conventional machine tools and serial kinematics robots, such as high flexibility, high stiffness, and high accuracy. PKM development is considered as a key technology for future robot applications in manufacturing industries. In this paper, we present various kinematics of PKM with their advantages and their disadvantages. Then, to achieve better accuracy and dynamic performance for both light machining and heavy machining and in order to reconfigure the machine quickly for task changes the static and dynamic performance must be better understood. So we study accuracy and dynamic performance and the impact during machining. We establish a model of static and dynamic properties, which can provide the tool for mechanical designers and the controller designers to evaluate the machine. This research work is related to industrial applications. What is considered in this work is the problem as encountered by the machine end user. The results obtained will benefit the machine end user in industry. The study of PKM in the production environment is a new topic in the PKM research area.

Key words: high-speed machining, parallel kinematics machines, dynamics, rigidity.

1.   INTRODUCTION

During the last ten years, the constraints of the automobile market have evolved. Today, the industrialization of an engine must take into account unguaranteed volumes of production, productivity requirements and rising quality standards, an increase in the number of versions of a product and rising environmental standards (pollution, safety, noise…).

D. Talabă and T. Roche (eds.), Product Engineering, 379–396.
© 2004 *Springer. Printed in the Netherlands.*

In the case of motorization, the market is very sensitive and the pollution-related norms have evolved every three years. Given these constraints, two types of industrialization exist, one for relatively fixed parts with high production (driving Block) on special and dedicated machines (transfer machine) and the other for evolutionary parts (cylinder heads) with lines containing standard machine tools (High Speed Machine tools).

In this article, we present HSM machine tools with a serial structure and their advantages and disadvantages and then we present the concept of machines with parallel architecture and we study how they can be integrated in the factory.

2. INDUSTRIALIZATION OF THE PARTS WITH STANDARD MACHINE TOOLS

2.1 High speed machining

To guarantee flexibility for the realization of cylinder heads, it is necessary to have a spindle embarked on a three axes module. In fact the objective is to increase production. These machines have existed for a long time but their low productivity did not allow high scale competitive industrialization in great production [1].

The design of HSM machine tools allowed for flexibility with significant productivity due to:
a) High speed machining with a spindle rotation of 15000 rpm to 40000 rpm.
b) Fast displacement and strong acceleration to minimize the dwell time machining.
c) A design dedicated to the high production to achieve a high level of reliability in spite of high sollicitations.

Table 1 shows the variation of requests in machines requests between a machine tool used in general mechanics and one in large-scale production. PCI has developed a range of HSM machine tools for this type of production. The machine tools have a structure "box in box" structure with kinematics chains optimized to have an excellent compromise between reliability and performance (Table 2).

Table 1. Annual sollicitations for the HSM machine tools (Figure 1)

Annual requests	HSM Machine tool	Conventional machine
Nb of tool changes per annum	1 000 000	50 000
Nb of rotations of axis B	250 000	25 000
Nb of parts produced	50 000	5 000

Table 2. Design features of the Meteor machine tool

	Meteor 5	Meteor 15
Spindle velocity	15000 rpm (24000)	same
Spindle power	25 KW (40 kW)	same
accelerations X, Y, Z	60 m/min	same
	γ= 6, 6, 12 m/s²	
workspace X, Y, Z (mm)	500, 630, 550	800, 800, 550
Length	1600 mm	1900 mm
Tool change duration	3,7 s	same

Figure 1. Machine tool "Meteor 5"

3. ANALYZING CONSTRAINTS OF THE AVERAGE FLEXIBLE DEVICES IN MASS PRODUCTION

3.1 Fundamental differences in design between the transfer lines and flexible machine tools

The most significant difference between the transfer lines and the flexible lines is that, in the first case, it consists of stations with simple machining in series and in the second of HSM machine tools in parallel [2, 3]. Table 3 makes it possible to compare the fundamental difference between the two types of industrialization. In the case of a flexible line, the number of machine tools is lower but on the other hand, they are sophisticated and more complex than the transfer machines. This creates lots of pressure on workshop logistics to use these machine tools.

Table 3. Comparative between line transfer and flexible line

1250 cylinder heads by day	Nb of machines of production	Unit cycle time
Transfer line	140 elementary machine tools with many sinpdles	1min
Flexible line	35 HSM four or five axis machine tools and one spindle	4 min

Today, the transfer lines provide increased quality control:
- strong insensitivity to variations in the environment (temperature, etc...) or parts (hardness, extra thickness, etc...).
- the means are well known over time.
- management of the controlled tools and few dynamic problems (balancing, frequency...) due to the spindle velocity is generally lower than 10000 rpm.
- immediate tracability of the parts due to the unity of the machining operations. If a hole is missing in a part, the problem station is immediately known, and intervention is rapid.
- production and maintenance engineers know very well this type of machines and the segmentation using simple machining stations simplifies diagnosis and improves maintainability.

3.2   Analyzing operation of the flexible lines

The use of a HSM machine tool is necessary on moving parts such as cylinder heads. On the other hand this means that production should be re-evaluated to guarantee high quality parts without adversely affecting the running costs or the production line [4].

3.2.1   Insensitivity to the means of variations in the environment

A machine tool is never in a stabilized cycle. In four minutes it can change tools 20 times with time loss for part or tool changes. This transitory mode generates possible defects which must be integrated:
a) Thermal deformations of the machines according to the room temperature in the factory (from 10 to 40°C). These possible drifts impose a rigorous regulation of the cutting fluid and its temperature change, the system of compensation or automatic measurement must be developed to readjust the machines;
b) Deformation of the parts during machining because if the part is at temperature, T1, and if the liquid is at temperature, T2, the few minutes of machining are sufficient to engender some dilatation of the part;
c) A fundamental point is the tool change. The tool change system must be able to guarantee that no chip slips between the attachment of the tool and the spindle because that can create a displacement of the tool.

3.2.2   Time reduction of flexible devices

The preventive maintenance of the HSM machine tool is heavier today than on conventional installations because:

- The HSM machine tools are light with high velocity, so some components wear out more quickly than a conventional machine. Moreover complexity of the trajectories can cause some damage to the components [5].
- The same machine carries out several different processes and whereas a run out of the spindle or a geometric defect is acceptable when tapping a hole, they no longer are when boring a hole.

To take into account these constraints, it is imperative to use robust solutions in mass production. The system must work even if the input data may vary by 20%. This reliability and quality aspect makes it possible to understand why HSM machine tools require specific design rules.

The principal rule is to have safety coefficients included in the integration of the processes or components. In all cases it is necessary to develop expertise which makes it possible to follow the general state of the machine tool.

3.2.3 Tool Management

The tools are fundamental to the machining quality chain:
1. Standard HSK does not impose any constraint on the cones of the tool holders. In mass production, a tool holder is used every three minutes with a clamping force of 2000 daN. It is imperative to use high-strength steels;
2. Balancing of the tools. It is necessary for the tools to be designed in such a manner that a change of plate or a drill does not deteriorate the original balancing;
3. Several replicas of the same tool exist with different lengths. The reliability of the information transfer between the machine tools is essential for part quality. An automatic process is recommended to avoid input errors in industrial facilities with hundreds of tools.

3.2.4 Parts traceability

With the merged strategy of a few machine tools, the same machining can be carried out on machines 1, 2 to N. The number of trajectory of the parts in the line is very significant, so it is necessary, when there are no independent controls, to have a software enabling final control of the line to know the trajectory of the part and to know which machine requires intervention.

3.2.5   Production and maintenance engineer

This point is very significant for the success of a project using HSM. The problems of the maintenance technicians are very different compared to a traditional workshop. Indeed a machine tool integrates a huge number of functions and the technical level required to make an evaluation needs competences in mechanics, hydraulics and electrical engineering.

If we make a parallel with aviation, we quickly realize that if a problem is too general for one man's memory functions, it is necessary to develop software to help in decision-making. That goes from a simple "check list" of the possible causes for a defect which one lists in the event of breakdown to using software to establish priorities for intervention of maintenance to optimize the quality or the productivity of the machine tool. This technology requires a higher level training for the operators.

The HSM machine tool allows flexibility of the production tool. Today with the constraints of the market on automobile products, this flexibility is imperative for the industrialization of various engine components and in particular the cylinder head. Equipment suppliers are also attracted towards this technology which makes it possible not to dedicate a production tool. In what follows, we present more particularly the modeling of the dynamic behavior of the machine tool Météor 5 and show its limits before presenting the technology of the machine tool with parallel kinematics [6].

4.   DETERMINATION OF THE RIGIDITY MAPPING FOR A HSM MACHINE TOOL

4.1   Introduction

In HSM machine tools, it is essential to ascertain static and dynamic rigidities, in particular to control the cutting parameters [7]. It is thus important to check in the design phase that the rigidity performances will be satisfied. The consistency check can be made by using a finite element model. The computing time to obtain a rigidity calculation being 6 minutes on a PC, it would take 46 hours on a CPU to obtain a suitable cartography. The originality in this work lies in the fact that a numerical experimental design is established to obtain in less than one hour of calculation, the rigidity cartography.

This work concerned a 3-axis HSM machine tool occupying a workspace of 500x500x550 mm^3 designed by our industrial partner who defined the design features of the machine tool.

4.2 Static rigidity of the machine tool obtained through numerical experimental design

The digital model of the machine tool is constructed with Ansys software. The structure is modelled classically using shell elements and we use global modeling for the connecting elements given by the manufacturers (Figure 2).

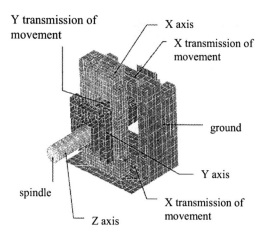

Figure 2. Finite element model of the machine tool

We wish to work out a model which will enable (starting from selected combinations), estimation of the answers for other combinations of the entry parameters. As no mathematical model is available to describe the behaviour of the physical system, the experimental designs offer an interesting approach. For the machine tool, we use an experimental design with three factors and on two levels and the three factors are the coordinates of the spindle in the workspace. They are the parameters that can be controlled and measured on the real machine tool.

In this case, the answer can be modelled by the following relations:

$$\hat{R} = \overline{R} + E_X X' + E_Y Y' + E_Z Z' + I_{XY} X' Y' + I_{XZ} X' Z' + I_{YZ} Y' Z', \quad (1)$$

in which, $\hat{R}$ is the estimated result; $\overline{R}$ is the average of all the answers; E_i is the effect of factor i on level +1; I_{ij} is the interaction of factors i and j on level +1; X', Y' and Z' are centered factors equal to +1 when the factor is tends towards the highest value and equal to −1 when it is at its lowest.

The centered factors are calculated with the centered formula:

$$X' = 2 \frac{X - \dfrac{X_{min} + X_{max}}{2}}{X_{max} - X_{min}}.$$ (2)

For each explored direction, an experimental design of this type is built. This leads us to the development of three models of rigidities for the machine tool each one corresponding to a machining direction. We use as factors the positions of the spindle within the workspace (X, Y and Z) and we vary from their minimal value to their maximum value. The plan is thus built starting from the computed values at the top of the workspace. The parameters selected (X, Y and Z) are those which are easily controllable in experiments. In addition, we use the notation of Yates who define that for each factor, the bottom grade is affected a value −1 and the high level a value +1. The experimental design is thus built according to the diagram in table 4. The rigidity values of the machine tool calculated at the points of the experimental design are given in table 5.

Table 4. Experimental design used for the modelling of the machine tool

	X	Y	Z
1	-1	-1	-1
2	-1	-1	+1
3	-1	+1	-1
1	2	3	4
4	-1	+1	+1
5	+1	-1	-1
6	+1	-1	+1
7	+1	+1	-1
8	+1	+1	+1

Table 5. Static rigidities calculated at the points of the experimental design

X (mm)	Y (mm)	Z (mm)	RX (daN/µm)	R_Y (daN/µm)	R_Z (daN/µm)
250	250	275	2,89	0,84	11,34
250	-250	275	3,09	0,86	12,08
-250	-250	275	3,18	0,86	12,29
-250	250	275	2,94	0,85	11,43
250	250	-275	7,68	2,42	12,54
250	-250	-275	9,28	2,42	13,40
-250	-250	-275	9,90	2,42	13,67
-250	250	-275	8,06	2,43	12,65

These values establish models of rigidities in the three directions according to positions X, Y and Z and their two degrees interactions. Table 6 contains the coefficients of the three built models.

Table 6. Averages, effects and interactions of the calculated models

Axe	R (daN/µm)	E_X	E_Y	E_Z	I_{XY}	I_{XZ}	I_{YZ}
X	5,87	-0,14	-0,48	-2,85	0,04	0,11	0,38
Y	1,64	0	0	-0,79	0	0	0,04
Z	12,43	-0,09	-0,44	-0,64	0,04	0	0,04

These three models build an approximate numerical rigidity chart for any workspace layout (Figure 3, 4). In each of the three rigidity models of the relations given below, the Z effect is the most influential of the effects retained for the models.

$$\begin{cases} \hat{R}_X = 5,87 - 0,48\ Y' - 2,85\ Z' + 0,38Y'Z' \\ \hat{R}_Y = 1,64 - 0,79\ Z' \\ \hat{R}_Z = 12,43 - 0,09\ X' - 0,44\ Y' - 0,64Z' \end{cases} \qquad (3)$$

These experimental designs can be estimated with a satisfactory degree of precision and a very low computing time compared to the calculations carried out by discretization of the spindle positions. They highlight the influence of the Z coordinate on the rigidity values. It is possible to make a study of the influence for the principal structural elements of the machine tool on the total rigidity and to deduce some rules from them for the design of the future machine tool or some means of improvement for the existing one [8].

From the work on these HSM machines, we decided to build a new one based on a parallel kinematics machine (PKM). The performances of the latter must be still higher than existing ones (speed 120m/min and acceleration 30 m/s^2) and for this PKM, we want to apply the same process for the rigidity study.

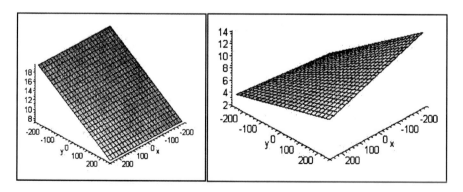

Figure 3. Numerical rigidity charts in directions X and Y

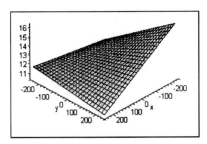

Figure 4. Numerical rigidity charts in direction Z

5.   DESIGN AND VALIDATION OF NEW PKM

5.1   Introduction

During the last few years, significant developments have occured in the field of HSM concerning cutting technology (increased spindle velocity) and machines structures [9, 10, 11]. The use of parallel kinematics machine for machine tools is relatively recent and is controlled only for some processes today. The goal is to reduce the moving masses and to increase accelerations and velocity. After a library search on the subject, we decided to lead a method of design for this type of machine with our partner. The final objective was to design a low cost machine tool with increased performance by integrating all the capitalized results.

5.2   Modelling of a new parallel kinematics machine

The machine selected has 3 degrees of mobility (figure 5). It is a hybrid kinematic architecture: Z axis is embarked on a parallel mechanism with 2 degrees of mobility. This mechanism ensures the movements in the plane (X,Y). This solution presents a reduced number of kinematic elements and connections. We have 5 elements and 6 connections [12, 13]. The whole of the elements is formed by 2 boxes, 2 arms, which carry the sliding Z axis. The two boxes are connected by a pivot to the frame. The 2 arms are bound by a rotation link and each one has a slide link with a box. The three joint slides are motorized by linear motors. The solution retained is protected by the French patent N° 99 13920.

We begin this new study from the experience previously gained, the development of the finite element model (figure 5). During the analysis of a complex structure, two aspects are to be considered. The first was created by finding a compromise between the precision and the complexity of the model. The second relates to the implementation of the method. Indeed, it is not possible to especially discretize by finite elements the components of joints when they contain rolling elements. However, the effect of their rigidity is essential in our analysis. The analyzed structure is a mechanism; so it is necessary to configure the relative positions of the elements and to adapt the grid according to the tool's position in the workspace. The analysis of the structure was carried out on Ansys software and the frame is not taken into consideration.

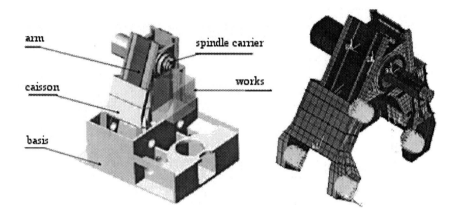

Figure 5. CAD Model and finite elements of the PKM

5.3 Rigidity analysis

We analyzed static rigidity at 27 points in the workspace. This enables us to obtain the space charts of rigidity using functions of approximation. Figures 6 and 7 represent the rigidity charts in the plane $z = 0$, with the origin of the reference mark as the centre of the workspace. For reasons of confidentiality, the values presented are adimensional. They are calculated compared to a reference rigidity $R_{réf}$. It is important to note that rigidity according to direction X is overall lower because of the railways position of the sliding Z axis. These results led us to propose the modification of this position for a better distribution of rigidity.

The results obtained enable us to better understand the behavior of this type of machine and to make the topological modifications necessary to the various elements. For this machine tool architecture, we can conclude that the rigidity map of the whole workspace evolved more than for a serial structure. So, our industrial partner has chosen an other concept for carrying out a new machine tool because it requires 5 or even 6 axis for machining.

Many PKM systems have been developed for this purpose. Theoretically, the PKM can be used to replace conventional machines. However many types of PKM can only perform very limited tasks due to the very limited workspace or high cost [14, 15].

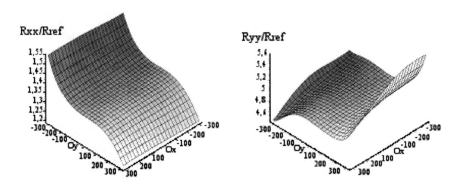

Figure 6. Charts of rigidity (z=0), $R_{xx} / R_{réf}$ and $R_{yy} / R_{réf}$

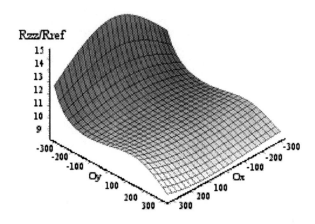

Figure 7. Charts of rigidity (z=0), $R_{zz} / R_{réf}$

5.4   Presentation of the machine based on a tricept robot

The HSM based on a tricept robot created by SMT Tricept AB is considered the most successful PKM application in manufacturing industry. The competitive edge for the tricept against other PKM's and serial machines are:
a) High dynamics for high machining productivity.
b) Full 5 axis (even 6 axis) machining capability for high flexibility.
c) Large workspace, 2400x2400x800 in XYZ, C(-360, +360) A(-10, +180).
d) Good static stiffness and high dynamic stiffness due to the Direct Measuring System technology.

e) Very good price/performance ratio.
f) A wide range of applications.
g) OEM-modules to build their own machine products.

The tricept robot is composed of a 3-DOF parallel structure and a 2-DOF spherical wrist. This machine architecture is a hybrid structure (Figure 8).

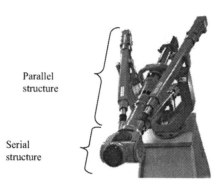

Parallel
structure

Serial
structure

Figure 8. Hybrid mechanism tricept robot

In order to evaluate the precision of the effector for a given acceleration and without the wrist, it is necessary to have prior knowledge of its static rigidity according to its position. The displacement of the mobile platform is studied according to the forces on the effector. Moreover, a hypothesis is made that only the legs can be deformed, thus one does not take into account the rigidity of the various motors and joints.

5.4.1   Modelisation of the legs

In a first approximation, the legs are modeled like an elastic solid composed of two full cylinders with various sections. The length of the sleeve is fixed, it is measured between the centre of the universal joint and the end of the sleeve. The leg's length is a function of the position of the platform centre. It is measured between the end of the sleeve and the center of the ball joint.

5.4.2   Determination of the rigidity in the workspace

When a force is used on the platform, each leg i is solicited in traction – compression mode (Figure 9).

This means that the displacement of the platform centre can be calculated during a drilling operation. This process is carried out along the axis $\vec{z}_f$ by the tricept robot. For the sake of simplification, the centre of the mobile platform is displaced in the same direction.

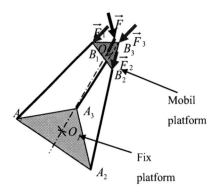

Figure 9. Forces on the platform

From this calculation, it is possible to use the same method as previous to determine the charts of rigidity for the tricept robot over the whole workspace. First, we study its rigidity in one plane for a given height (Figure 10). The results are presented in Figure 11.

The results show that during a drilling operation, the rigidity is very high in the fixed platform direction. The evolution throughout the whole workspace is more different than in the other two machine tools. We can see that the position of the part has great importance according to the process. In future work, we will study this influence with the dimensional precision in milling to optimize the dynamical behaviour.

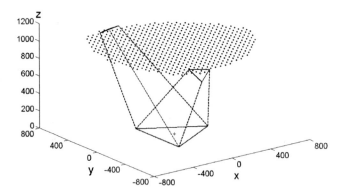

Figure 10. Points for rigidity determination in the workspace at z=1000

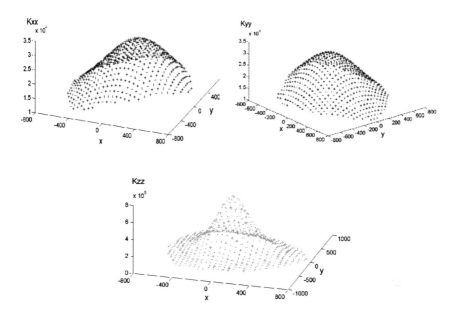

Figure 11. Charts of rigidity in the three directions

Other parameters must be taken into account in order to optimize the pocketing process, for example the minimization of the machining time.

6. TOOL PATH OPTIMIZATION FOR HIGH SPEED MACHINING

The pocketing form is obtained by the movement of a tool which must meet certain constraints such as mechanical ones during cutting efforts. To calculate a pocketing tool path, an algorithm based on the Voronoi diagram make its possible to calculate a helical tool path from the inside to the pocket outside. But the result presents a set of discontinuities in tangency located at the radial tool path linking and at the connections between elementary tool paths.

So, to reduce the machining time, it is possible either to decrease the length of the tool path, or to increase the instantaneous feed rate of the tool [16, 17, 18]. Each machining operation generates trajectories with a significant number of discontinuities in tangency or curvature, which disturb the behaviour of the machine tool during the trajectory follow up. In fact, each machining direction change during machining involves:

a) A reduction in the real feed rate of the tool in order to respect the contour error;

b) Solicitation of the machine tool structure during decelerations and accelerations;

c) Variation in the cutting conditions due to variations in the feed rate;
d) An increase in the machining time due to the reduction in the feed rate.

The purpose of the pocket machining optimization is to find the trajectory which minimizes the machining time, among all the possible trajectories which respect the various geometrical and mechanical constraints. A parallel study of the instantaneous feed rate and the radial depth of cut of the tool at a corner clearing is proposed.

In order to validate this method, we compare a tool path calculated using our method to tool paths generated by a Computer Aided Manufacturing system (Figure 12). We test these tool paths on the HSM machine tool "Meteor 5".

The gain is equal to 10%, using linear/circular interpolation format. The polynomial interpolation format allows a gain of about 15% (Table 7).

The proposed method brings down machining times even if the percentage of the tool path machined at the programmed feed rate is lower. We are today able to generate a tool path where the feed rate is equal to the programmed one or to the maximized feed rate over 73% of the tool path. All the experiments show that the benefits generated by our method are about 15% to 25% in terms of machining time depending on the pocket geometry whatever the machine tool used.

Table 7. Experimental machining time and average feed rate

	Path length (mm)	Machining time (s)			Average feed rate (m/min)		
		G1	G1G2G3	Bpline	G1	G1G2G3	Bspline
CAM System	2831	28,5	25,9	/	5,9	6,5	/
Our method	2793	26	23,2	25,4	6,5	7,2	6,6

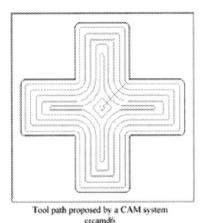

Tool path proposed by a CAM system
crcamd6

Tool path adapted with our method
cradpd6

Figure 12. Tool paths obtained by a CAM system and our method

7. CONCLUSIONS

In this paper, we have presented several architectures of HSM machine tools and we have seen that for a serial machine tool, the evolution of the rigidity in the whole workspace is more intuitive than for a PKM machine tool. Actually, it is very difficult to manufacture parts with the second one because we must possess more information about them and it's important to know how to clamp the part in order to respect the dimensional, form deviation and surface roughness specifications according to the rigidity.

The PKM machine tools are very important for industry because their use leads to greater flexibility in production. Then, for productivity requirements, we can impose a minimization of machining time for pocket machining. The method is based on the computation of corner radii according to the kinematic behaviour of the machine tool, and the radial depth of cut variation.

Another path of investigation is lubrication during machining. Today, machine tools must use less cutting fluid to avoid pollution problems. Some works are being conducted on minimal lubrication which makes it possible to have a small quantity of lubricant directly directed at the contact point between the tool and the part. The goal is to obtain a more ecological process which leads to lower waste and less residues.

This research work is related to a direct industrial application and the result will benefit the machine end user in industry. The study of PKM in the production environment is a new topic in the PKM research area and there remains a lot of work to do in many related subjects both in mechanics but also in mechatronics.

REFERENCES

1. O. VIDAL, *Démarche de conception d'un centre d'usinage grande vitesse pour la production de pièces grande série*. (3èmes assises Machines et Usinage Grande Vitesse, mars 2004, Clermont Ferrand, pp. 9-20).
2. F. JOVANE, *Present and future of flexible automation*. (The 4th Chemnitz Parallel Kinematics Seminar, PKS 2004 proceedings vol. 24, Planery paper, pp. 1-20).
3. F. JOVANE, *Turning manufacture into manufuture*. (opening session, CIRP 53rd, General Assembly, Montreal, Canada, 2003).
4. N. HENNES, D. STAIMER, *Application of PKM in aerospace manufacturing-high performance machining centers ECOSPEED, ECOSPEED-F and ECOLINER*. (The 4th Chemnitz Parallel Kinematics Seminar, PKS 2004 proceedings vol. 24, pp.557-567).
5. R. NEUGEBAUER, *Structure design an optimization of parallel kinematics*. (The 4th Chemnitz Parallel Kinematics Seminar, PKS 2004 proceedings vol. 24, pp. 15-45).
6. J. TLUSTY, *Fundamental comparison of the use of serial and parallel kinematics for machines tools*. (annals of the CIRP, vol. 48, 1999, pp. 351-356).

7. R. LAUROZ, P. RAY, G. GOGU, *From classical machining to high speed machining :
    comparative analysis.* (Proceedings of the International Seminar on Improving Machine
    Tool Performance, San Sebastian, Spain, July 1998, pp. 105-116).
8. R. LAUROZ, P. RAY, G. GOGU, O. VIDAL, *Approche de conception d'une machine de
    travail et d'usinage à grande vitesse.* (IDMME'98, 2ème Conférence Internationale sur la
    Conception et la Fabrication Intégrées en Mécanique, Compiègne, France, mai 1998, pp.
    877-884).
9. G. PRITSCHOW, *Parallel kinematic machines (PKM) – limitations and new solutions.*
    (annals of the CIRP, vol. 49, 2000, pp. 275-280).
10. J. ASSARSSON, *Simulation and analysis of parallel kinematic machines.* (Thesis, Lund
    University, 2001).
11. G. PRITSCHOW, C. EPPLER, T. GARBER, *Influence of the dynamic stiffness on the
    accuracy of PKM.* (3rd Chemnitz Parallel Kinematic Seminar, Chemnitz, mai 2002, pp.
    313-333).
12. B. C. BOUZGARROU, P. RAY, G. GOGU, C. BARRA, O. VIDAL, *Static and dynamic
    characterization of new parallel high speed machine tool.* (Proceedings of the II
    International Seminar on Improving Machine Tool Performance, La Baule, France, 3rd –
    5th July 2000, actes sur CD Rom, 12 pages).
13. B. C. BOUZGARROU, P. RAY, G. GOGU, *Design and optimization of new high speed
    machine tool with hybrid kinematics.* (3ème conférence internationale sur la coupe des
    métaux et l'usinage grande vitesse, Metz, Juin 2001, 5 pages).
14. F. CACCAVALE, G. RUGGIERO, B. SICILIANO, L. VILLANI, *On the dynamics of a
    class of parallel robots.* (Advances in robot kinematics, Kluwer, 2000, pp187-196).
15. J. CHEN, W. HSU, *Design and analysis of a tripod machine tool with an integrated
    cartesain guiding and metrology mechanism.* (Precision engineering, Elsevier, 2004, 24
    pages).
16. V. PATELOUP, E. DUC, P. RAY, *Corner optimisation for pocket machining.*
    (International journal of Machine tools & Manufacture, 2004, 12 pages).
17. V. PATELOUP, E. DUC, P. RAY, *Corner optimization for pocket machining.* (IDMME
    2004, 5ème Conférence Internationale sur la Conception et la Fabrication Intégrées en
    Mécanique, Université de Bath, 10 pages).
18. M. TERRIER, A. DUGAS, J. Y. HASCOET, *Parallel kinematics machine tools and high
    speed milling.* (IDMME 2004, 5ème Conférence Internationale sur la Conception et la
    Fabrication Intégrées en Mécanique, Université de Bath, 10 pages).
19. G. AMILIEN, *La microlubrification en usinage.* (3èmes assises Machines et Usinage
    Grande Vitesse, mars 2004, Clermont Ferrand, pp. 75-79).

HIGH DEGREE ACCURACY MODELLING AND CALIBRATION OF SERIAL ROBOTS WITH LARGE ERRORS

M. Neagoe[1], G. Gogu[2] and D. Diaconescu[1]
[1]*Transilvania University of Brasov, Romania;* [2]*Laboratory of Mechanics and Engineering, University Blaise Pascal and French Institute of Advanced Mechanics, France*

Abstract: The paper deals with the improvement of accuracy modelling and calibration process, in the case of serial robots with large errors. Beginning from the first degree kinematic modelling, a general third-degree modelling of errors in serial robot chains is generated and applied for a RRR Puma type serial robot-structure. The model is based on the use of error Jacobian matrices, namely first, second and third-degree error Jacobian, respectively. Through the high-degree error model, the end-effector positioning can be more accurately estimated. Furthermore, a general calibration algorithm based on the third degree error model is presented, with application for the same RRR serial manipulator. Finally, a numerical example of the proposed algorithms is presented and conclusions.

Key words: serial robots, accuracy modelling, high degree error models, high degree kinematic calibration.

1. INTRODUCTION

A robot is a complex system, on which many error factors act; therefore it has an actual behaviour that deviates from the desired (commanded) one, established on the basis of the *nominal robot model*. Robot accuracy can be enhanced, without structural or constructive improvements, through *calibration*, which allows the use in the command process of an *actual (correct) model*, more accurately related to the real robot. Calibration attempts to identify the most important error factors, to model theirs influences on the robot accuracy and to obtain the actual (real, correct) values of the modelling parameters by analysing the experimental data from

397

D. Talabă and T. Roche (eds.), Product Engineering, 397–408.

accuracy testing. Thus, the nominal models can be optimised and the actual models (used in command/control process) ensure the increase of robot accuracy. General considerations about robot calibration are published by many authors e.g. Mooring [5], Hollerbach [4] and Bernhardt [1].

The *accuracy models*, which play a central role in the *calibration process*, establish *the end-effector errors* depending on the source of the errors. Several methods are used in precision analysis [1–8]; most of them are based on the homogenous operators and Denavit-Hartenberg convention, in order to model the *infinitesimal* errors. With the aim to increase the efficiency of the precision models in the case of large errors (*finite errors*), the authors developed a new approach about the *high degree* modelling of serial robot precision and its application in robot calibration.

In this paper, the authors present a general method used in high degree error modelling of serial robots (section 2), applied to a RRR Puma type robot structure (section 3); based on the obtained error models, an algorithm used in high degree calibration is developed in section 4. Finally, the identification algorithm efficiency is outlined based on a numerical example and the final conclusions are presented (section 5).

2.   ON THE HIGH DEGREE MODELLING OF ROBOT ERRORS

The robot positioning – orientation errors $\Delta \mathbf{X}^1$ due to $\Delta \mathbf{q}$ deviations are modelled in literature [1–5] based on the robot Jacobian:

$$\Delta \mathbf{X}_e = \mathbf{J}_e \cdot \Delta \mathbf{q}, \tag{1}$$

where $\Delta \mathbf{X}_e = [d_{ex}\ d_{ey}\ d_{ez}\ \delta_{ex}\ \delta_{ey}\ \delta_{ez}]^T$, $\mathbf{J}_e$ – the robot Jacobian, $\Delta \mathbf{q}$ – the vector of variable errors, e – the frame in which the Jacobian $\mathbf{J}_e$ and the error vector $\Delta \mathbf{X}_e$ are reduced[2].

This method allows obtaining a simple error model (*linear model*); its disadvantage is the modelling inaccuracies induced by the use of only first degree terms from power series of *sin* and *cos* functions. Hence, error modelling can be improved introducing the *high degree* terms. The error-Jacobians[3] used in *high degree* modelling can be obtained using an algorithmic process based on the homogenous operators [6, 7].

---

[1]Through **aldine** symbols vectors or matrixes are symbolized and through *italic* symbols - the scalars.

[2]By **reducing** the infinitesimal displacements vectors and the corresponding Jacobian in a known frame, we understand: a) *reducing* the infinitesimal vector in the origin of the known frame; b) *expressing* the new obtained vector in the frame considered

According to the relation (1), a robot can be modelled as a linear transformation of errors from *variable (generalised) space* to *operational space* (Figure 1).

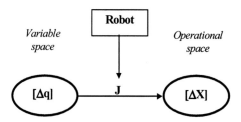

Figure 1. First degree error–model

An *m*–degree error model of a robot with *n* degrees-of-freedom takes into account all the deviations up to those with the maximum degree *m*. Two categories of deviations, and implicitly of Jacobians, interfere in such a model:

a) *k*–degree *decoupled deviations*, $1 \le k \le m$, which are described by the vector of the deviations at power *k*: $[\,\Delta q_1^k \ \Delta q_2^k \ \cdots \ q_n^k\,]^T$. The influence model of these deviations is made through *k–degree decoupled error Jacobians*, further marked with $\mathbf{J}^k$.

b) *k*–degree *coupled deviations*, $2 \le k \le m$, which model the coupling effects of two or more source-errors through some specific Jacobian's, further named *coupled error Jacobian's*. In the second–degree error model the influences of all coupled deviations as $\Delta q_j \cdot \Delta q_k$, with $j \ne k$ are modelled. The *m*–degree error model extends the $(m-1)$ degree model by taking into account all coupled terms as $\Delta q_{t_1}^{x_1} \cdot \Delta q_{t_2}^{x_2} \cdots q_{t_m}^{x_m}$, in which the subscripts t_j are different and $\sum x_i = m$. The *m*–degree error model is based on the diagram from Figure 2 and uses all error Jacobians up to the *m*–degree, with at least 2 non-zero exponents x_i. In this case, the operational errors are:

$$[\Delta \mathbf{X}_e] = [\mathbf{J}_e^{\mathrm{I}}] \cdot [\Delta \mathbf{q}] - \frac{1}{2}[\mathbf{J}_e^{\mathrm{II}}] \cdot [\Delta \mathbf{q}^2] - \frac{1}{6}[\mathbf{J}_e^{\mathrm{III}}] \cdot [\Delta \mathbf{q}^3] + \ldots \pm \frac{1}{m!}[\mathbf{J}_e^m] \cdot [\Delta \mathbf{q}^m] -$$

$$- \frac{1}{2}[^c\mathbf{J}_e^{\mathrm{II}}] \cdot [\Delta \mathbf{q}_{t_1} \Delta \mathbf{q}_{t_2}] - \frac{1}{6}[^c\mathbf{J}_e^{\mathrm{III}}] \cdot [\Delta \mathbf{q}_{t_1}^{x_1} \Delta \mathbf{q}_{t_2}^{x_2} \Delta \mathbf{q}_{t_3}^{x_3}] + \ldots \pm \frac{1}{m!}[^c\mathbf{J}_e^m] \cdot [\Delta \mathbf{q}_{t_1}^{x_1} \ldots \Delta \mathbf{q}_{t_m}^{x_m}],$$

$$(2)$$

[3]In high degree error–modelling we use some Jacobian–type matrix, namely *first degree error-Jacobian, second degree error-Jacobian,* and respectively *third degree error-Jacobian.* In the case of joint errors, the first degree error–Jacobian is the robot Jacobian used in velocities modelling.

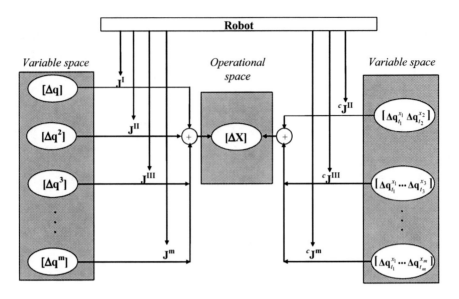

Figure 2. High degree error–model

In the relation (2) J_e^m and $^cJ_e^m$ represent *the decoupled* and respectively *coupled m–degree error Jacobian.*

The algorithms used to describe the two categories of error Jacobians (*decoupled* and *coupled*), for $m = 3$, will be further presented below.

3.   THIRD DEGREE MODELLING OF SERIAL CHAIN ACCURACY

In order to illustrate the algorithm application in modelling the third-degree errors, a Puma type manipulator chain is considered with three degrees of freedom ($R_z \perp R_x \parallel R_x$ – Figure 3), which is characterized by the following parameters: $a_1 = 1500$ mm, $a_2 = 150$ mm, $b_2 = 1000$ mm, $a_3 = 150$ mm, and $b_3 = 1000$ mm. For simplicity and a better understanding, the following assumptions are considered: a) the *source–errors* are only the *joint–errors* and b) only the end-effector's positioning errors are measured.

In order to establish the manipulator deviations, the following steps are covered:

1. The frame "e" in which the deviations are described and, respectively, the error Jacobians are established. In this example it is considered that $\mathfrak{R}_e = \mathfrak{R}_3 = \mathfrak{R}_P$ (Figure 3, b).
2. *The modelling of the finite displacements (the geometrical model) of the* manipulator by using the method TCS [2, 3] (considering $\mathfrak{R}_e = \mathfrak{R}_P$):

$$\mathbf{A}_{01} = \mathbf{R}^z(q_1) \cdot \mathbf{T}^z(a_1), \tag{3}$$

$$\mathbf{A}_{12} = \mathbf{R}^x(q_2) \cdot \mathbf{T}^x(a_2) \cdot \mathbf{T}^y(b_2), \tag{4}$$

$$\mathbf{A}_{23} = \mathbf{R}^x(q_3) \cdot \mathbf{T}^x(a_3) \cdot \mathbf{T}^y(b_3). \tag{5}$$

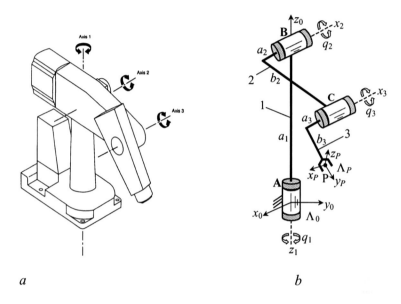

a	b

Figure 3. The RRR Puma type serial manipulator (a) and its parameterisation (b).

3. *Establishment of the first, second and third degree decoupled error Jacobians.* Applying the relations presented in [6, 7], the expressions (6)–(8) of the error Jacobians $\mathbf{J}^I$, $\mathbf{J}^{II}$ and $\mathbf{J}^{III}$ are obtained:

$$\mathbf{J}^I = \begin{bmatrix} -b_2 cq_2 - b_3 c(q_2+q_3) & 0 & 0 \\ (a_2+a_3)c(q_2+q_3) & b_2 sq_3 & 0 \\ -(a_2+a_3)s(q_2+q_3) & b_3+b_2 cq_3 & b_3 \end{bmatrix}; \tag{6}$$

$$\mathbf{J}^{II} = \begin{bmatrix} a_2+a_3 & 0 & 0 \\ b_2 cq_2 c(q_2+q_3)+2b_3 cq_2 cq_3 c(q_2+q_3)+b_3(sq_2^2-cq_3^2) & b_3+b_2 cq_3 & b_3 \\ -b_2 cq_2 s(q_2+q_3)-2b_3 cq_2 cq_3 s(q_2+q_3)+b_3(sq_2 cq_2+sq_3 cq_3) & -b_2 sq_3 & 0 \end{bmatrix}; \tag{7}$$

$$\mathbf{J}^{III} = \begin{bmatrix} -b_2 cq_2 - b_3 c(q_2 + q_3) & 0 & 0 \\ (a_2 + a_3)c(q_2 + q_3) & b_2 sq_3 & 0 \\ -(a_2 + a_3)s(q_2 + q_3) & b_3 + b_2 cq_3 & b_3 \end{bmatrix}.$$

$$(8)$$

4. The *second* and *third degree coupled error Jacobians* ($^c\mathbf{J}^{II}$ and $^c\mathbf{J}^{III}$ – rel. 9 and 10) can be obtained applying the algorithm presented in [6, 7]:

$$^c\mathbf{J}^{II} = \begin{bmatrix} -2b_2 sq_2 - 2b_3 s(q_2 + q_3) & 2b_3 s(q_2 + q_3) & 0 \\ 0 & 0 & -2b_3 \\ 0 & 0 & 0 \end{bmatrix};$$

$$(9)$$

$$^c\mathbf{J}^{III} =$$

$$\begin{bmatrix} -6b_3 c(q_2 + q_3) & 0 & -3b_2 cq_2 - 3b_3 c(q_2 + q_3) \\ 0 & \begin{array}{c} 3b_2[sq_3 - cq_2 s(q_2 + q_3)] + \\ + 3b_3[-2cq_2 cq_3 s(q_2 + q_3) + sq_2 cq_2 + sq_3 cq_3] \end{array} & 0 \\ 0 & \begin{array}{c} 3b_2[cq_3 - cq_2 c(q_2 + q_3)] + \\ + 3b_3[-2cq_2 cq_3 c(q_2 + q_3) + cq_2^2 + cq_3^2] \end{array} & 0 \end{bmatrix}$$

$$\begin{bmatrix} 0 & -3b_3 c(q_2 + q_3) & 0 & 0 \\ -\dfrac{3}{2}b_3 s(2q_2 + 2q_3) & 0 & 0 & 0 \\ \dfrac{3}{2}b_3[1 - c(2q_2 + 2q_3)] & 0 & -3b_3 & -3b_3 \end{bmatrix}$$

$$(10)$$

where $cq_i = \cos(q_i)$, $sq_i = \sin(q_i)$, $i = 1...3$.

5. *The robot positioning deviations* (described in the frame $\Re_P$) are established with the relation (11) in the *first-degree modeling*, relation (12) – *second-degree modeling* and relation (13) – *third-degree modeling*.

$$\Delta\mathbf{X} = \mathbf{J}^I \cdot \begin{bmatrix} \Delta q_1 \\ \Delta q_2 \\ \Delta q_3 \end{bmatrix} ; \quad \Delta\mathbf{X} = \mathbf{J}^I \cdot \begin{bmatrix} \Delta q_1 \\ \Delta q_2 \\ \Delta q_3 \end{bmatrix} - \frac{1}{2}\mathbf{J}^{II} \cdot \begin{bmatrix} \Delta q_1^2 \\ \Delta q_2^2 \\ \Delta q_3^2 \end{bmatrix} - \frac{1}{2}{}^c\mathbf{J}^{II} \cdot \begin{bmatrix} \Delta q_1 \Delta q_2 \\ \Delta q_1 \Delta q_3 \\ \Delta q_2 \Delta q_3 \end{bmatrix} \quad (11), (12)$$

$$\Delta X = J^I \cdot \begin{bmatrix} \Delta q_1 \\ \Delta q_2 \\ \Delta q_3 \end{bmatrix} - \frac{1}{2} J^{II} \cdot \begin{bmatrix} \Delta q_1^2 \\ \Delta q_2^2 \\ \Delta q_3^2 \end{bmatrix} - \frac{1}{6} J^{III} \cdot \begin{bmatrix} \Delta q_1^3 \\ \Delta q_2^3 \\ \Delta q_3^3 \end{bmatrix} - \frac{1}{2}{}^c J^{II} \cdot \begin{bmatrix} \Delta q_1 \Delta q_2 \\ \Delta q_1 \Delta q_3 \\ \Delta q_2 \Delta q_3 \end{bmatrix} - \frac{1}{6}{}^c J^{III} \cdot \begin{bmatrix} \Delta q_1 \Delta q_2 \Delta q_3 \\ \Delta q_1^2 \Delta q_2 \\ \Delta q_1 \Delta q_2^2 \\ \Delta q_1^2 \Delta q_3 \\ \Delta q_1 \Delta q_3^2 \\ \Delta q_2^2 \Delta q_3 \\ \Delta q_2 \Delta q_3^2 \end{bmatrix}$$

$$(13)$$

After numerical substitutions in relations (11)–(13), the values of the positioning effector errors ($\Delta r = \sqrt{\Delta x^2 + \Delta y^2 + \Delta z^2}$ and its components Δx, Δy, Δz), reduced in $\mathfrak{R}_P$, are systematized in Table 1, considering large errors: $\Delta q_1 = \Delta q_2 = \Delta q_3 = 1°$, for a configuration defined by $q_1 = 60°$; $q_2=45°$; $q_3= 30°$.

Table 1. $\Delta q_1 = \Delta q_2 = \Delta q_3 = 1°$ ($q_1 = 60°$; $q_2 = 45°$; $q_3 = 30°$)

Puma robot positioning errors – reduced in $\mathfrak{R}_P$				
	First degree model	Second degree model	Third degree model	Exact model
Δx [mm]	-16.85858600	-16.10040590	-16.09491824	-16.09508918
Δy [mm]	10.08181961	9.302604381	9.303908176	9.30398500
Δz [mm]	44.96400395	45.18226427	45.16788881	45.1678460
Δr [mm]	49.06747049	48.85896564	48.84411204	48.84414341

The comparative numerical analysis of the inaccuracies of error models (deviations from exact model, Figure 4) highlights that the second degree modelling improves significantly the estimation of the positioning errors; introducing the third degree modelling, a non-significant difference relative to the exact model (finite model, which establishes the positioning errors as finite displacements) is obtained. In this way, the error estimation is highly improved and a more accurate error model is obtained.

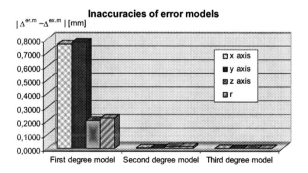

Figure 4. The inaccuracies of error models ($\Delta^{er.m}$ –errors established with error models, $\Delta^{ex.m}$ – errors established with exact models).

4.   HIGH DEGREE KINEMATIC CALIBRATION OF SERIAL MANIPULATORS

The *parameter identification* step, as part of *calibration process*, supposes to obtain the *actual (real, correct) values* of the *modelling parameters* by processing the experimental data from accuracy testing. Thus, the nominal models can be optimised and the new correct models (used in command / control process) allow the robot accuracy to increase.

In order to illustrate the algorithm used in high degree calibration, the same Puma type manipulator is considered (Figure 3). Also, the following assumptions are considered: a) *ideal links*, the geometrical parameters are exactly known; b) only the joint variable (q_1, q_2, q_3) are affected by constant errors; c) only the position error (Δx, Δy, Δz) measurement are available; d) the *third degree kinematic calibration* is applied.

In this case, the precision model becomes a three input – three output system; hence, a single positioning measurement is enough to identify the source-errors (Δq_1, Δq_2, Δq_3).

With these specifications, the *third degree identification algorithm* consists in the following steps:

– Identification of the first degree solution beginning from first degree error model (see rel. 11):

$$\begin{bmatrix} \Delta x \\ \Delta y \\ \Delta z \end{bmatrix}_P = [\mathbf{J^I}] \cdot \begin{bmatrix} \Delta q_1 \\ \Delta q_2 \\ \Delta q_3 \end{bmatrix} => \begin{bmatrix} \Delta q_1 \\ \Delta q_2 \\ \Delta q_3 \end{bmatrix} = [\mathbf{J^I}]^{-1} \cdot \begin{bmatrix} \Delta x \\ \Delta y \\ \Delta z \end{bmatrix}_P , \tag{14}$$

where $\mathbf{J^I}$ is the *first–degree error Jacobian* and P – the frame $\Re_P$ where the errors are reduced.

– The first degree model solution (Δq_1, Δq_2, Δq_3) has to be corrected with (ε_1, ε_2, ε_3) values in order to solve the second degree error model (see rel. 12). Thus, the following relation is obtained:

$$\begin{bmatrix} \Delta x \\ \Delta y \\ \Delta z \end{bmatrix}_P = [\mathbf{J^I}] \cdot \begin{bmatrix} \Delta q_1 + \varepsilon_1 \\ \Delta q_2 + \varepsilon_2 \\ \Delta q_3 + \varepsilon_3 \end{bmatrix} - \frac{1}{2} [\mathbf{J^{II}} \cdot {}^c\mathbf{J^{II}}] \cdot \begin{bmatrix} (\Delta q_1 + \varepsilon_1)^2 \\ (\Delta q_2 + \varepsilon_2)^2 \\ (\Delta q_3 + \varepsilon_3)^2 \\ (\Delta q_1 + \varepsilon_1)(\Delta q_2 + \varepsilon_2) \\ (\Delta q_1 + \varepsilon_1)(\Delta q_3 + \varepsilon_3) \\ (\Delta q_2 + \varepsilon_2)(\Delta q_3 + \varepsilon_3) \end{bmatrix} . \tag{15}$$

Neglecting the terms $\varepsilon_i \cdot \varepsilon_j$, $i, j = 1...3$, a linear system (rel. 16) in

unknowns (ε_1, ε_2, ε_3) is obtained and solved.
- Because the second degree error solution ($\Delta q_1 + \varepsilon_1$, $\Delta q_2 + \varepsilon_2$, $\Delta q_3 + \varepsilon_3$) is an approximation of the real solution, a new correction (δ_1, δ_2, δ_3) is introduced in order to solve the third degree error model (rel. 17).

$$\left(\left[\mathbf{J}'\right]-\frac{1}{2}\left[\mathbf{J}^{II}\cdot{}^{c}\mathbf{J}^{II}\right]\cdot\begin{bmatrix} 2\Delta q_1 & 0 & 0 \\ 0 & 2\Delta q_2 & 0 \\ 0 & 0 & 2\Delta q_3 \\ \Delta q_2 & \Delta q_1 & 0 \\ \Delta q_3 & 0 & \Delta q_1 \\ 0 & \Delta q_3 & \Delta q_2 \end{bmatrix}\right)\cdot\begin{bmatrix} \varepsilon_1 \\ \varepsilon_2 \\ \varepsilon_3 \end{bmatrix}=\frac{1}{2}\left[\mathbf{J}^{II}\cdot{}^{c}\mathbf{J}^{II}\right]\cdot\begin{bmatrix} (\Delta q_1)^2 \\ (\Delta q_2)^2 \\ (\Delta q_3)^2 \\ \Delta q_1\cdot\Delta q_2 \\ \Delta q_1\cdot\Delta q_3 \\ \Delta q_2\cdot\Delta q_3 \end{bmatrix}, \quad (16)$$

$$\begin{bmatrix} \Delta x \\ \Delta y \\ \Delta z \end{bmatrix}_P=\left[\mathbf{J}'\right]\cdot\begin{bmatrix} \Delta q_1+\varepsilon_1+\delta_1 \\ \Delta q_2+\varepsilon_2+\delta_2 \\ \Delta q_3+\varepsilon_3+\delta_3 \end{bmatrix}-\frac{1}{2}\left[\mathbf{J}^{II}\cdot{}^{c}\mathbf{J}^{II}\right]\cdot\Delta^{II}-\frac{1}{6}\left[\mathbf{J}^{III}\cdot{}^{c}\mathbf{J}^{III}\right]\cdot\Delta^{III}.$$

$$(17a)$$

$$\Delta^{II}=\begin{bmatrix} (\Delta q_1+\varepsilon_1+\delta_1)^2 \\ (\Delta q_2+\varepsilon_2+\delta_2)^2 \\ (\Delta q_3+\varepsilon_3+\delta_3)^2 \\ (\Delta q_1+\varepsilon_1+\delta_1)(\Delta q_2+\varepsilon_2+\delta_2) \\ (\Delta q_1+\varepsilon_1+\delta_1)(\Delta q_3+\varepsilon_3+\delta_3) \\ (\Delta q_2+\varepsilon_2+\delta_2)(\Delta q_3+\varepsilon_3+\delta_3) \end{bmatrix},$$

$$\Delta^{III}=\begin{bmatrix} (\Delta q_1+\varepsilon_1+\delta_1)^3 \\ (\Delta q_2+\varepsilon_2+\delta_2)^3 \\ (\Delta q_3+\varepsilon_3+\delta_3)^3 \\ (\Delta q_1+\varepsilon_1+\delta_1)(\Delta q_2+\varepsilon_2+\delta_2)(\Delta q_3+\varepsilon_3+\delta_3) \\ (\Delta q_1+\varepsilon_1+\delta_1)^2(\Delta q_2+\varepsilon_2+\delta_2) \\ (\Delta q_1+\varepsilon_1+\delta_1)(\Delta q_2+\varepsilon_2+\delta_2)^2 \\ (\Delta q_1+\varepsilon_1+\delta_1)^2(\Delta q_3+\varepsilon_3+\delta_3) \\ (\Delta q_1+\varepsilon_1+\delta_1)(\Delta q_3+\varepsilon_3+\delta_3)^2 \\ (\Delta q_2+\varepsilon_2+\delta_2)^2(\Delta q_3+\varepsilon_3+\delta_3) \\ (\Delta q_2+\varepsilon_2+\delta_2)(\Delta q_3+\varepsilon_3+\delta_3)^2 \end{bmatrix},$$

$$(17b)$$

$$\left(\left[\mathbf{J}'\right]-\frac{1}{2}\left[\mathbf{J}^{II}\cdot{}^{c}\mathbf{J}^{II}\right]\cdot\mathbf{E}^{II}-\frac{1}{6}\left[\mathbf{J}^{III}\cdot{}^{c}\mathbf{J}^{III}\right]\cdot\mathbf{E}^{III}\right)\cdot\begin{bmatrix} \delta_1 \\ \delta_2 \\ \delta_3 \end{bmatrix}=-\left[\mathbf{J}'\right]\cdot\begin{bmatrix} \varepsilon_1 \\ \varepsilon_2 \\ \varepsilon_3 \end{bmatrix}+\frac{1}{2}\left[\mathbf{J}^{II}\cdot{}^{c}\mathbf{J}^{II}\right]\cdot\Delta_*^{II}+\frac{1}{6}\left[\mathbf{J}^{III}\cdot{}^{c}\mathbf{J}^{III}\right]\cdot\Delta_*^{III},$$

$$(18)$$

$$
\mathbf{E}^{II} = \begin{bmatrix}
2(\Delta q_1 + \varepsilon_1) & 0 & 0 \\
0 & 2(\Delta q_2 + \varepsilon_2) & 0 \\
0 & 0 & 2(\Delta q_3 + \varepsilon_3) \\
\Delta q_2 + \varepsilon_2 & \Delta q_1 + \varepsilon_1 & 0 \\
\Delta q_3 + \varepsilon_3 & 0 & \Delta q_1 + \varepsilon_1 \\
0 & \Delta q_3 + \varepsilon_3 & \Delta q_2 + \varepsilon_2
\end{bmatrix},
$$

$$
\mathbf{E}^{III} = \begin{bmatrix}
3(\Delta q_1 + \varepsilon_1)^2 & 0 & 0 \\
0 & 3(\Delta q_2 + \varepsilon_2)^2 & 0 \\
0 & 0 & 3(\Delta q_3 + \varepsilon_3)^2 \\
(\Delta q_2 + \varepsilon_2)(\Delta q_3 + \varepsilon_3) & (\Delta q_1 + \varepsilon_1)(\Delta q_3 + \varepsilon_3) & (\Delta q_1 + \varepsilon_1)(\Delta q_2 + \varepsilon_2) \\
2(\Delta q_1 + \varepsilon_1)(\Delta q_2 + \varepsilon_2) & (\Delta q_1 + \varepsilon_1)^2 & 0 \\
(\Delta q_2 + \varepsilon_2)^2 & 2(\Delta q_1 + \varepsilon_1)(\Delta q_2 + \varepsilon_2) & 0 \\
2(\Delta q_1 + \varepsilon_1)(\Delta q_3 + \varepsilon_3) & 0 & (\Delta q_1 + \varepsilon_1)^2 \\
(\Delta q_3 + \varepsilon_3)^2 & 0 & 2(\Delta q_1 + \varepsilon_1)(\Delta q_3 + \varepsilon_3) \\
0 & 2(\Delta q_2 + \varepsilon_2)(\Delta q_3 + \varepsilon_3) & (\Delta q_2 + \varepsilon_2)^2 \\
0 & (\Delta q_3 + \varepsilon_3)^2 & 2(\Delta q_2 + \varepsilon_2)(\Delta q_3 + \varepsilon_3)
\end{bmatrix},
$$

Δ_*^{II} and Δ_*^{III} are obtained from relation (17b) imposing $\delta_1 = \delta_2 = \delta_3 = 0$.

Finally, neglecting the third degree terms $\delta_i \cdot \delta_j \cdot \delta_k$, $i, j, k = 1...3$ and second degree terms $\delta_i \cdot \delta_j$, $i, j = 1...3$, a linear system (rel. 18) with the unknowns $(\delta_1, \delta_2, \delta_3)$ is obtained and solved.

The improved solution $(\Delta q_1 + \varepsilon_1 + \delta_1, \Delta q_2 + \varepsilon_2 + \delta_2, \Delta q_3 + \varepsilon_3 + \delta_3)$ obtained from the third degree error model allows a good estimation of the real solution.

A single set of the positioning end-effector's errors (reduced in $\mathfrak{R}_P$) is measured, considering the following robot joint configuration: $q_1 = 60°$; $q_2 = 45°$; $q_3 = 30°$; the „measured"end-effector's errors are established from the geometrical model, considering large errors i.e. $\Delta q_1 = \Delta q_2 = \Delta q_3 = 1°$:

[Δx, Δy, Δz] = [-16.09508918, 9.303985, 45.1678460] [mm].

Applying the third degree algorithm already presented, the following results are obtained:

a) First degree source-error solution:

[Δq_1, Δq_2, Δq_3]^I = [0.9547°, 0.9179°, 1.1517°],

with the least-squares deviation from exact solution: 0.17838°.

b) Second degree source-error solution:

[ε_1, ε_2, ε_3] = [0.0449575°, 0.0820505°, -0.15253447°] =>

[Δq_1, Δq_2, Δq_3]^{II} = [0.99966°, 0.99995°, 0.99922°],

least-squares deviation: 0.00084621°.

c) Third degree source-error solution:

$[\delta_1, \delta_2, \delta_3] = [0.000341343°, 0.000056891°, 0.0007645277°] =>$
$[\Delta q_1, \Delta q_2, \Delta q_3]^{III} = [1.0000106°, 1.00000706°, 0.9999872°]$,
least-squares deviation: 0.0000180197°.

In conclusion, the *high degree identification algorithm* allows a successive improvement of the solution; the third degree solution is closed to the real solution even for large values of source-errors (Figure 5).

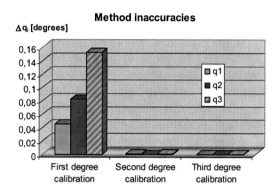

Figure 5. The inaccuracies of the calibration models.

5. CONCLUSIONS

Based on the precision modelling and numerical simulations made on a RRR serial structure, the following conclusions could be drawn:

- Starting from the limits of the existing accuracy models (first degree models), a high degree modelling of robot errors and its application is presented in first part of the paper.
- The third degree error model is proposed to be applied in the premise of a large RRR manipulator (with large values of link lengths) and large deviations of error sources (joint variables).
- The decoupled and coupled error Jacobians, used in the high degree error modelling, can be easily described applying an algorithm proposed by the authors. This algorithm can be also applied to compute both positioning and orienting end effector's errors.
- The high degree error models can be used to calculate with high accuracy the influence of the source-errors with large values on the serial robot operational errors.

The numerical analysis of the positioning errors highlights the following aspects:

1. The first–degree modelling describes with high deviations (Table 1, *first degree model*) the real state of errors (Table 1, *exact errors*), due to the

method imprecision's in the case of high values of the source–errors.

2. The second error model assures a consistent improvement of all components (Δx, Δy and Δz) of the end effector's errors.

3. By introducing the third degree modelling, a non-significant difference relative to the exact model is obtained.

The high degree error models can be applied in calibration of robots (in the *identification step*) as well, with positive effects on the solution accuracy. The general aspects about calibration and the results for the RRR serial structure, presented in this paper, allow drawing the following conclusions:

a) The accuracy (precision) model is usually the kernel of the calibration process. This model yields to the good calibration results only in the case of small (infinitesimal) errors. For this reason, the authors propose in the paper a high degree precision model, which allows obtaining accurate identification results even for large errors.

b) The numerical results of the third degree calibration highlight the efficiency of this approach in the case of large errors: the classical (first degree) calibration model yields to a relatively inaccurate solution (deviation from exact solution: $0.17838°$). Instead, the solution of the third degree model is much closer to the exact solution (deviation: $0.0000180197°$).

REFERENCES

1. R. BERNHARDT, S. L. ALBRIGHT, *Robot calibration* (Chapman &Hall, 1993).
2. F. DUDITA, D. DIACONESCU, G. GOGU, *Linkages: inventics and kinematics in a phylogenetic approach*" (in Romanian), Ed. Tehnica, Bucuresti, 1989.
3. G. GOGU, P. COIFFET, A. BARRACO, *Représentation des déplacements des robots* (Ed. Hermes, Paris, 1997).
4. J. HOLLERBACH, CH. WAMPLER, *The calibration index and taxonomy for robot kinematic calibration methods* (International Journal of Robotics Research, vol. 15, No. 6, 1996, pp. 573-591).
5. B. MOORING, Z. ROTH, M. DRIELS, *Fundamentals of Manipulator Calibration* (John Wiley&Sons Inc., New York, 1991).
6. M. NEAGOE, *Contributions to the study of industrial robot precision* (in Romanian, PhD Thesis, Transilvania University of Braşov, 2001).
7. M. NEAGOE, C. JALIU, N. CRETESCU, *High degree modelling of errors in robot chains* (Third International Conference "Research and Development in Mechanical Industry" RaDMI 2003, 19-23 September 2003, Herceg Novi, Serbia and Montenegro, pp. 1389-1407, 2003).
8. R. PAUL, *Robot manipulators: Mathematics, programming, and control* (The MIT Press, 1981).

APPLICATION OF IMAGE ANALYSIS FOR THE STUDY OF STRUCTURAL MODIFICATIONS IN FLOWING EMULSIONS
Theoretical approach and practical application

A. G. Pocola and D. O. Pop
Technical University of Cluj-Napoca, Romania

Abstract: The behaviour of flowing emulsions is influenced by dispersion, concentration and type modifications that can occur during and due to the flow associated hydrodynamic processes. The modifications of the dispersion degree, concentration or even emulsion type have a direct influence over the viscosity, lubricating capacity, stability and thermodynamic properties of the flowing emulsion. Consequently, in order to control the characteristics of a flowing process for an emulsion we must have more information regarding dispersion, concentration, appearance of the inversion points. This paper presents an image analysis based method with which one can study the modification of the above-mentioned parameters. The proposed method can also be extended for other processes with associated structural modifications, which can be studied through image analysis.

Key words: flowing emulsions, hydrodynamic instability, image analysis.

1. ASPECTS CONCERNING FLOWING EMULSIONS

There are many instances when fluid mixtures have to be transported by means of pipes [1, 2, 3 and 4]. One of the most important aspects concerning the transportation of such mixtures lies in their structural stability during the flowing process. The most often used fluid mixtures are the oil-water emulsions.

Starting from the behaviour during flow of such emulsions, this paper proposes a research method based upon image analysis and processing, a method that can be later on applied to other types of fluid mixtures.

409

D. Talabă and T. Roche (eds.), Product Engineering, 409–424.
© 2004 *Springer. Printed in the Netherlands.*

Following experiments related to oil-water mixes, Nädler and Mewes [1] noticed that when a liquid mixture flows through a pipe, no matter if the supply output is constant, the areas occupied by each of the phase in some two sections of the pipe are different (Figure 1).

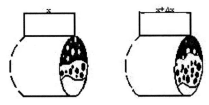

Figure 1. The structure variation of a flowing liquid mixture

According to the experiment mentioned earlier, Nädler and Mewes classified the flowing cases into seven important groups (Figure 2).

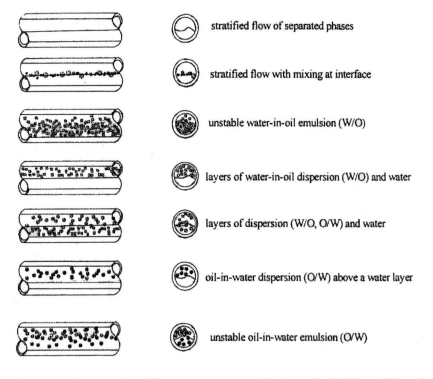

Figure 2. Oil-water mixture flow systems representation (according Nadler and Mewes): stratified flow of the two phases, stratified flow with two-interface mixture, unstable water in oil emulsion flow type, stratified water flow and emulsion of water in oil, stratified water flow, emulsion of oil in water and emulsion of water in oil, stratified water flow and oil in water emulsion, unstable oil in water emulsion type flow.

Nädler and Mewes's research confirm the presence of the segregation phenomenon (pointed out by the stratified flow) observed by Sigre and Silberberg [5] as well as the disappearance of this phenomenon and increase of emulsification in the case of eddy flow.

The conditions for one or other flow systems to exist are presented in Figure 3, as dependent on the oil-water mix. The straight lines represent constant ratios of the volume outputs corresponding to the two phases.

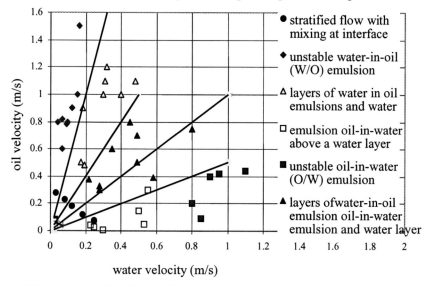

Figure 3. Dependence of the oil-water flow systems on the component surface rates (according Nadler and Mewes)

In most cases, the flow is achieved by pumping and the motion of the fluid mixture is determined by the pressure difference.

Yan and Kuroda [6] proved that two fluids of different viscosities and making part of a fluid mixture do not possess equal flowing rates, though subjected to the same pressure difference.

The expression of the rate difference of the two components of the fluid mixture is given by the equation:

$$v_d - v_c = \frac{\phi_c(\mu_d + \mu_{dc}) - \phi_d(\mu_c + \mu_{cd})}{2(\mu_d \cdot \mu_c - \mu_{dc} \cdot \mu_{dc})} \cdot [h \cdot y - y^2] \cdot \frac{\partial p}{\partial x}, \qquad (1)$$

where μ_d, μ_c, μ_{cd} and μ_{dc} are the four partial viscosities of the emulsion [4]; μ_d is „the effect" of the dispersed phase viscosity upon its own shearing stress (own effect), μ_c is „the effect" of the continuous phase viscosity upon its own shearing stress (own effect), μ_{dc} is „the effect" of the continuous phase viscosity upon the dispersed phase shearing stress, μ_{cd} is „the effect" of the dispersed phase viscosity upon the continuous phase shearing stress, ϕ_d and ϕ_c are volume fractions of the dispersed, respectively continuous phase, while h represents the thickness of the fluid film formed between the two surfaces of the pipe.

As one can see from Eq. (1), the rate difference between the components of the fluid mixture while flowing depends upon the mixture concentration, the components viscosities, the position versus the pipe walls and the pressure gradient that characterizes the flow.

The motion of the two liquids at various velocities leads to different concentration areas, and, hence, inside the fluid mixture there can come up structural modifications due to coagulation, coalescence, inversion and breaking up. These modifications are not desirable as they influence the properties of the flowing fluid mixtures [6, 7, 8].

2. THE IMAGE ACQUISITION AND PROCESSING METHOD APPLIED TO THE STRUCTURAL MODIFICATIONS OF THE FLOWING FLUID MIXTURES, DURING THE EXPERIMENT

The aspects presented above are significant for the behaviour of the mixtures. This is the reason why we propose here an experimental research concerning the structural modifications that occur inside a fluid mixture, during and due to the flowing process. The method is based upon the image acquisition inside the flowing mixture (Figure 4) and image analysis and processing in a second stage [9].

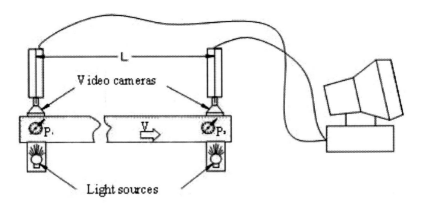

Figure 4. Image acquisition for a flowing emulsions

2.1   Image acquisition

For the image acquisition, we propose the Lab View-IMAQ type hardware produced by National Instruments. The system can be used with computer systems of the type PCI, PXI and Compact PCI [10].

The system contains a 1408 acquisition board for the black and white image acquisition. The 1408 acquisition board is compatible both with the standard video systems RS-170, CCIR, NTSC, PAL and the non-standard ones. It also contains four video channels and four input/output modules (I/O).

The dedicated software package NI-IMAQ that was specially designed for the IMAQ hardware control was used. The NI-IMAQ software package incorporates the virtual instrumentation module (VIs) required to control and give parameters to the data acquisition process.

Three acquisition modes are available in the LabView-IMAQ software package: snap, grab and sequence. These modes of image acquisition allow obtaining either separate image photos or a film with successive images continuous running.

2.1.1   Image acquisition type

a) Snap-type image acquisition. The snap-type image acquisition mode is based upon the storage of only one image in the buffer. When this mode is initiated, the image acquisition system stores the image frame to be subsequently subjected to this procedure in a buffer. The individual acquisition mode is suitable to the applications carried out at low speed or that are static.

b) Grab-type image acquisition.The grab-type image acquisition mode is based on the continuous storage of more successive images in the same buffer. This function produces an image acquisition in a continuous loop, with three distinct phases: structuring, acquisition, storage and transfer, and cycle finish. Structuring is the operation that occurs only once at the beginning of each continuous image acquisition process and has the role of initiating the process under discussion. Acquisition, storage and transfer are operations that takes place for each current image taken over by the acquisition system and stored in the buffer.

c) Sequence-type image acquisition. The sequence-type image acquisition is controlled by a sequence duration variable and a transfer lag to multiple buffer variables. This type of acquisition is recommended for the processes where multiple images have to be taken over. Mention should be made that images can be taken over frame by frame or that the number of frames that can be ignored can be specified. Similar to the grab-type, this type of acquisition is also characterized by structuring operations that take place of the process beginning. During this operation, a string containing the number of frames to be ignored between successive image acquisitions is initiated.

The function that defines the manner in which acquisition is carried out can be synchronous or asynchronous. When it is synchronous, the image is not transferred until the whole sequence is acquired. In the case of asynchronous acquisition, the transfer occurs immediately after the image acquisition.

The image acquisition modes to be set up in the image processing software must be correlated with the features of the process studied and implicitly with the video camera performance.

2.1.2   Camera selection

At present, three variants of technique are implemented in the video cameras for image acquisition, namely:
–   Frame by frame image acquisition. This technique relies on the successive stage integration of the even lines in the image fields, followed by their downloading, and the integration of the odd lines in the image fields. The resulting image in the sensor field following scanning will contain either odd, or even only lines. When the download process is complete, the successive images will be mixed.
–   Field image acquisition. In this scanning technique, both even and odd lines are taken over by the image sensors. For instance, lines 1 and 2 of the image shall be mixed to form line number 1 in the resulting image. Similarly, lines number 2 and 3 shall be mixed to form line 2, the process going on until the final configuration of the resulting image. According to the description above, the scanning techniques mentioned require a definite image processing time. Moreover, in case of movement in the scanning field, the image will be blurred. In order to remove the effect, more modern cameras, use the progressive scanning technique.
–   Progressive scanning. Progressive scanning technique cameras transfer the whole image frame as received by the sensors with no combination of frames. The obvious advantage of this type of scanning lies in the capacity to detect motion, but on the other side it is incompatible with TV systems, as it requires too large a frequency band to transmit signals. However, image analysis applications do not require the rough image to be straightforwardly processed by the video camera. More precisely, the video camera is used both as a sensor and it is connected to a data acquisition board with an incorporated video board. Progressive scanning cameras became more popular due to the applications related to image

acquisition. They possess a direct digital output that removes part of the process components that are time consuming so hat the processing stages in the TV systems are no more needed.

2.1.3 Highlighting

The highlighting of the area of the image to be acquired is very important, mainly for the applications in which details appearing after acquired image processing should be as accurate as shade difference.

On the other hand, highlighting can help the distinction between the field to be analysed and the adjacent visual fields. While the visual field of interest should be properly lighted, the neighbouring fields should not be lighted and thus the studied area can be made distinct and saving in memory and transfer resources can be made.

In this context, it is desirable to have a strong contrast between the area under scrutiny and the areas to be ignored. The powerful lighting of the first area can contribute to decrease the importance of noise signals due to light radiations from the surrounding environment, ordinary lighting, reflected light beams etc.

The highlighting technique can be adapted to the process features whose running should be analysed with image acquisition. Lighting can be: direct, inverse, stroboscopic, with dark field, diffuse etc.

In the case of direct lighting, both the image receiver and the light source are on the same side of the visual field of study. The one-direction direct lighting is detrimental mainly because of ordinary lights, double image or reflections. Generally, in this kind of lighting the light beams are partially collimated.

To avoid errors and unclear images, mainly due to double and blurred images, diffuse light is used for the direct lighting of the visual field. The diffuse light is not collimated and it can be obtained by passing the light fascicle through a semi-transparent plastics or glassy material.

To point out shapes and contours, the inverse (from backwards) lighting is used. In this kind of lighting, the image receiver and the light source are distinct from the visual field, as they are on each of the two sides of it. It is characterized by very strong contrasts, by highlighted shapes and separation outlines, but some of the details inside the image are also affected too.

When very high velocity images are acquired, stroboscopic lighting is recommended. This is defined by a succession of extremely short length light impulses (microseconds). Due to this type of lighting, the image field already taken over by the receiver can be processed and transmitted without having to avoid the danger of an overlapped image field. This kind of lighting diminishes the chance of blurring and double contours in the case of moving images.

2.2   Processing and conditioning the acquired images

An image can be described by a function f(x, y) representing the space distribution of the light intensity in a visual field. Function f(x, y) is the digitised form of the actual image, the accuracy with which it is reproduced depending on the sensitivity and processing capacity of the video camera. For example, a one 8 bit-colour video camera can detect, digitise and transmit 256 shades of colour in the black and white range.

A point of some co-ordinates (x, y) is called pixel and the number of pixels contained in an image define the image resolution. Each pixel is characterized by a certain shade of color numbered from 0 to 255. By 8-bit process digitizing, 0 will be allocated the row 00000000, while 255 will be allocated the row 11111111.

In the conditioning of an image operations that aim at facilitating the further processing operations are included. The conditioning operations are applied to the source image to point out some of the aspects in it and to remove things that can be ignored. Linear or Boolean arithmetical operators, Laplace, Gauss, gradient, damping filter operators can be applied to a source image as well as other operations such as: the separation of the image areas in groups, or the grouping of neighbouring areas, the deletion of small-sized areas etc.

To perform conditioning operations, a virtual or real matrix type image should be interposed between the source image formed of the same number of pixels and the resulting image formed of similar number of pixels. The interposed image has the same number of pixels as the source, respectively resulting image. A conditioning operation is applied between each pixel belonging to the source image and the equivalent pixel of the virtual image. As a consequence, the features of the pixel in the resulting image are reached.

As mentioned before, the conditioning operations are performed at the level of each pixel, so that the whole image is conditioned in the procedure called "pixel by pixel".

If A is a source image of NxM pixels (ps) and B is an operational image of NxM pixels (pop), then a conditioning operation can be applied between the two images, and the result is the image C that also has NxM pixels. (pr). The general form of a conditioning operation is given by:

$$p_r(x, y) = p_s(x, y) \mathbf{Op} p_{op}(x, y). \tag{2}$$

On the basis of this general formula, arithmetic or Boolean conditioning operations can be defined. These operations can be used to mix, separate and compare images, their most frequent applications being those related to the comparison of two images received in different time intervals. The search for joint elements in two or more images, the comparison with a model image the arithmetic or Boolean operators also serve at modifying the contrast, or taking off brightness of an image respectively.

2.3 Structural and dimensional evaluation of the elements in the acquired image field

After an image field is acquired and processed to highlight specific and significant research aspects (outline, colour, gradients etc) a certain number of structural and dimensional evaluation functions can be applied to the image.

Table 1 presents part of the functions to be defined in the Lab View IMAQ software package to perform the structural and dimensional evaluation of the elements found in the images taken over for the experimental purposes [10].

By means of structural evaluation, one can get information related to the emulsion type, its dispersion degree, structural kinematics e.g. time modification of the dispersion degree, segregation, coalescence and breaking up.

This structural evaluation can be direct or comparative. Individually analysing the processed images and extracting data on the shape and size of drops with reference to a standard, the dispersion degree or contrast represents the direct evaluation.

Comparative evaluation is performed by comparing information from more successive images taken in the same experiment (structural kinematics evaluation), or in experiments conducted on various development conditions (evaluation of the significance of initial conditions related to the process development).

The structural emulsion evaluation during the design stage and evaluation of the modifications coming up during and due to the flowing process consider the following aspects:

a) As an evaluation of the emulsion degree of dispersion during its design stage, the dispersion degree of an emulsion can be assessed by the size and uniformity of the drops in the disperse stage and by the uniformity of drop settlement in the continuous phase mass. These appreciations can be made by direct assessment of the taken images, stored and possibly processed.

b) For the evaluation of the liquid phase drops deformation degree in a flowing emulsion, the starting point is that with the initially forming emulsion, the dispersed liquid drops are spherical in shape. The drops deformation degree is assessed versus the spherical shape they had in the initially forming stage emulsion.

c) Analysis of the oil deposits formed during and because of the flowing process shows that the instability of an emulsion generally manifests in a succession of events, such as coagulation, coalescence, breaking up or inversion. For an "oil-in-water" emulsion, coagulation leads to the crowding of the oil drops in a certain area (dispersion degree modified), while coalescence leads to their joining together and forming of larger drops or even oil deposits.

Table 1. Measurement parameters returned by IMAQ complex measure

Function	Description of function application outcome
Area (pixels)	Particle individual and summed up area computation (in pixels)
Area (calibrated)	Particle individual area computation (in the unit associated to a pixel area)
Number of holes	Number of areas of a different shade inside particles and distinct from them by an closed contour, called holes (incorporated areas)
Hole's area	Holes area computation (in the unit associated to a pixel area)
Total area	Total area computation (particles and incorporated areas)
Scanned area	Computation of the area of the entire image analysed at a certain moment (visualised)
Area/ Scanned Area [%]	Ratio between the sum the sum of the particles area and the visualized area
Area/ Total Area [%]	Ratio between the sum the sum of the particles area and the total area
Centre of mass (X)	X co-ordinate of the mass centre of a particle
Centre of mass (Y)	Y co-ordinate of the mass centre of a particle
Longest segment length	The length of the longest horizontal segment to be drawn inside a particle
Perimeter	The length of the exterior contour related to a particle
Hole's perimeter	The length of the exterior contour related to an incorporated area
Mean chord X	The average of the horizontal segments length to be drawn inside a particle
Mean chord Y	The average of the horizontal segments length to be drawn inside a particle
Max intercept	The longest segment length to be drawn inside a particle
Particle orientation	Direction of the longest segment [angle]
Ellipse minor axis	Length of the equivalent ellipse minor axis whose perimeter and surface are equal to that of a given particle
Ellipse major axis	Length of the equivalent ellipse major axis whose perimeter and surface are equal to that of a given particle
Ratio of equivalent ellipse axes	Ratio between the equivalent ellipse axes
Hydraulic radius	The area of the particle surface versus its perimeter
Waddel disk diameter	Diameter of the circle having the same surface as the particle

The occurrence of this phenomenon is extremely important in the framework of the research on the structural modifications taking place in the emulsions during and owing to flow processes.

d) Observation on the points of inversion and breaking up. After oil deposits appear, the balance in that area of the emulsion becomes unsure and affects its structural stability.

The instability of the emulsion leads either to the emulsion breaking up and to the coming up of some open separation contours between the emulsion phases (Figure 5.a), or to inversion, when some continuous phase (incorporated areas) appear inside the disperse phase, their contours being closed this time [11] (Figure 5.b).

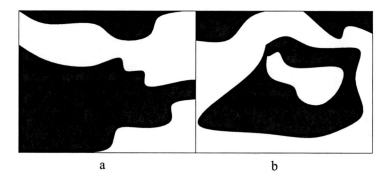

a   b

Figure 5. Manifestations of emulsion instability in lubricating films: a) breaking up, b) inversion.

Part of the elements subjected to structural evaluation by direct visual analysis can also be assessed quantitatively (the deformation degree is assessed by the size ratios, the dispersion degree by the variation of the drop size and oil deposits dimension, the instability degree by the number of inversion points).

The elements in an image field are characterized by linear dimensions, area of surface, orientation and position. Besides these one must consider also the shape, colour, contrast, uniformity, distribution and others.

While the size of the first group characteristics (structural evaluation) can be assessed visually following an earlier image processing (filtering, relieving etc.), the second group characteristics can be assessed (dimensional evaluation) only with some specific functions.

e) Determining the emulsion concentration modification degree. One of the most important aspect noticed in almost all the experiments related to emulsion transportation was the modification of the concentrations of the emulsions during and due to the specific flow process. The experimental research made so far has been based on a large variety of evaluation methods of this parameter, e.g. the modification of the viscosity, the lubricating film assessment and direct image analysis.

In the experimental method proposed in this paper, it is recommended to evaluate the modification of the emulsion concentration by using typical functions of the image assessment. These specific functions can be applied both to the images taken over at various time moments and in various flow path points.

f) Evaluation of the structural kinematics of the flowing emulsions. The structural kinematics of the flowing emulsions represents a synthesis of the behaviour of the emulsion during the flow process specific conditions. Structural kinematics is a type of analysis inserted in this paper to help the global description of an emulsion during and due to the flow process.

Due to the fact that the flow process develops in a finite time interval, the structural modifications inside the flowing fluid mixtures should be expressed versus this time interval.

Practically, the modification of any of the parameters described above is a proof the structural kinematics that will or will not be followed by an effect at the macroscopic level of the flow process.

3.   APPLICATION

To demonstrate the above-mentioned theoretical and methodological aspects an oil-in-water emulsion image acquisition application will be presented below.

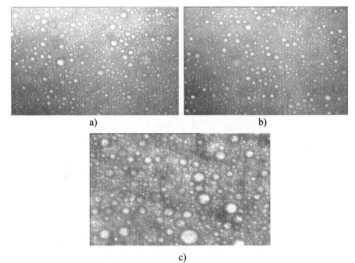

a)   b)

c)

Figure 6. Images of the states of a flowing "oil-in-water" emulsion

Figures 6 a, b and c show the images of three different states of a flowing emulsion.

The rough images can be further processed to "extract" as easily as possible the information needed in the analysis process. Figure 7 presents a sequence of the processing of the acquired image.

Thus, a series of conditioning operations are applied to a rough image (the emulsion in Figure 6.a).

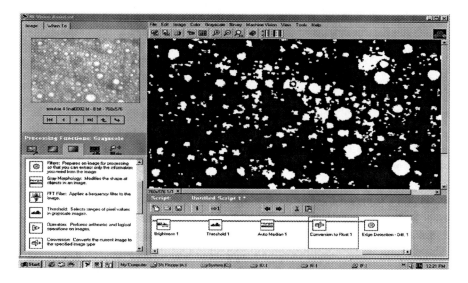

Figure 7. A sequence from the image processing process

Following the processing, simpler forms of images such as the ones in Figures 8 and 9 can be found, and the separation interface between the components of the fluid mixture is more obvious.

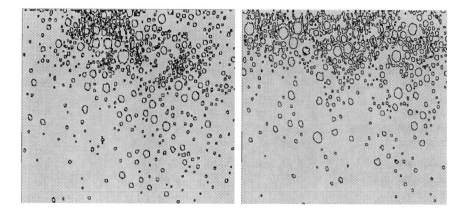

Figure 8. Flowing oil-in water emulsion processed images

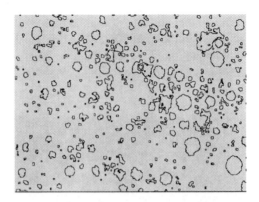

Figure 9. Flowing oil-in water emulsion further processed image

The processed images can be further subjected to a structural and dimensional evaluation process. After this evaluation, data with reference to the mixture structural kinematics (inversion point occurrence, modified dispersion degree, oil deposits setting etc.) is available. Figure 10 presents a detail from the processed image of a flowing emulsion.

To each oil-occupied region an identification number is allotted. Table 2 presents the numerical results of the application of one of the specific evaluation functions outlined in Table 1.

Table 2

Object	210	225	239	252	345	410	429
Centre of Mass X	300	410	350	246	375	364	409
Centre of Mass Y	102.5	103.7	137.9	135	180	196	224
Holes' Perimeter	0	0	160	0	0	0	96.36
Equivalent Ellipse Major Axis	19.76	2.98	82.35	36.97	55.55	4.9	112.6
Equivalent Ellipse Minor Axis	5.54	1.71	19.77	22.42	7.79	2.34	13.86
Hydraulic Radius	1.89	0.52	6.8	6.78	2.73	0.75	4.86
Waddel Disk Diameter	10.46	2.26	40.35	28.79	20.81	3.39	39.51
Area	86	4	1279	651	340	9	1226
Holes' Area	0	0	187	0	0	0	95
Particle & Holes' Area	86	4	1466	651	340	9	1321
Image Area	4.42E+05						
Number of Holes	0	0	10	0	0	0	9
%Area/(Particle & Holes' Area)	100	100	87	100	100	100	92
Ratio of Equivalent Ellipse Axes	3.57	1.75	4.16	1.65	7.13	2.09	8.13
Elongation Factor	2.69	1.96	1.64	1.25	1.79	2.53	1.4
Compactness Factor	0.6	0.67	0.45	0.75	0.41	0.6	0.42

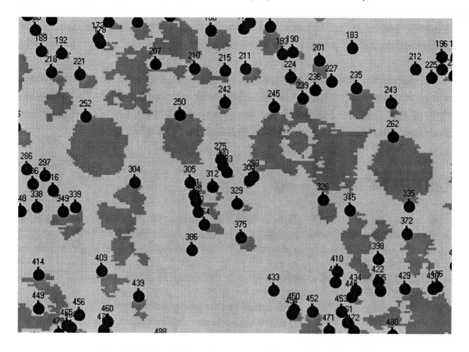

Figure 10. Detail from the processed image from Figure 7

It is easy to visualize the inversion points that appear in two of the regions (Object 239, 429), highlighted by the function "Holes".

4. CONCLUSIONS

The modifications of the dispersion degree, concentration or even emulsion type have a direct influence over the viscosity, lubricating capacity, stability and thermodynamic properties of the flowing emulsion. Consequently, in order to control the characteristics of a flowing process for an emulsion one must have enough information regarding dispersion, concentration, appearance of the inversion points

An image analysis based method for the study of the modification of the flowing emulsion parameters has been presented. To this aim, an experimental research concerning the structural modifications that occur inside a fluid mixture, during and due to the flowing process has been proposed. The method is based upon the image acquisition inside the flowing mixture and image analysis and processing in a second stage.

The image analysis is based on the use of Lab View-IMAQ type hardware produced by National Instruments, which can be used with computer systems of the type PCI, PXI and Compact PCI and the dedicated software package NI-IMAQ that was specially designed for the IMAQ hardware control to provide the necessary parameters to the data acquisition process.

To demonstrate the theoretical and methodological aspects an oil-in-water emulsion image acquisition application was also presented. The proposed method can also be extended for other processes with associated structural modifications, which can be studied through image analysis.

REFERENCES

1. M. NADLER, D. MEWES, *Flow induced emulsification in the flow of two immiscible liquids* (International Journal Multiphase Flow, vol. 23, No.1, 1997, pp. 55-68).
2. J. L. ZAKIN, a.o., *Transport of oils and oil-in-water emulsions* (Journal Fluid Engineering, no. 101, pp. 100-104).
3. R. PAL, *Pipeline flow of unstable and surfactant-stabilized emulsion* (Journal AIChE no 39, pp. 1754-1764).
4. A. PILEHVARI, a.o., *Oil-in-water emulsions for pipeline transport of viscous crude oils* (Paper SPE 18218 SPE Annual Technical Conference and Exhibition, Houston, 1988).
5. G. SIGRE, A. SILBERBERG, *Radial particle displacements in Poiseuille flow of suspensions* (Nature, nr. 21, ian.1961, pp 209-210).
6. S. YAN, S. KURODA, *Lubrication with emulsion: first report: The extended Reynolds equation* (Wear, no 206, pp. 230-237).
7. A. POCOLA, D. POP, *Consideration concerning the emulsions concentration due to the velocities differences between phases* (Annals of MTM 2001, Proceeding of 5th International MTM Symposium, 4-6 October 2001 UTCN).
8. T. NAKAHARA, a.o., *Observations of liquid droplet behavior and oil film formation in O/W type emulsion lubrication* (Transaction of A.S.M.E., vol. 110, pp. 348-353).
9. H. M. YASUHIKO, a.o., *Cinemicrophotographic study of boiling of water-in-oil emulsions* (International journal multiphase flow, vol. 6, pp. 255-266).
10.***, *LabVIEW Machine Vision and Image Processing* (Austin, 1998).
11.A. POCOLA, *Contributions to the lubrication with emulsions oil-water* (PhD Thesis, Technical University of Cluj-Napoca, 2003).

THE MANUFACTURE TECHNOLOGY OF TUBE MODELS BY SELECTIVE LASER SINTERING

C. Petra
"Petru Maior" University of Targu Mures, Romania

Abstract: The paper presents some specific aspects of the technology tube models realized by Selective Laser Sintering (SLS) process. Selective laser sintering is ideal for applications that require functional testing. These prototypes are durable, rugged and able to withstand the harshest testing environments.

Key words: tube models, laser sintering.

1.   INTRODUCTION

Selective Laser Sintering (SLS®, registered trademark by DTM™ of Austin, Texas, USA) is a process that was patented in 1989 by Carl Deckard, a University of Texas graduate student. The selective laser sintering process is a free-form fabrication method to create components by precise thermal fusing (sintering) of powdered materials. Parts of complex geometries are built in successive layers that define subsequent cross sections of the component.

Materials most often used in selective laser sintering are polycarbonate and nylon. For polycarbonates, densities of 75-92% of the standard injection-molded material can be obtained; for nylon even better values, 87-93%, have been obtained. Another material that is well suited for the process is investment casting wax. This allows the direct construction of wax patterns for foundry use.

Thermoplastic powder is spread by a roller over the surface of a build cylinder (Figure 1). The piston in the cylinder moves down one object layer thickness to accommodate the new layer of powder. The powder delivery system is similar in function to the build cylinder. Here, a piston moves upward incrementally to supply a measured quantity of powder for each

D. Talabă and T. Roche (eds.), Product Engineering, 425–432.
© 2004 *Springer. Printed in the Netherlands.*

layer. A laser beam is then traced over the surface of this tightly compacted powder to selectively melt and bond it to form a layer of the object. The fabrication chamber is maintained at a temperature just below the melting point of the powder so that heat from the laser need only elevate the temperature slightly to cause sintering. This greatly speeds up the process. The chamber is also filled with nitrogen to prohibit the oxidation of the materials at the elevated temperature. The process is repeated until the entire object is fabricated.

The typical layer thickness for the SLS® process is between 0.004" and 0.006".

After the object is fully formed, the piston is raised. Excess powder is simply brushed away and final manual finishing may be carried out. No supports are required with this method since overhangs and undercuts are supported by the solid powder bed. That's not the complete story, though. It may take a considerable length of cool-down time before the part can be removed from the machine. Large parts with thin sections may require as much as two days of cooling time.

The Sinterstation® 2500 System Process Chamber

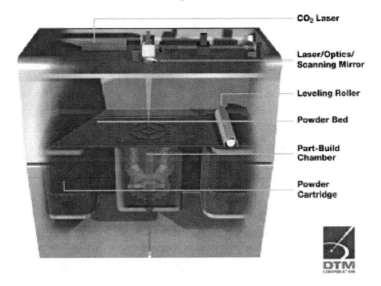

Figure 1. The working principle of SLS

SLS offers the key advantage of making functional parts in essentially final materials. However, the system is mechanically more complex than stereolithography and most other technologies. A variety of thermoplastic materials such as nylon, glass filled nylon, and polystyrene are available.

Surface finishes and accuracy are not quite as good as with stereolithography, but material properties can be quite close to those of the intrinsic materials. The method has also been extended to provide direct fabrication of metal and ceramic objects and tools.

1.1   Advantages

a) Capable of producing the toughest part compared with other process.
b) Selective laser sintering provides exact representations of complex designs in just days. This means that without delay, one could receive a superior design communication tool. Using the physical prototype, you can detect errors early and correct them in advance.
c) Large variety of material can be used, including most engineering plastic, wax, metal, ceramic, etc.
d) Parts can be produced in short time, normally at a rate of up to 1 inch per hour.
e) No post curing of parts is required.
f) During the building process, the part is fully supported by the powder and no additional support is required.
g) Parts can be built on top of others.

1.2   Disadvantages

- The powder material requires being heat up to the temperature below the melting point before the building process which takes about 2 hours. After building the parts, it also takes 5 to 10 hours to cool down before removing the parts from the powder cylinder.
- The smoothness of the surface is restricted to the size of the powder particles and the laser spot resulting that the surface of the part is always porous. Smooth surface can only be obtained by post processing.
- The process chamber requires continuous supply of nitrogen to provide a safe environment for the sintering process to be taken place resulting expensive running cost of the process.
- Toxic gases will be generated from the process which leads to an environmental issue.
- Process using different material require different license.

2.   THE PRECISION STUDY OF TUBE MODELS

A primary requirement that any client has from the producers of goods, be they industrial or consumer goods, is quality. An important quality component is precision.

It is very important for a model, a part in general, to have the necessary precision, (tolerance field) because of the attraction of some unwanted elements, such as:

a) Poor functioning of the assembly to which the part belongs.
b) The appearance of some costs (expenses) derived from the necessity to perform some subsequent operations to improve precision.
c) Technical flaws of some tools-machines because of imprecisely made parts.

In this case, knowing the destination of the model and the environment it will work in, it is interesting to see the precision achieved. The obtained information will characterize not only the quality of the surface of the detail "ELBOW A4" but the entire fabrication process by selective laser sintering.

The precision study of the model made by SLS will be done by experimental research on the machine ECLIPSE 500. Basically, measurements will be made on the real part and they will be compared to the parameters of the CAD model, which is considered the ideal model of reference. The difference between the two values will be the actual deviation of the real surface from the nominal one.

Stages:

a) Introducing the 3D model of the prototype in ECLIPSE 500.
b) Setting, fixing and orientation of the part on the measuring table of the machine.
c) Overlapping the system of coordinates of the machine to that of the model.
d) Using the "Umess" soft to find out possible deviations of shape. In the present case there are no such deviations as there are no factors to influence this e.g. the wear of the tool, an elastic system machine-tool-part and the cutting forces.
e) Using the second software "HOLOS" to determine deviations of position. Before the proper measurement a tolerance field will be chosen to compare the found values.
f) Interpreting the results.

For a better observation of the phenomena the Sinter Station model was placed on exact measuring table, in the exact position it had while being made, that is on the plane side of it.

For the result to correspond to reality it is necessary to make many measurements. That is why, the entire surface of the model will be full of as many virtual measuring points as possible. On "ELBOW A4" 390 measurements were made (Figure 2). The values that appear in the measuring protocol are as follows:

–   ISTMASS – value measured on the physical model;

- NENNMASS – value existing in 3D model;
- O.TOL – superior deviation of the tolerance field;

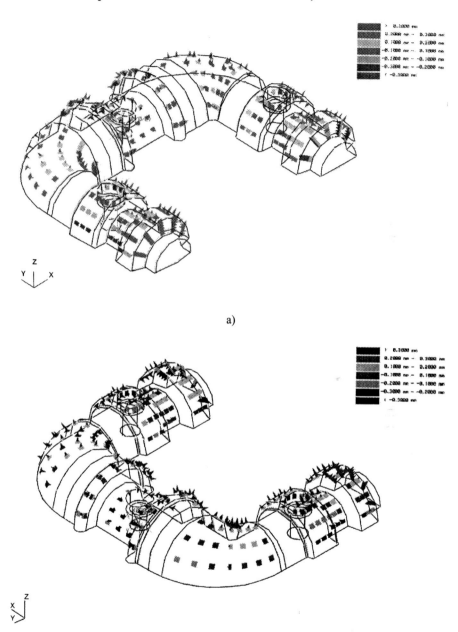

a)

b)

Figure 2. Distribution of errors on the entire surface

- U.TOL – inferior deviation of the tolerance field;
- ABW – effective deviation of the point measured to line 0;
- UEB – value of exit from the tolerance field;
- X, Y and Z represent the coordinates of the measured points and D is the distance between point ISTMASS and point NENNMASS.

At a preliminary measurement it was noted that the real points are moved from the nominal ones in an interval between –0.322 and +0.509; in some regions there is an excess of material and in others a lack of it.

It is interesting to notice the distribution of these errors on the entire surface of the part. For this reason, the part has been divided into 16 zones. They represent some simple surfaces (2 plane surfaces and 14 curved surfaces).

3.   CONCLUSIONS

Although in the graphical representations made after the measurements everything is described on an amplified scale, the tolerances are within the normal limit for such a procedure. Indeed, the SLS process is not very precise (Table 1, 2, 3, 4 and Figures 3, 4) in comparison with other classic technologies, but this can be overlooked taking into consideration the short time it takes to make these prototypes.

Table 1. Distribution of errors according to different tolerance fields (total number of measurements : 390): Nr of points in the interval

Tolerance field	Nr of points in the interval	Percentage(%)
-0,1…+0,1	149	38,21
-0,2…+0,2	307(158)	78,71(40,52)
-0,3…+0,3	372(65)	95,38(16,66)
over 0,3	390(18)	100(4,61)

Table 2. Distribution of errors according to different tolerance fields: Nr of X values in the interval

Tolerance field	Nr of X values in the interval	Percentage(%)
-0,1…+0,1	294	75,39
-0,2…+0,2	367(73)	94,1(18,72)
-0,3…+0,3	390(23)	100(5,89)

Table 3. Distribution of errors according to different tolerance fields: Nr of Y values in the interval

Tolerance field	Nr of Y values in the interval	Percentage(%)
-0,1…+0,1	315	80,77
-0,2…+0,2	374(59)	95,89(15,14)
-0,3…+0,3	387(13)	99,23(3,33)
over 0,3	390(3)	100(0,76)

Table 4. Distribution of errors according to different tolerance fields: Nr of Z values

Tolerance field	N of Z values in the interval	Percentage(%)
-0,1...+0,1	256	65,64
-0,2...+0,2	363(107)	93,07(27,44)
-0,3...+0,3	390(27)	100(6,92)

In this case, if we compare the values of the deviations that resulted from the measurements on the part, we can see that out of the total of 390 – 149 fit the tolerance field chosen for comparison, in percentage 38.20%.

Although the average deviations that Sinter Station 2000 works with are of ±0.2 mm it is noticed that in the tolerance field only 78.71% of the values fit, whereas normally, over 90% should. This is because of the requirement that all measured points need to meet: at the same time and on no direction they are allowed to exceed the maximum established value.

Best precision is obtained on Oy axis. An explanation is that the palpate on ECLIPSE 500 was imposed the measuring pace. Deviations because of the system are almost nonexistent and thus few erroneous values will result. Deviations on Ox are very close in value and number to those on Oy, i.e. the distribution of errors in the xOy plane is uniform.

From the date, one can see that as compared to Ox and Oy, there are more deviations on Oz that exit the tolerance field. The part manufactured by SLS grows by successive adding of layers. Due to this principle an inevitable scale effect arises, that is visible with the naked eye. This can be diminished by reducing the thickness of the layers (to a certain extent, naturally) but this is time consuming.

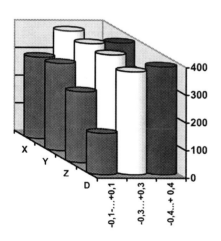

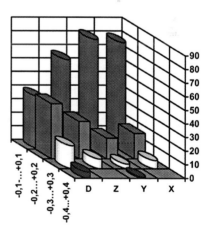

Figure 3. Distribution of errors according to different tolerance field

Figure 4. Number of values in the new tolerance field (%)

Comparing a model made using conventional technology to one made using SLS the quality – price ratio (the effort to achieve this) is superior to SLS. Precision has to be there where and when it is asked for. That is why, sometimes trying to obtain more precise parts than necessary is useless and nor recommended.

In practice, parts can be positioned and oriented in a variety of ways. The chosen one is that which ensures a superior surface quality, in that region of the part which is imposed. In the case of horizontal and vertical surfaces (unlike the curved ones) the scale effect will be zero and thus the quality of the surfaces will be very good.

REFERENCES

1. C. PETRA, *Fabricarea rapidă a prototipurilor prin Sinterizare Selectivă cu Laser folosind drept material DuraFormul* (lucrare de disertaţie).
2. P. BERCE, N. BÂLC, a.o., *Fabricarea rapidă a prototipurilor* (Bucureşti, Editura Tehnică, 2000).
3. www.atirapid.com
4. www.alexdenuoden.nl

MOBILE ROBOT SYSTEM CONTROLLED THROUGH MOBILE COMMUNICATIONS

T. Butnaru, F. Gîrbacia, F. Tîrziu and D. Talabă
Transilvania University of Brasov, Romania

Abstract: A relatively new research direction in mobile robotics concerns the communication technology for the control mobile robots in conjunction with other devices or between a team of mobile robots. This paper describes an approach for the remote control of a mobile robot using Bluetooth communication. Aspects regarding Bluetooth technology, architecture and components of the mobile robot developed are presented. Software for Bluetooth communication and a virtual environment in which the operator may create a path for robot are also presented.

Key words: mobile robots, Bluetooth, virtual environment, wireless communication.

1.   INTRODUCTION

A relatively new research area in mobile robotics concerns the communication technology for the control of the mobile robots in conjunction with other devices or between a team of mobile robots. A good communication system between the user and the robot will expand the capability and versatility of mobile robots. In this context, wireless communication allows a mobile robot to be used as a part of more complex systems or to interact with the user using common equipment (e.g. mobile phone, Pocket PC).

Nowadays, these aspects risen significant interest because many new types of applications become thus possible, especially in the domestic area. In [4], a wireless local area network is created based on the IEEE 802.11 standard for a team of mobile autonomous robots. Some problems of this communication protocol are identified, e.g. high degree of message losses and limited reach of messages. For the communications between a mobile

D. Talabă and T. Roche (eds.), Product Engineering, 433–442.

robot and a Palm Pilot M500 an infrared interface is used in [9]. Palm Pilot M500 is mounted on the robot platform and the distance between the transmitter and the receiver is very small. It was shown that the infrared communication has some disadvantages such as: infrared receiver can be at an angle of maximum 30 degrees from the infrared transmitter, the distance between the transmitter and the receiver can be of maximum 10 meters and the transmission of the signal is not allowed trough the walls. The infrared communications seems to be not secure, accurate and reliable and finally not a very good method of wireless communications.

This paper present an approach based on a system which uses Bluetooth communication, a relatively recent wireless radio communication standard. The tested system was made up using the following components: wheeled mobile robot, software for control using PC and a virtual environment that enable the user to create various paths for the robot. These components are described in the following sections.

2.   THE MOBILE ROBOT ARCHITECTURE

The architecture chosen for the mobile robot is of type "three wheeled robot". Two of the wheels are driving wheels and one is a steering wheel.

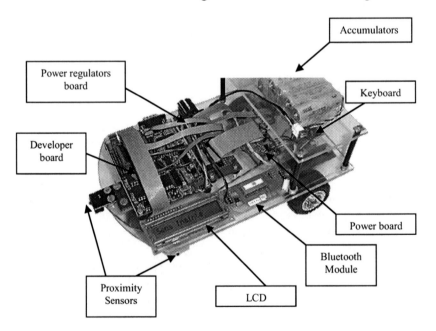

Figure 1. Components of the mobile robot

The mobile robot program was developed using the assembler language provided by Intel (MSC51 family). The data received from the Bluetooth device is similar with the data received from the RS232 interface therefore, the way it is handled is also similar. In order to build the mobile robot the following components have been used (Figure 1):

a) System developer board based on a PHILIPS 80C552 microcontroller. This is an 8 bit CISC microcontroller included in MSC51 family developed by the Intel.

b) Wireless Bluetooth module used to transmit data from/to PC or other mobile devices (e.g. mobile phone, Pocket PC).

c) For the motor alimentation a standard power board was used.

d) A board with voltage regulators for conversion power input to the TTL level.

e) An LCD display unit to display messages on the mobile robot control unit.

f) A keyboard and four proximity sensors.

g) Two stepper motors used for the robot propulsion.

h) High capacity electric accumulators of 19.2V/1300mA.

i) Mechanical support board for the components.

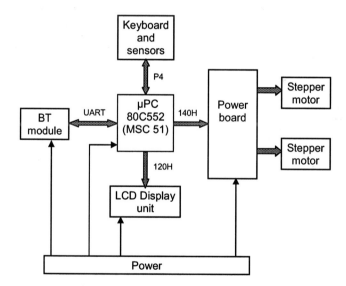

Figure 2. Hardware schematic connection of mobile robot.

The hardware diagram for communication between peripherals and the microcontroller is presented in Figure 2. The command signals are received by the microcontroller using the UART interface (to which the Bluetooth module is connected), or from the sensors and the keyboard connected to the port 4 of the microcontroller. These signals are decoded and transformed in commands for the stepper motors and the LCD display. On the port 0 of the microcontroller, other devices are connected: the power board used to supply stepper motors, the LCD display device and the external memory. The command signal of the microcontroller is transmitted to the corresponding device using the 74LS138 address decoder.

The commands signals transmitted to the LCD display and stepper motors are 120H respectively 140H (Figure 2). The stepper motors are connected to the microcontroller using the gains model ULN 2003. This gain convert the TTL signals to output voltage needed to insure a proper functionality of the stepper motors. The mechanic power is transmitted from the motors to the wheels through two stage geared transmissions with the ratio of gear equal with 5.625 (Figure 3). These features generate enough power to transport a weight of up to 5 kg.

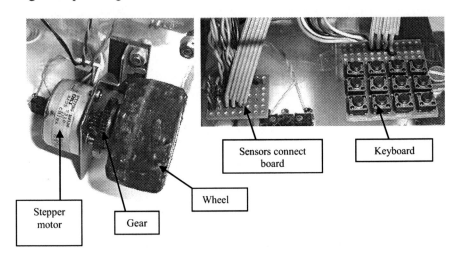

Figure 3. Preview the transmission system and the keyboard.

The keyboard (Figure 3) is used to transmit the following commands:
–   move the robot to the left, right, forward or backward,
–   start/stop,
–   switch between virtual and normal mode of the robot control.

These commands are displayed on the LCD display and also transmitted to the PC using the Bluetooth interface.

The keyboard is organized in form of 4 x 4 matrix connected to pins 1-4 of the P4 port [2].

The proximity sensors (Figure 3) have the same functionality, as the keyboard commands.

The LCD is based on the connection scheme presented in Figure 4. The LCD's data lines are directly connected to the controller's data bus. The

LCD is accessed as a port, involving 4 different addresses (usually the address lines A0 and A1 are connected, via the latch circuit 74373, to the lines RS and R/W), while the address decoder 74138 activates the access line E of the LCD. The routines written for the LCD device are meant to perform various operations (e.g. write/read commands, write data, initialization of LCD).

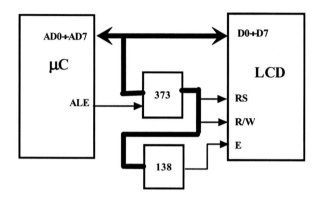

Figure 4. Connection LCD display to microcontroller

3. THE MOBILE ROBOT COMMAND USING BLUETOOTH TECHNOLOGY

3.1 Bluetooth technology

The Bluetooth technology was created by Ericsson in 1994 and is used to replace the cables in the office, in laboratories or at home. As a result a new concept of personal area network has emerged and a new kind of network was introduced.

The devices used within this technology are small, low-power, short range and need cheap radio chips. The power of the signal is around 1mW, compared to 3W for mobile phones. They have a short range of action (about 10 meters) and thus the probability to interact with other devices in the area is reduced. This is also due to the transmission method Frequency-Hopping Spread-Spectrum.

A simple Bluetooth network is called pico-net and can comprise up to 8 devices. There is only one master device in such a network, the other devices being slaves. The master device is established during the connection and it is the device that initiates the communication.

For the communication module of this application an embedded Bluetooth module from Merlin Systems Corp. Ltd was used and an USB

Bluetooth dongle from MSI. The two components provide a channel for transferring commands from the computer to the mobile robot.

The architecture of the system as shown in Figure 5 consists in two main parts: a Bluetooth enabled PC and a mobile robot carrying the embedded Bluetooth module. The connection established between them is a point-to-point connection.

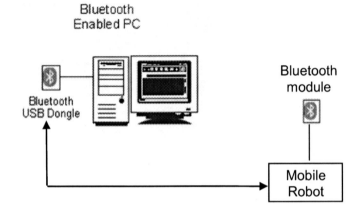

Figure 5. Architecture of the system with Bluetooth

Bluetooth has a wide range of applications such as: mobile phones, PDA, automotive, cameras and scanners, etc. In order to use Bluetooth in such applications, different kind of data transfer between devices is provided and established by the so called "profiles". The profiles have been developed in order to describe how the implementations of the user models have to be implemented. The user models describe a number of user scenarios where Bluetooth performs the radio transmission.

Among the most used Bluetooth profiles one can note:
a) SDAP (Service Discovery Application Profile) used to detect the services provided by another Bluetooth device
b) SPP (Serial Port Profile) used to create a virtual serial port over Bluetooth and establish a serial RS-232 connection.

For our application we have used SDAP in the initialization phase and then SPP in order to connect the PC to the mobile robot and transmit the commands.

3.2   The robot command system

The Mecel SDK – Demo Kit has been used to develop the software for the communication with the mobile robot. The software has a functional part and a graphical user interface (Figure 6).

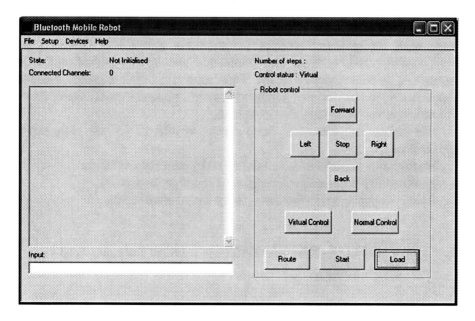

Figure 6. Mobile robot command program from PC using Bluetooth

In order to command the robot from the PC (Figure 7) some few steps have to be followed:

a) First a security mode is initialized (this aspect concerns the level of security established for the communication).

b) The next step is the setup of the transport layer.

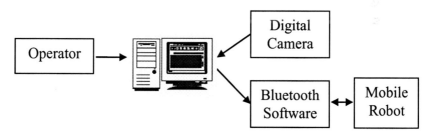

Figure 7. Direct mobile robot control using Bluetooth technology.

c) Search for the Bluetooth devices that are in range; once a device is found, its address is displayed; at this point a problem of the used technology was noticed with the connection time which is slow, because of the radio delay which appears when other Bluetooth module are negotiating the protocols.

d) The next step is the search for the available services for the devices that are meant to connect to.

After these stages, the communication between the Bluetooth devices can begin using the identified services. For the robot control two methods have been implemented, such as the commands can be transmitted either in Normal or Virtual Control mode. Thus, one can select between Normal control mode to transmit the commands one by one and Virtual control mode to transmit a file containing the commands.

The following problems have been identified in the considered application:
– the communication is reliable but in a short range (up to 30 m),
– the reliability is generally decreasing with range (expected),
– the connection time is slow (sometimes over 5 seconds).

4.   VIRTUAL ENVIRONMENT FOR THE ROBOT CONTROL

To evaluate system behavior a virtual environment has been developed (Figure 8) that would enable the operator of the robot to navigate through a series of obstacles. The obstacles foreseen are defined both in the virtual word and in the real world. The operator drives the virtual robot trough the obstacles and generate a file containing a path for the real robot. The instructions generated in the virtual environment are read from the generated file and transmitted to the robot using the Bluetooth communication.

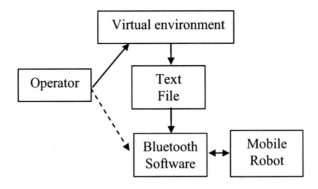

Figure 8. Virtual aided mobile robot control using Bluetooth technology

The virtual vehicle displayed in the VR environment is implemented as a computer simulation which incorporates a model of the vehicle's dynamics. For the implementation of this environment GLScene has been used, an OpenGL based 3D library for Delphi that is distributed under the Mozilla Public License. This library combines the use of Delphi's VCL to give higher-level management of the 3D scene. GLScene provides a hierarchical object structure, movement functions, and procedural objects to provide a 3D engine. Collision between the virtual robot and virtual obstacles is enabled by a function provided by this library. To control the virtual robot, the operator can use a joystick or the keyboard.

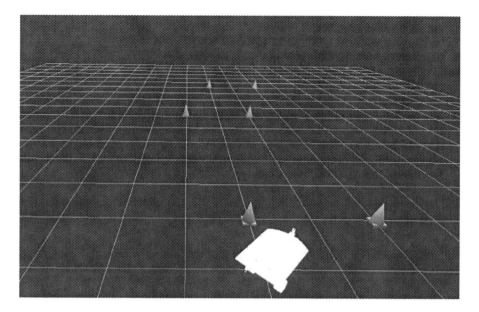

Figure 9. Virtual Environment scene for the robot control.

5. CONCLUSIONS

The system presented in this paper provides a reliable real-time wireless communication for application that consists in mobile robot systems using Bluetooth technology. A virtual environment was created that enables the user to create various paths for the mobile robot.

The proposed approach can be applied for the communication between team of mobile robot system as well.

REFERENCES

1. M. MILLER, *Discovering Bluetooth* (Sybex Incorporated, July 2001).
2. A. DUMITRIU, T. BUTNARU, *Solution for human-machine interfaces in systems with microcontrollers.* (The VITH International Conference on Precision Mechanics and Mechatronics COMEFIM-6, 10-12 October 2002, Brasov, pp. 67-72).
3. A. DUMITRIU, *Techniques for Information Processing* (Transilvania University of Brasov, 1996).

4. E. NETT, S. SCHEMMER, *An architecture to support cooperating mobile embedded systems* (Conf. Computing Frontiers 2004, pp. 40-50).
5. O. FEGER, *Mc-Tools2. Die 8051-Mikrocontroller Familie.Einführung in die Software.* (Hardware+Software Verlags OHS, Traunstein, Germany, 1991).
6. GLScene, 2004 – OpenGL Solution for Delphi. Founded and previously developed by LISCHKE M., currently developed and maintained by GRANGE E., www.glscene.org, last visited on 10 July 2004.
7. http://www.mecel.com, MECEL BLUETOOTH™ SDK - DEMO KIT
8. http://www.palowireless.com/infotooth/tutorial/profiles.asp
9. S. GOVINDAN, *Creating a Wireless Interface between a Robot and Handheld PDA,* (2003).
10. A. KEDDAR, N. CISLO, D. TALABĂ, J. G. FONTAINE, *Several mechanical architectures for telepresence with Semi-Legged Mobile Robot* (The 9th World Congress IFTOMM on Theory of Machines and Mechanisms, Milano, 1995, pp. 2348-2352).
11. D. TALABĂ, J. G. FONTAINE, S. BUCHHEIT, E. H. GAMAH, *RAMSES-Walking platform with pneumatic actuators.* (Buletinul Simpozionului Național cu Participare Internațională PRASIC'94, Brașov, 1994, pp.215-223).

Part 5

GREEN ENERGY

1.   INVITED LECTURES

2.   CONTRIBUTIONS

OPTIMIZATION OF THE WIND GENERATION: COMPARISON BETWEEN TWO ALGORITHMS

G. Tapia and A. Tapia Otaegui
Systems Engineering and Control Department, University of the Basque Country, Spain

Abstract: Wind energy generation is increasing its participation in energy distribution and has to compete with other energy sources that are not so variable in terms of generated active power. It is important to consider that nowadays, the active power demand can vary quite rapidly and different sources of electricity generation must be available. In the case of wind energy, wind speed predictions are an important tool to help producers make the best decisions when selling the energy produced. These decisions are crucial in the electricity market, because of the economic benefits for producers and consequently their profitability, depends on them. Hence, the optimisation of wind energy production and consequently the economic benefits derived from its connection to the grid becomes one of the most important problems to be solved in the very close future. This paper presents two control strategies developed for the active power regulation of wind farms made up with double fed induction generators, in order to obtain the maximum active power from the wind hitting the blades of a mill. The dynamic performance of a real 660kW generator is analysed when both algorithms are applied.

Key words: wind power generation, double fed induction machine, production optimisation, sliding mode control.

1. INTRODUCTION

Wind energy is becoming one of the most important renewable energy sources in several countries all over Europe. But even if it is a clean source and it is zero fuel cost, there are some problems when trying to connect this kind of distributed generation to the electric grid.

On the other hand, the geographical location of wind farms is another fundamental question. Because wind energy is disperse, intermittent and not very regular in terms of the amount of generated power it is important to

445

D. Talabă and T. Roche (eds.), Product Engineering, 445–472.

consider that when a "good location" is identified and used, a generator can work 6000 hours per year and the amount of energy produced can be similar to the energy produced by a generator working at full power for 2200 hours per year. Therefore, before a wind farm is built the local wind conditions must be evaluated exactly and the predictions analysed.

Consequently, several research works related to the regulation of wind farms have been published [1, 2, 3, 4, 5, 6, 7]. These papers present research results of voltage and frequency control algorithms developed to obtain the maximum economic profits under the best control conditions. The generator analyzed in this paper (the most used one in the recently built wind farms) is based on a double fed induction machine (DFIM), which dynamic performance can be described by the following features:

a) Its rotor speed can vary from subsynchronous to supersynchronous speed.
b) It generates a constant frequency active power.
c) The generated active and reactive power can be controlled in an independent way.

On the other hand, it is important to consider that the economic benefits of the exploitation of a wind farm are proportional to the active power generated in the farm. Additionally, in a DFIM, the generated active power depends on the power coefficient, C_p, which is different for each generator and is directly related to the proportion of power extracted from the wind hitting the blades of a windmill. This means that, for each instantaneous wind speed in a mill, there is a turbine rotation speed which delivers to the grid the maximum active power generated by the above mentioned mill. Therefore, if the maximum generation point for each wind speed can be established, following the trajectory described by all the maximum generation points, the economic benefits of the exploitation of wind energy will be maximum too.

Finally, this kind of double fed induction generator can be controlled in order to generate or absorb reactive power, so that, the voltage level of the electric grid where it is connected to, can be maintained within the permitted limits [7, 8]. As the control of the active and reactive power can be developed in an independent way, two different kind of references must be established:

- Tracking of the trajectory of the maximum active power generation points for the best economic exploitation conditions.
- Reactive power constant reference for the voltage level control.

The second type of control, the reactive power control, is not going to be analyzed in this paper and the only important subject under analysis is the economic profits of the exploitation of a wind farm. So, the main objective of this paper is the development of a control law to ensure that the best trajectory for the active power generation is followed and to compare the obtained dynamic performance with the typical PI controller performance.

2. PROBLEM DESCRIPTION

This paper tries to describe the control algorithms developed for the active power control in order to find a solution both for wind farm installers and for utilities. The above mentioned solution is related to the generation of electricity under the best conditions, which can be classified as:

1. Optimisation of the active power level obtained from the wind hitting the blades of the windmill which leads to important economic benefits.
2. Possibility of establishing different references for active power generation in order to deal with situations of deregulated electric market, that means, constant references fixed by the electricity buyer.

Taking into account the actual and real configuration of wind farms (electronic components, generator, protection systems, etc.), the developed solution must be simple, effective and robust so as to be installed in these real systems. Thus, the methodology used to complete the work is based on the following objectives:

a) Analysis of the dynamic performance of a 660 kW double fed induction generator (DFIG) in order to calculate its active and reactive power generation capability.
b) Development of a precise mathematical model for a DFIG and validation of the model with real data.
c) Design of a traditional PI and another type of control-law to ensure the optimisation of the active power generation in a DFIG. In this way, the trajectory described by the maximum generation for each wind speed will be followed.
d) Implementation of the designed control laws on the developed and validated model and analysis of their dynamic performance.
e) Possibility of installing the designed laws on a real generator: advantages and problems.

3. DOUBLY FED INDUCTION GENERATOR ELECTRICAL MODELLING

The first step of the work developed in order to obtain the wind farm model was to analyse the double fed induction generator and obtain a correct model, [9]. When trying to model a double fed induction machine (DFIM) under different wind speeds, it is important to consider that this kind of wound rotor machine has to be fed both from stator and rotor sides, Figure 1.

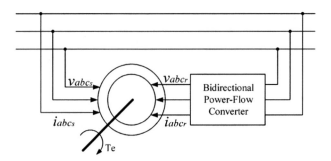

Figure 1. Double fed induction machine

Normally, the stator is directly connected to the grid and the rotor side is interfaced through a variable frequency power converter. In order to cover a wide operation range from sub synchronous to super synchronous speeds, the power converter placed on the rotor side has to operate with power flowing in both directions. This is achieved by means of a back-to-back PWM inverter configuration. The operating principle of a DFIM can be analysed using the classical theory of rotating fields and the well-known d-q model, as well as both three-to-two and two-to-three axes transformations, [10].

Then, to deal with the machine dynamic behaviour in the most realistic possible way, both stator and rotor variables are referred to their corresponding natural reference frames. In other words, the stator side current and voltage components are referred to a stationary reference frame, while the rotor side current and voltage components are referred to a reference frame rotating at rotor electrical speed, ω_r. When aiming to express the induction machine electrical model in the above-mentioned reference frames, it is first necessary to perform the Clarke's transformation, [10], from the three-phase to the d-q current and voltage system.

In this way, taking the general three-phase model of the electric machine dynamic performance as a starting point, the "Quadrature-Phase Slip-Ring" model, [10], for the DFIM might be expressed through the following matrix equation:

$$\begin{bmatrix} u_{sD} \\ u_{sQ} \\ u_{r\alpha} \\ u_{r\beta} \end{bmatrix} = \begin{bmatrix} R_s + pL_s & 0 & pL_m\cos\theta_r & -pL_m\sin\theta_r \\ 0 & R_s + pL_s & pL_m\sin\theta_r & pL_m\cos\theta_r \\ pL_m\cos\theta_r & pL_m\sin\theta_r & R_r + pL_r & 0 \\ -pL_m\sin\theta_r & pL_m\cos\theta_r & 0 & R_r + pL_r \end{bmatrix} \begin{bmatrix} i_{sD} \\ i_{sQ} \\ i_{r\alpha} \\ i_{r\beta} \end{bmatrix}, \tag{1}$$

where, L_s and L_r are the stator and rotor side inductances, L_m is the magnetising inductance, R_r and R_s are the stator and rotor side resistances, θ_r, is the rotor angle (variable) and p corresponds to the derivative operation, p = d/dt. The roles of current and voltage components in (1) need to be interchanged so as to treat voltages as independent variables - system inputs - and currents as dependent variables - system outputs - as corresponds to a voltage fed doubly fed induction generator (DFIG).

Figure 2 represents a typical DFIM control structure that works with two cascaded control loops. The inner loop (stator-flux oriented vector control) task consists in controlling independently the rotor current direct - i_{rx} - and quadrature – i_{ry}- components expressed according to the reference frame fixed to the stator-flux linkage space phasor. Two identical PI controllers are typically used to implement this inner control loop. The outer one governs both the stator side active and reactive powers.

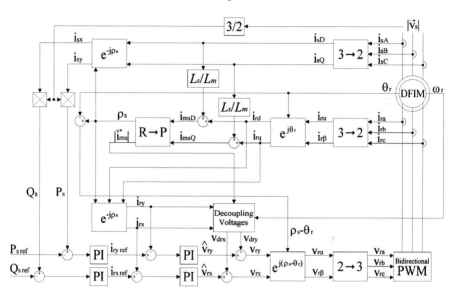

Figure 2. Control structure for the DFIM

3.1 Operating Conditions

In order to develop a correct model of the wind generator (a 660 kW real generator), some operative restrictions must be considered. First, the reactive power generation/absorption capacity for each generated active power of the modelled generator must be considered; that means that the typical P/Q curves depending on the ambient temperature have to be evaluated.

On the other hand, the currents that can be driven by the double-sided PWM must be analysed to obtain its operative conditions. From the tests performed in a real test bed, the curves appearing in Figure 3 have been obtained. The three inner curves correspond to the P/Q curves (active/reactive power curves) of the generator and the most restrictive one is related to the hottest ambient temperature.

It can be seen that the outer curve, corresponding to the limitations imposed by the inverter are less restrictive than those fixed by the generator itself. Consequently, it must be ensured that the operation range of each generator(P_i, Q_i), is always maintained inside the most restrictive curve ($P_{i\ max}$, $Q_{i\ max}$), the one obtained for the hotter ambient temperature, Figure 4.

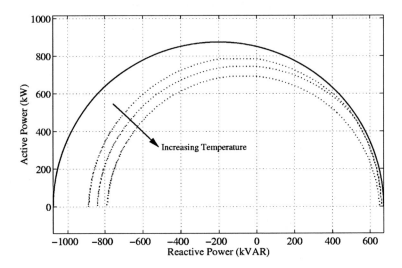

Figure 3. P/Q curves for the generator + inverter

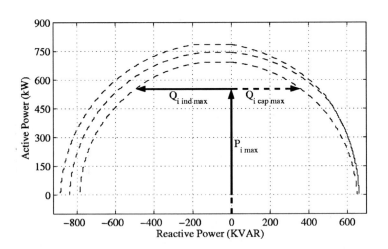

Figure 4. Limitation for the P/Q operation in the DFIM generator

3.2   Experimental validation of the model

Before starting with the analysis of different active power controllers, the validity of the developed DFIM model must be demonstrated.

Table 1 presents the main parameters, referred to a 20° C ambient temperature, of a typical double fed induction machine of 660 kW, used in the wind farm installation selected for the optimisation algorithm validation.

Table 1. DFIG Electric Parameters

R_s, stator resistance per phase	0.0067 Ω
X_{sl}, stator leakage reactance per phase	0.0300 Ω
n, general turns ratio	0.3806
X_m, mutual reactance	2.3161 Ω
R_r, rotor resistance per phase	0.0399 Ω
X_{rl}, rotor leakage reactance per phase	0.3490 Ω

The model presented in section 3 has been implemented under the MATLAB/SIMULINK simulation package. Figures 5 and 6 show respectively, the dynamic behaviour of the simulated and the real 660 kW generator when, it is generating 300 kW as a result of the instantaneous wind speed with a cosφ=1, and a power factor set-point change takes place. This new set point corresponds to a cosφ=0.857 capacitive power factor. Generated active and reactive powers are considered to be positive. Since wind conditions remain constant during the whole trial, it can be seen how the generated active power is not changed, while the generated reactive power varies rapidly so as to track the new power factor set point.

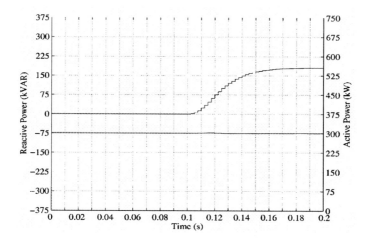

Figure 5. Generator model performance under step changes in the power factor

From figures 5 and 6, it can concluded that the model dynamic performance corresponds exactly with the real generator performance and so, it will be used to validate different control-laws. It has to be mentioned that, a correct model of a DFIM can be very useful in the near future because of the continuous development that this kind of generator has in Spain, related to the increasing promotion of the use of wind energy.

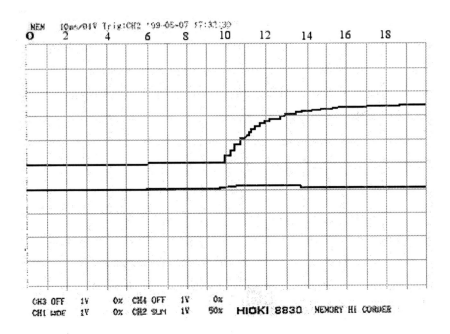

Figure 6. Real generator performance under step changes in the power factor

4.   ACTIVE AND REACTIVE POWER REFERENCES

Actually the most widely used algorithms for the control of DFIM based generators establish the reactive power generation reference equal to zero. This is due to the fact that no voltage regulation (through reactive power control) is required in wind farms, even if this control can be developed when using DFIGs, [7].

On the other hand, as it has been mentioned above, the generation of active power is directly related to economic benefits for a wind farm owner and consequently, the definition of a correct reference for the active power generation becomes a crucial part of the whole process. As this is the main objective of the present work, the first step that is going to be analysed is to set a correct active power reference.

4.1 Active power reference

To start with the active power reference establishment it is important to consider the mechanical power, P_m, developed by a wind turbine which depends directly on the blade radius, the power coefficient and the wind speed hitting the blades of the generator,

$$P_m = \pi/2\,C_p(\lambda, \beta)\rho r^2 v^3$$
$$\lambda = \omega_r\, r/v \qquad (2)$$

where r = blade radius; C_p = power coefficient; V = wind speed; ω_r = rotational speed; λ = tip speed ratio; β = pitch angle; ρ = air density.

For the analysis developed in this paper, the power coefficient of a 660 kW generator has been considered, Figure 7. There it can be seen that, for any given wind speed, there is a rotation speed, ω_r' which generates a maximum power $P_m = P_{opt}$.

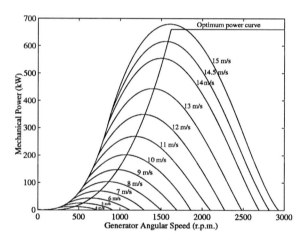

Figure 7. Curve of the optimum Cp for the 660 kW machine.

Additionally it is important to consider that when wind speed changes, the rotation speed varies so as to follow the change. But as a reliable measurement of the wind speed hitting the blades of the generator can not be ensured, the equation for obtaining the target power, P_{opt}, can be modified eliminating the dependence on wind speed.

$$P_m = P_{opt} = \left(\pi/2\,C_{p\,opt}\,\rho r^2\,(r/\lambda_{opt})^3\right)\omega_r^3 . \qquad (3)$$

It can be seen in (3) that the generated power is a function of the rotor speed and follows a cubic law:

$$P_{opt} = k\,\omega_r^3 \qquad\qquad\qquad (4)$$

But, even if the generated power can be theoretically unlimited, it is really limited because wind turbine components can be operated only within their mechanical and electrical restrictions. Consequently, when the reference value for P_{opt} is greater than the nominal value of the electric machine, P_{Nom}, the pitch control is activated (to generate the nominal value) and the optimum active power reference is not followed till $P_{opt} \le P_{Nom}$.

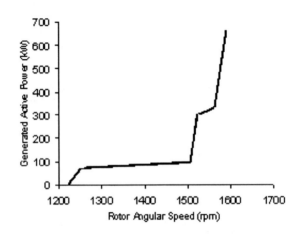

Figure 8. Linearization of the cubic curve in several sections

Taking into account the developed analysis, some implementation aspects must be considered. The execution of a cubic algorithm for trajectory tracking can cause some problems, due to the actual hardware characteristics of wind generators and the memory needed to develop the mathematical calculations of the algorithm. Thus, some linear approximations must be developed to implement the **P$_{opt}$** cubic curve in the hardware of the generators:
– Linear curve in sections, Figure 8.
– Linear curve in sections, considering hysteresis, Figure 9.
Both approximations are almost equal but the hysteresis has been introduced in the second one, Figure 9, not to damage the machine because when sudden changes on the wind speed occur, the reference curve changes from one linear section to another one and consequently, the rotor angular speed changes quite rapidly.

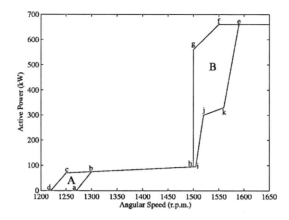

Figure 9. Linearization with hysteresis

4.2   Active Power Control: different situations

To start with the design of the control laws that have to be applied for active power regulation, it is necessary to consider that the active power generation in a wind mill can be analysed under two different scenarios:
a) tracking of the cubic curve, P_{opt} curve (increasing economic benefits),
b) tracking of a constant reference that can be modified depending upon the needs of the "energy buyer".

In the first case, a special algorithm must be developed to follow a variable reference in order to obtain always the P_{opt} value and consequently, important economic benefits. The control algorithm applied for this purpose must ensure the correct tracking of the reference curve. In the second one, the active power reference is constant and defined by the "buyer", normally by the utility which is the owner of the electric grid where the wind farm is connected to. In this case, the reference is transmitted to the wind farm and to each mill in the farm using its communication network and the set point modifications are quite slow (approximately every five minutes). Therefore, a PI controller would be enough to ensure the reference maintenance in this case.

5.   DESIGN OF THE CONTROL ALGORITHM FOR THE TARGET POWER TRACKING

As the main objective of this paper is to deal with the best situation from the economic point of view, only the first scenario of active power control is going to be considered and two different control algorithms will be designed and implemented in the model of the wind generator developed in section 3.

The first one is a traditional PI controller (the same as the used in the second scenario with constant references) and the second one is based on a sliding-mode controller. The results obtained from the implementation of both algorithms will be compared in order to detail some conclusions for future real applications.

The first algorithm described in this section is a PI controller because its implementation is quite simple, it evaluates a small amount of data and its performance is very robust. On the other hand, its performance for trajectory tracking purposes has proved not to be very exact.

The second one, the sliding mode controller, is very suitable for trajectory tracking problems but its implementation becomes more complicated than the one in the first case. Its performance can be as robust as the obtained with the PI algorithm and consequently, it has been considered as an interesting option to analyse.

5.1   Design of the PI controller

When controlling the active power generated by a DFIG, two PI controllers must be tuned, Figure 2. The structure of the PI controllers can be better explained in Figure 10. The first one corresponds to the inner loop (vector control) and the outer one, to the controller for the trajectory tracking purpose. The PI controllers used in both cases present a especial structure [11], with a b parameter in the reference signal, in order to develop a pole assignment trying to assign at the same time, the closed loop poles and the resultant zero. In this way, the desired system dynamic performance (defined for the pole assignment), is not significantly disturbed

$$u(t) = k_p (by_{ref} - y) + k_p/T_i \int_0^t (y_{ref} - y)dt . \qquad (5)$$

Once the active power reference has been established (cubic curve or linear approximation of the curve), it is necessary to measure both the rotor angular speed, ω_r, and the actually generated active power, P_s. Taking into account the measurements of these two data and the value of the instantaneous active power reference, $P_{s\,ref}$, the obtained error, $P_{s\,ref} - P_s$, can be evaluated.

Then, the outer control loop can be analysed in order to develop the set point for the i_{ry} current component (through a PI controller), needed in the inner control loop (vector control)

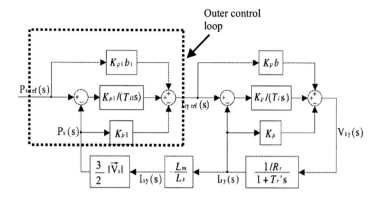

Figure 10. Active power control loop using a PI controller

$$i_{ry\,ref} = -K_{p1}\left[\left(b_1 P_{s\,ref} - P_s\right) + \frac{1}{T_{i1}}\int_0^t \left(P_{s\,ref} - P_s\right)dt\right].$$ (6)

And the transfer function of the closed loop relating the active power target (the optimum one), $P_{s\,ref}$, with the really obtained one, is evaluated as

$$\frac{P_s(s)}{P_{s\,ref}(s)} = \frac{K_g\left(s + \dfrac{1}{bT_i}\right)\left(s + \dfrac{1}{b_1 T_{i1}}\right)}{s^3 + a_2 s^2 + a_1 s + a_0},$$ (7)

where,

$$K_g = \frac{3L_m\left|\vec{v}_s\right|K_p bK_{p1} b_1}{2L_s L_r'} \quad ; \quad a_0 = \frac{3L_m\left|\vec{v}_s\right|K_p K_{p1}}{2L_s L_r' T_i T_{i1}};$$

$$a_2 = \frac{R_r + K_p}{L_r'} + \frac{3L_m\left|\vec{v}_s\right|K_p bK_{p1}}{2L_s L_r'};$$ (8)

$$a_1 = \frac{K_p}{L_r' T_i} + \frac{3L_m\left|\vec{v}_s\right|}{2L_s L_r'}\left(\frac{K_p K_{p1}}{T_i} + \frac{K_p bK_{p1}}{T_{i1}}\right)$$

Due to the fact that the obtained closed loop system is a third order system, it is not possible to develop a pole assignment technique to obtain the proportional and integral constant values ($\mathbf{K_{p1}}$ and $\mathbf{T_{i1}}$) of the PI controller. Consequently, an approximation of the inner loop system will be considered: an overdamped second order system dynamic performance is quite similar to the performance of a first order system, with the same t_s, settling time. Thus, it is possible to say that the inner loop can be expressed as

$$\frac{I_{ry}(s)}{I_{ry\,ref}(s)} \approx \frac{1}{1+\frac{t_s}{4}s}, \tag{9}$$

and the closed loop transfer function can be represented by,

$$\frac{P_s(s)}{P_{s\,ref}(s)} = \frac{\dfrac{6L_m\left|\vec{v}_s\right|K_{p1}b_1}{L_st_s}\left(s+\dfrac{1}{b_1T_{i1}}\right)}{s^2 + \dfrac{4L_s+6L_m\left|\vec{v}_s\right|K_{p1}}{L_st_s}s + \dfrac{6L_m\left|\vec{v}_s\right|K_{p1}}{L_st_sT_{i1}}} \tag{10}$$

Now, the operating requirements of the closed loop have to be established in order to tune the PI controller of the outer loop. For this especial double fed induction generator, an over damped response is required, with a ts1=70 ms settling-time. Considering these objectives, the following controller parameters are obtained,

$$K_{p1} = \frac{L_s(5.8t_s - 2t_{s1})}{3L_m\left|\vec{v}_s\right|t_{s1}}, \tag{11}$$

$$T_{i1} = \frac{\xi^2 t_{s1}(5.8t_s - 2t_{s1})}{16.82t_s}.$$

Additionally, an anti wind-up system must be introduced in the algorithm not to apply to the converter a current value greater than the maximum one it can deliver; that means, $i_{r\,ref} \leq I_{MAX}$, Figure 11. Consequently, if the optimum active power generation has to be ensured, the $i_{rx\,ref}$ current component must be limited (the generated reactive power is limited too), while the $i_{ry\,ref}$ current component, related to the generated active power, is maintained in the value obtained from the developed PI controller.

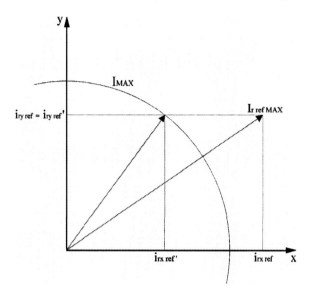

Figure 11. PI controller algorithm with an anti wind-up

5.1.1   Simulation results

Some simulation tests have been performed to analyse the validity of the PI control algorithm. One of these tests appears in Table 2, where it can be seen that important wind speed changes have been considered.

The results obtained for active power, voltages applied to the generator rotor and the curve ωm, angular speed versus Ps, generated active power, appear in Figures 12a, 12b, 13 and 14.

Table 2. Wind speed time evolution and power factor references

Time Period (s)	Wind Speed (m/s)	Reference cos φ_{ref}	
		Value	Type
0 – 1	15	0.98	Capacitive
1 - 3	6	0.98	Capacitive
3 - 5	6	0.95	Capacitive
5 - 7	12	0.95	Capacitive
7 - 9	12	1.00	-------
9 - 11	8	0.90	Inductive
11 – 12	18	0.97	Inductive

The generated active power and its set point are displayed in Figure 12a. A detail of the active power performance appears in Figure 12b. It can be seen that when sudden changes occur in the wind speed the tracking of the reference is not very good because there are some delays between the reference change and the obtained response.

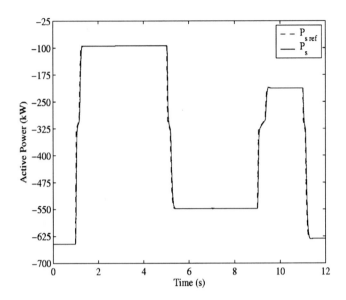

Figure 12a. Generated active power

On the other hand, the voltages applied to the rotor, Figure 13, are capable of being really applied with the used converter. In Figure 14, the ω_m versus P_s graphic appears. It can be concluded that there are some tracking problems that can be improved, even if the whole performance of the algorithm can be defined as quite correct.

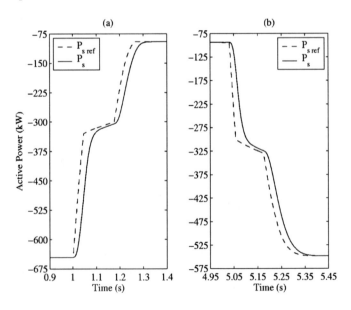

Figure 12b. Detail of Figure 12a

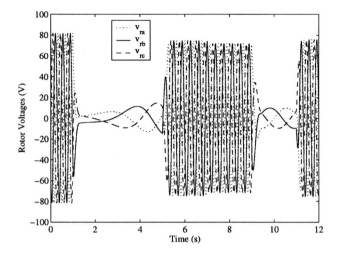

Figure 13. Voltages applied to the rotor

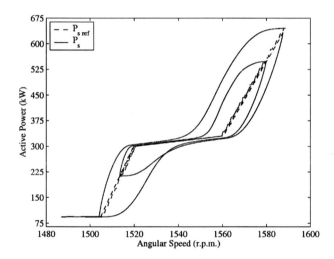

Figure 14. ω_m versus P_s curve

5.1.2 Experimental results

The PI controller previously described and analysed by simulation tests, has been applied in some real generators of a wind farm located in Navarra (northern part of Spain). The used generators, based on the 660 kW electric machine modelled in section 3, and the controller, appear in Figures 15a and 15b respectively.

Figure 15a. Double fed induction generator located in the wind mill (INDAR)

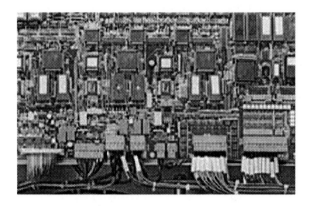

Figure 15b. Generator controller (EHN)

It is important to consider that the hardware installed in those generators is not capable of using the cubic reference curve and a linear approximation with hysteresis has been applied as the optimum active power generation target function.

The results obtained with the application of this type of linear reference and the implementation of the designed PI controller can be observed in Figure 16. As it was proved in the simulation tests, the reference tracking is quite good, but the hysteresis introduced in the reference signal can cause some problems for the controller behaviour. The active power data have been measured during one hour using a sampling period of one second and considering a test developed with variable wind speed. Therefore, it can be concluded that the PI controller presents a correct performance under not important changes in the wind speed and can be used for this type of optimum active power tracking but at the same time, when sudden changes in the wind speed occur, the reference tracking can be improved using other type of control algorithms.

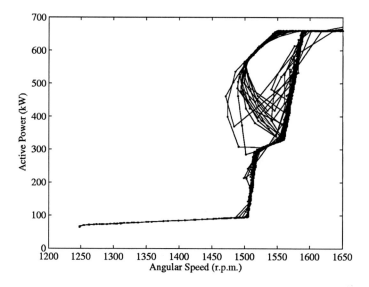

Figure 16. Tracking of a linear reference in a 660 kW DFIG

5.2 Sliding mode control

Once the PI controller has been implemented and checked (both by simulations and real tests), a sliding mode controller is developed trying to obtain a better tracking of the target optimum active power generation curve. The sliding mode control [12] has been selected because its performance is as robust as the one obtained with the PI controller and the data needed to implement it on the generator hardware is not so excessive.

The selected controller is based on the Slotine and Li [3] structure. The first step of the described process [13] is to substitute the PI controller of the outer loop in Figure 10 by the sliding mode control and therefore, to obtain the expression that relates the variable to be controlled, P_s, and the control signal, $i_{ry\ ref}$

$$\frac{P_s(s)}{I_{ry\ ref}(s)} = \frac{-\dfrac{3L_m\left|\vec{v}_s\right|K_p}{2L_sL_r'T_i}(bT_is+1)}{s^2+\left(\dfrac{K_p+R_r}{L_r'}\right)s+\dfrac{K_p}{T_iL_r'}}. \tag{12}$$

If the system zero disappears, b=0, the expression is simplified and the dynamic performance of the generator is not even affected, so that,

$$\left[s^2 + \left(\frac{K_p + R_r}{L_r^{'}} \right) s + \frac{K_p}{T_i L_r^{'}} \right] P_s(s) = -\frac{3L_m \left| \vec{v}_s \right| K_p}{2L_s L_r^{'} T_i} I_{ry\, ref}(s)$$

$$\frac{d^2 P_s}{dt^2} = -\left(\frac{K_p + R_r}{L_r^{'}} \right) \frac{dP_s}{dt} - \frac{K_p}{L_r^{'} T_i} P_s - \frac{3L_m \left| \vec{v}_s \right| K_p}{2L_s L_r^{'} T_i} i_{ry\, ref} \qquad (13)$$

Then, before defining the sliding surface [13], the function f and the parameter b can be expressed as,

$$f(\frac{dP_s}{dt}, P_s) = -\left(\frac{K_p + R_r}{L_r^{'}} \right) \frac{dP_s}{dt} - \frac{K_p}{L_r^{'} T_i} P_s, \quad b = -\frac{3L_m \left| \vec{v}_s \right| K_p}{2L_s L_r^{'} T_i} \qquad (14)$$

and the second order equation in (13) can be redefined

$$\frac{d^2 P_s}{dt^2} = f(\frac{dP_s}{dt}, P_s) + bi_{ry\, ref} . \qquad (15)$$

Thus, the sliding surface can be described as,

$$s = \left(\frac{dP_s}{dt} - \frac{dP_{s\, ref}}{dt} \right) + \lambda (P_s - P_{s\, ref}) = 0 . \qquad (16)$$

In this way, if the system dynamics is maintained over the sliding surface, $s = 0$, according to (16), the generated active power error, $P_s - P_{s\, ref}$, will decrease exponentially with a T time-constant, $T = 1/\lambda$

$$P_s - P_{s\, ref} = e_{P_s}$$
$$e_{P_s}(t) = e_{P_s}(t = 0)e^{-\lambda t} . \qquad (17)$$

Once the sliding surface has been defined it is necessary to analyse the maximum and minimum values of **f** and **b** and to obtain the control signal $i_{ry\, ref}$ that has to be applied to the internal loop of the system (vector control). These upper and lower limits for **f** and **b** depend directly on the generator parameters and consequently, their variations must be accounted.

The first parameters to analyse are the ones related to the PI controller parameters (K_p and T_i) of the internal loop which, once evaluated, are maintained constant during the generator operational life. L_m, the magnetising inductance decreases slowly as the generator becomes older and can reach 80% its initial value. R_r, the rotor resistance, is not very variable because the speed operation area is quite narrow; consequently, the variation can be considered as not significant. The feeding voltage of the system can be modified between 5% and 10% its nominal value. Considering that n, L_{ls} and L_{lr} (general turns ratio, stator and rotor leakage inductances respectively) remain almost constant, the L'_r (rotor transient inductance) variation must be considered,

$$L'_r = \sigma L_r = \left(1 - \frac{L_m^2}{L_s L_r}\right) L_r = L_r - \frac{L_m^2}{L_s} =$$

$$L_{lr} + \frac{L_m}{n} - \frac{L_m^2}{nL_m + L_{ls}} =$$

$$\frac{n^2 L_m L_{lr} + nL_{ls} L_{lr} + nL_m^2 + L_m L_{ls} - nL_m^2}{n^2 L_m + nL_{ls}} = \tag{18}$$

$$\frac{n^2 L_m L_{lr} + nL_{ls} L_{lr} + L_m L_{ls}}{n^2 L_m + nL_{ls}},$$

if $L_s = L_{ls} + nL_m$, then

$$L'_r L_s = L_m (nL_{lr} + L_{ls} / n) + L_{lr} L_{ls}$$

$$\tag{19}$$

$$\frac{L_m}{L'_r L_s} = \frac{L_m}{L_m (nL_{lr} + L_{ls} / n) + L_{lr} L_{ls}}$$

If L_m decreases, L'_r decreases too in a lower proportion and the same thing occurs when L_m increases its value. The relation $L_m / L_s L'_r$ varies in the same way but less than L'_r.

For each P_s and dP_s/dt, taking into account R_r and L'_r parameter values and the sings of the above mentioned P_s and dP_s/dt, f_{max} and f_{min} can be obtained. Thus, if $P_s < 0$ (generating active power, the most common case) and $dP_s/dt > 0$, then

$$f_{max} = -\frac{1}{L_r^{'min}}\left[(K_p + R_r)\frac{dP_s}{dt} + \frac{K_p}{T_i}P_s\right] \quad if \quad \frac{K_p}{T_i}|P_s| > (K_p + R_r)\frac{dP_s}{dt},$$

$$f_{max} = -\frac{1}{L_r^{'max}}\left[(K_p + R_r)\frac{dP_s}{dt} + \frac{K_p}{T_i}P_s\right] \quad if \quad \frac{K_p}{T_i}|P_s| < (K_p + R_r)\frac{dP_s}{dt},$$

$$f_{min} = -\frac{1}{L_r^{'max}}\left[(K_p + R_r)\frac{dP_s}{dt} + \frac{K_p}{T_i}P_s\right] \quad if \quad \frac{K_p}{T_i}|P_s| > (K_p + R_r)\frac{dP_s}{dt}$$

$$f_{min} = -\frac{1}{L_r^{'min}}\left[(K_p + R_r)\frac{dP_s}{dt} + \frac{K_p}{T_i}P_s\right] \quad if \quad \frac{K_p}{T_i}|P_s| < (K_p + R_r)\frac{dP_s}{dt} \quad (20)$$

Consequently, for any value of P_s and dP_s/dt it can be concluded that

$$\hat{f} = \frac{f_{max} + f_{min}}{2} = -\frac{L_r^{'min} + L_r^{'max}}{2L_r^{'min}L_r^{'max}}\left[(K_p + R_r)\frac{dP_s}{dt} + \frac{K_p}{T_i}P_s\right],$$

$$F = \left|\hat{f} - f^{min}\right| = \left|\hat{f} - f^{max}\right| = \left|\frac{L_r^{'min} + L_r^{'max}}{2L_r^{'min}L_r^{'max}} - \frac{1}{L_r^{'min}}\right|\left|(K_p + R_r)\frac{dP_s}{dt} + \frac{K_p}{T_i}P_s\right|, \quad (21)$$

$$F = \left|\frac{L_r^{'max} - L_r^{'min}}{2L_r^{'min}L_r^{'max}}\right|\left|(K_p + R_r)\frac{dP_s}{dt} + \frac{K_p}{T_i}P_s\right|.$$

Then, defining

$$\frac{L_m}{L_s L_r'} * \frac{K_p}{T_i} = fraction$$

,

$$b_{max} = -\frac{3}{2}\left|\overrightarrow{v_s}\right|^{max} fraction^{max},$$

$$b_{min} = -\frac{3}{2}\left|\overrightarrow{v_s}\right|^{min} fraction^{min} \quad (22)$$

and,

$$\hat{b} = \sqrt{b_{max} \cdot b_{min}} = -\frac{3}{2} \sqrt{\left|\overrightarrow{v_s}\right|^{min} fraction^{min} \left|\overrightarrow{v_s}\right|^{max} fraction^{max}}, \qquad (23a)$$

$$\beta = \sqrt{\frac{b_{max}}{b_{min}}} = \sqrt{\frac{fraction^{max} * \left|\overrightarrow{v_s}\right|^{max}}{fraction^{min} * \left|\overrightarrow{v_s}\right|^{min}}}. \qquad (23b)$$

As it is mentioned in [13], when the existence condition is applied $\dot{s} = 0$, an estimation of $i_{ry\ ref}$ is obtained,

$$\dot{s} = \left(\frac{d^2 P_s}{dt^2} - \frac{d^2 P_{s\ ref}}{dt^2}\right) + \lambda \left(\frac{dP_s}{dt} - \frac{dP_{s\ ref}}{dt}\right) =$$

$$(24)$$

$$= f + b * i_{ry\ ref} - \frac{d^2 P_{s\ ref}}{dt^2} + \lambda \left(\frac{dP_s}{dt} - \frac{dP_{s\ ref}}{dt}\right) = 0,$$

$$\hat{i}_{ry\ ref} = \frac{\hat{f} + \dfrac{d^2 P_{s\ ref}}{dt^2} - \lambda \left(\dfrac{dP_s}{dt} - \dfrac{dP_{s\ ref}}{dt}\right)}{\hat{b}}. \qquad (25)$$

and applying the η-reachability condition,

$$s * \dot{s} \le -|s| * \eta \qquad (26)$$

the control signal $i_{ry\ ref}$ that has to be applied to the internal loop, to the vector control, is achieved

$$i_{ry\ ref} = \hat{i}_{ry\ ref} - \frac{k}{\hat{b}} sign(s), \qquad (27)$$

where,

$$k \ge \beta(F + \eta) + (\beta - 1)\left|\hat{b}\hat{i}_{ry\ ref}\right|. \qquad (28)$$

5.2.1   Design of the control algorithm

When the sliding mode control has been analysed, the algorithm to implement it on the generator hardware must be developed. The steps that have to be followed can be described as:

d) Definition of the active power reference, $P_{s\,ref}$.
   – Measurement of the rotor angular speed, ω_r.
   – Calculation of P_{sref}, Figure 7.
e) Measurement of the generated active power, Ps.
f) Calculation of f, b, fmin, fmax, bmin and bmax.
g) Calculation and approximation of dPs/dt, dPs ref/dt and d2Ps ref/dt2.
h) Calculation of $\hat{i}_{ry\,ref}$ and the current $i_{ry\,ref}$ through (17) and (19).

5.2.2   Simulation results

Using this algorithm some simulations have been developed in order to test the dynamic performance of the control system described above. The simulation test is the same as the one used for the PI controller (Table 2), considering that the power factor references are transformed into reactive power references through,

$$\cos \varphi_{ref} = P/\sqrt{(P^2 + Q_{ref}^2)} . \tag{29}$$

In Figures 17, 18 and 19 some simulation results can be observed. Figure 17a represents the generated active power through the test during 12 seconds. In Figure 17b, a detail of the active power is described. It can be seen that the generated power tracks almost perfectly the active power reference. Comparing these results with the ones obtained from a PI implementation (Figures 12a and 12b), it can be seen that the sliding mode control performance, in terms of trajectory tracking, is quite better than the obtained with the PI controller. Anyway it is interesting to analyze the control signal derived from this control law, that means, the voltages that must be applied to the rotor, Figure 18. These voltages are quite similar to the ones obtained from the PI algorithm and it can be concluded that, the application of the sliding mode control should be really done. In Figure 19 the ω_m versus P_s curve can be seen. In this Figure, the correct tracking of the reference trajectory can also be seen.

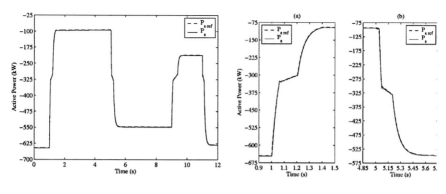

a. Generated active power   b. Detail.

Figure 17 Generated active power

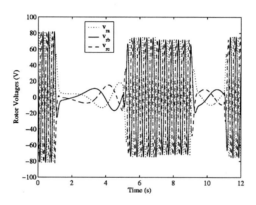

Figure 18. Voltages applied to the rotor

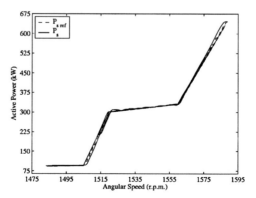

Figure 19. ω_m versus P_s curve

Finally, in Figures 20, 21 and 22, some comparisons between the two controllers performance are presented. The dotted lines correspond to the PI controller and the continuous lines, to the results obtained with the sliding mode controller. In Figure 20, the control signal, $i_{ry\,ref}$, is analysed.

Both signals (PI and sliding mode control) are quite similar and consequently the possibility of applying that law in the real system can be again derived. In Figure 21 the mechanical angular speed is presented and in Figure 22, the obtained reactive power can be observed. Again, the results obtained from both controllers are quite similar and therefore, the validity of the sliding mode control for real applications can be derived. The only special needs are related to the hardware characteristics of the control system in the generators.

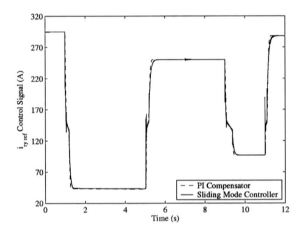

Figure 20. $i_{ry\,ref}$ control signal in both tests

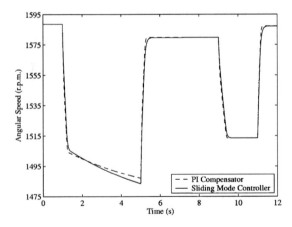

Figure 21. Mechanical angular speed with both controllers

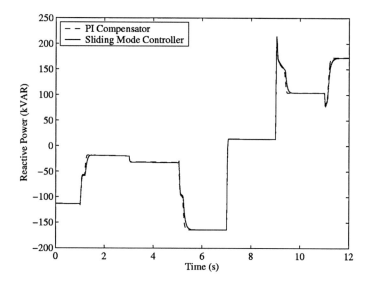

Figure 22. Obtained reactive power

6.   CONCLUSIONS

One of the most important conclusions of the presented work is that the tracking of the optimum active power generation curve, P_{opt} curve, has been developed through two different control algorithms, always looking for simplicity and robustness.

The algorithm based on a traditional PI controller can be used under every circumstance without variations on the control hardware of the actual wind generators. The one based on a sliding mode control presents a better performance for trajectory tracking applications, with error minimization characteristics but with the need of more computational operations.

On the other hand, even if both algorithms present a correct dynamic performance in the developed tests, only the PI controller has been really implemented in a wind farm. Thus, it would be interesting to continue with the started work analyzing the real implementation of the sliding mode control. It has to be mentioned that, in this case, a slightly different hardware of control must be used due to the computational needs of the algorithm.

Finally, as new power regulation systems related to the renewable energy sources are being applied in different countries, some research in this field is needed, concerning the generated active power and its "quality" together with economic aspects of wind farm exploitation.

ACKNOWLEDGMENT

The authors are especially grateful to R. Criado (from IBERDROLA) for his helpful comments in the research work development. The authors are also grateful to the induction generator manufacturing company, INDAR, and the control systems company, INGETEAM, for their contribution in providing the electric machine characteristics and all the test results needed for the research development.

This work has been supported by the research program of the University of the Basque Country (UPV/EHU).

REFERENCES

1. S. BHOWMIK, R. SPEE, J. H. R. ENSLIN, *Performance optimization of doubly-fed wind power generation system.* (IEEE Trans. on Industry Applications 1999; 35(4), pp 949-958).
2. C.- M. LIAW, C.- T. PAN, Y.-C. CHEN, *Design and Implementation of an Adaptive Controller for Current-fed Induction Motor.* (IEEE Trans. on Industrial Electronics 1988; 35(3), pp 393-401).
3. J.J.E. SLOTINE, W. LI, *Applied Nonlinear Control.* (Englewood Cliffs, New Jersey: Prentice Hall, 1991).
4. R. SPEE, S. BHOWMIK, *Wind turbines. Encyclopedia of Electrical Engineering.* (New York, Wiley, 1999).
5. A. TAPIA, G. TAPIA, X. OSTOLAZA, J. R. SÁENZ, R. CRIADO, J. L. BERASATEGUI *Reactive Power Control of a Wind Farm made up with Doubly Fed Induction Generators (I).* (Proceedings of the IEEE Porto Power Tech'2001, September 2001. (Vol. IV), DSR1-054).
6. A. TAPIA, G. TAPIA, X. OSTOLAZA, I. ZUBIA, J. R. SÁENZ, *Reactive Power Control of a Wind Farm made up with Doubly Fed Induction Generators (II).* (Proceedings of the IEEE Porto Power Tech'2001, September 2001. (Vol. IV), DSR1-055, Oporto).
7. G. TAPIA, A. TAPIA, J. L. BERASATEGUI, J. R. SAENZ, *Voltage Regulation of Distribution Networks through Reactive Power Control* (IFAC Publications – Elsevier Science Ltd., ISBN 008 044184, Proceedings of the 15th IFAC World Congress 2002, March/April 2003, pp 377-382).
8. A. TAPIA, G. TAPIA, X. OSTOLAZA, *Reactive power control of wind farms for voltage control applications.* (Renewable Energy 2003, in press).
9. A. TAPIA, G. TAPIA, X. OSTOLAZA, J. R. SÁENZ, *Modelling and Control of a Wind Turbine Driven Doubly Fed Induction Generator* (IEEE Trans. on Energy Conversion 2003; 18(2), pp 194-204).
10. P. VASS, *Vector control of AC Machines.* (New York: Oxford University Press, 1991).
11. K. J. ÅSTRÖM, T P. HÄGGLUND, *Controllers: Theory, Design and Tuning.* (USA: Instrument Society of America, 1995).
12. A. SABANOVIC, F. BILALOVIC, *Sliding Mode Control for AC Drives.* (IEEE Trans. on Industry Applications 1989; 25(1): pp 70-75).
13. J.J.E. SLOTINE, W. LI, *Applied Nonlinear Control.* (Englewood Cliffs, New Jersey: Prentice Hall, 1991).

DESIGN AND DETERMINATION OF THE MOST COST EFFECTIVE PV CONFIGURATION SYSTEMS TO MEET THE LOADS OF A HOUSEHOLD

S. N. Kaplanis
Technological Educational Institute of Patra, Greece

Abstract: A methodology to determine the most cost effective solution for an integrated hybrid PV system to meet the loads of a household is presented. The PV hybrid system consists of a photovoltaic generator backed up by a Diesel generator or by wind power. Such systems show greater reliability to meet the loads and it, often, represent the most effective solution, whenever wind potential is not adequately available. The primary objectives are: a) to size a PV hybrid system which meets the load demand of a household, b) to present a methodology to determine the most cost effective solution amongst the possible configurations and the various types of system's components.

Different scenarios about the loads have been studied and analyzed. A wide variety of tools, ranging from simple rules of thumb to sophisticated software packages, exist for the analysis and dimensioning of PV systems. The development of a friendly package to determine the most cost effective configuration out of several possible PV systems has been a matter of investigation in this paper. The approach followed considers the technical characteristics and power performance of the PV generator and its associated components, as well as the total daily load and each load profile. The sizing is based on the daily performance and the daily load profile requirements, introducing all possible corrections to PV performance and power transmission. The economic analysis takes into account the life cycle of the components, as well as the inflation rate and gives the most cost effective solution.

Key words: photo-voltaics, PV-sizing, Ah and Wh method.

D. Talabă and T. Roche (eds.), Product Engineering, 473–508.
© 2004 *Springer. Printed in the Netherlands.*

1. GENERAL DESCRIPTION OF PHOTOVOLTAIC SYSTEMS

The basic element in the PV market is the PV cell, the module or panel and finally the PV generator. PV cells and modules are rated on the basis of the power delivered under Standard Testing Conditions (STC) defined as:
a) 1 kWh/m² solar intensity on it and sunlight spectrum , AM1.5
b) PV cell temperature of 25°C.

The power output measured under STC is expressed in terms of peak power or "Watt peak" or W_p, when the PV system operates at the M.P.P. see Figure 1 [1-4].

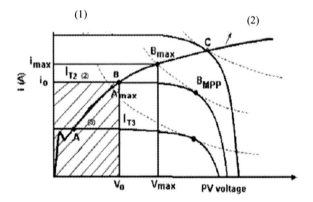

Figure 1. The characteristic I, V of a PV system (1) and the characteristic I, V of a motor (2).

They intersect at C (MPP) which is the operating point of the system PV generator and motor for intensity level I_{T1} and at B for intensity level I_{T2}. The power from the PV generator transferred to the motor, in the latter case is given by the shaded area. While the system operated at the MPP, I_T would be much greater corresponding to the area under the dotted line. Such a case exhibits a mismatch between the PV generator and the load, which here is represented by the motor.

PV modules are integrated into systems i.e. PV-arrays or PV generator designed as to meet the loads of specific applications. The components added to the PV modules constitute the **"Balance Of System" or BOS**. Balance of system components can be classified into four categories:

1st. Batteries – to store the electrical energy and provide power on demand at night or on overcast days.

2nd. Inverters (DC/AC) – required to convert the DC power produced by the PV modules into AC power.

3rd. Controllers – that allow the management of the energy storage to the battery system and deliver power to the load.

4th. Other components such as cabling, electronics, meters and sensors.

5th. Structural frame – required to mount or install the PV modules.

Not all systems require all those components. For example, in systems where no AC load is present an inverter is not required. For, on grid systems, the utility grid acts as the storage medium and therefore batteries are not required. Some other systems, also, require other components which are not strictly related to photovoltaic. Some PV systems, for example, include a fossil fuel generator that provides electricity, when the batteries become depleted. All these cases are part of the general configuration to be shown later in the Figures 11 and 16.

Three main types of PV-cells based on Si are developed so far.

- The single or mono-crystalline Si, Figure 2.
- The poly-crystalline, pc-Si, PV-cell, Figure 3.
- The amorphous, a-Si, or thin film PV-cell, Figure 4. They exhibit flexibility but lower efficiency.

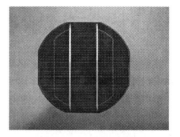

Figure 2a. PV-module from monocrystalline silicon mounted on the south wall of the R.E.S. Lab. of the T.E.I. Patra. This PV system powers house appliances which simulate a household, as installed within the space of the laboratory.

Figure 2b. PV monocrystalline cell

Figure 4. PV-modules from poly-crystalline silicon and a zoom to the surface of such a cell

Figure 4. PV-modules from amorphous silicon

2.   APPLICATION OF PV SYSTEMS

The applications of PV systems cover the domestic, the industrial and the agricultural sectors, but also the telecommunications and the space sector, as the Figures 5, 6, 7 and 8 below highlight.

Figure 5. Satellite that covers its energy needs with PV modules mounted on its surface

Figure 6. A PV generator used for AC and DC lighting purposes. To the left the DC/AC inverter is shown.

Figure 7. PV generator in the T.E.I. of Patra, Greece. Iinstalled power, 2.5 KW$_p$

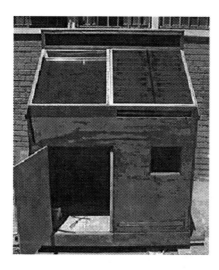

Figure 8. A solar test cell to study the performance of energy roofs with embedded PV panels. The design is such as to run the PV powered roof under the PV + thermal concept. The south wall painted greenish functions under the passive solar concept as explained below.

2.1   The PV and Thermal (combined) design

Following the policy to achieve the maximum use of any element of the system, increasing its usability, either this is expressed in efficiency or in yield (kWh/kW$_p$ for a period of time), PV panels may cover part or the whole roof or similarly for the façade. Such a design has the following effects:

a) Decreases the building's structural cost as compared to the case where the PV elements are placed upon the building shell.

b) The final installation is more aesthetic

c) Less cost for installation as several frame parts are not required

d) A stream of air is allowed to slide past the PV panel's back surface. As the PV surface develops medium temperatures, around 60-70^{0}C, (depending on the solar intensity on the PV cell and the ambient temperature) the stream of air may convey heat off the PV panel and heat up the space it is finally driven to. The same concept is followed when ducts with eco-Freon are into thermal contact with the PV panel's back surface. A double phase flow is then developed within the inclined tubes. Freon gas is driven up to a heat exchanger to provide finally hot water in a tank. This stored heat is used for either hot domestic water use or in floor heating or other needs. Such a system is shown in Figure 8.

e) A similar approach is followed in facades; see façade in Figure 8. In cases 4 and 5 the PV generator provides enough thermal energy to meet the space heating loads.

3.   SIZING A PV SYSTEM

Sizing a photovoltaic system is a very important task in the system's design process. In the sizing process one has to take into account four basic factors [1-4]:

a) the solar radiation on the site and generally the METEO data [5,6],

b) the daily power consumption (Wh/day) and specifically the load profile see Figure 9, and the types of electric loads,

c) the storage system to contribute to system's energy independence for a certain period of time,

d) the B.O.S. elements and their specifications.

If the PV generator is oversized it will have a big impact to the final cost and to the price of the power produced.

If, on the other hand, the PV-generator is undersized, problems might occur in meeting the power demand at any time, while in addition to that, the battery system will soon be disabled as it is doubtful if the system can charge it at its nominal capacity.

The sizing should be carefully planned, examining various possible PV system configurations and various models or types of B.O.S. components, in order to get to a cost effective and reliable PV system.

3.1   Solar radiation data

The amount of solar radiation at a site at any time, either it is expressed as solar intensity (W/m^2) or solar radiation in MJ or Wh/m^2. Those data are primarily required to provide an answer to the amount of the power to be produced by the PV generator.

The amount of electrical energy produced by a PV-array depends primarily on the radiation at a given location and time.

Data on solar radiation are usually given in the form of global radiation that is beam, direct and diffuse radiation over a horizontal surface; see Meteonorm Database [5].

3.2   Load Data

As it concerns the loads, one may get the proper information on data according to the appliances to be powered by the PV generator.

These appliances or otherwise consumers or loads could be domestic appliances like: TV sets, lights, refrigerator, kitchen, vacuum cleaner, washing machine, coffee machine etc.

The load estimation can be done using a table based methodology as suggested through Table 1.

Thus the load profile is determined as in Figure 9, which provides the daily load in **Wh/day** or **Ah per day** at a voltage **Vs,** and the power that is transferred.

*Table 1.*Table for load profile estimation

Appli-ance	Qt	AC or DC	P Y/N	Run Hours [W]	Hours /Day	Days /Week	W-hours /Day	Surge [W]	Ph-L Y/N	Type Load	Oper. Period
1	2	3	4	5	6	7	8	9	10	11	12
Total Daily Wh			Max. AC Wattage								

Notes about the columns:

Column 1: Appliance type

The appliances which are to be used/connected are registered hereto.

Column 2: Items Quantity

It provides the quantity of the above appliances or consumers or loads for the application studied. An example of multiple identical appliances is the lighting lamps. There is no need to list every light bulb in the house separately.

Column 3: AC or DC

The column distinguishes the consumers as DC or AC. This has a significant impact to the PV configuration and the efficiency. Remember that the DC output of a PV generator to be inverted to an AC, consumes power by just being on. That is even the AC loads is not on the inverter in stand by mode consumes some power.

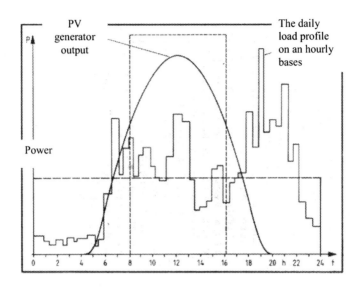

Figure 9. Daily load on an hourly basis and the PV generator output

Column 4: Inverter Priority

The purpose of this column is to provide a feeling for the normal operating wattage of the inverter. For example, if an appliance is on for a good deal of time, or if the appliance has to have access to the inverter power, then this is to be an inverter priority load.

Any appliance that turns itself on and off must be an inverter priority load. However, some loads are operated infrequently. That implies there is freedom to decide what other appliance is allowed to operate at the same time. That is, the PV-generator has room to accommodate this load's power consumption. Such loads are not inverter priority ones.

This column helps to choose the size of the DC/AC inverter. It also, helps to determine the inverter's average operating efficiency.

The inverter's efficiency depends on the ratio P_L/P_N where P_L is the load power and P_N the inverter nominal power; see Figure 10.

Column 5: Operating Power (Watts)

In this column the power consumption of each load is registered. Electrical appliances display their power data along with other specifications.

The noted watts value represents the highest case scenario, i.e. the highest power that the appliance will ever draw e.g. when the full range of the consumer is used, the case of a radio. Generally, for usual cases one has to reduce the rated wattage on the sticker by about 25% e.g. lamps.

Column 6: Hours per Day

This column gives info on the time period each consumer is on per day. In some ways this information is easy to figure out. Example is the washing machine which takes about twenty minutes to complete a cycle. As it concerns the light bulbs one has to estimate how much time per day each light is turned on. That depends on the life style or services style in the premises.

Some appliances turn themselves on and off automatically.

Refrigerators start up when the temperature inside gets lower than preset. They run until they are cooled down to a certain temperature. Then, they turn themselves off. This is called a **"duty cycle"** and can be estimated by direct observation.

To determine the energy consumed, the time figure of column 6 is multiplied with the power figure of column 5.

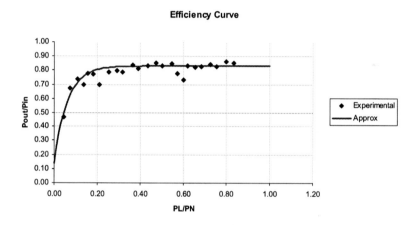

Figure 10. Inverter efficiency curve as the experiment gave

Column 7: Days per Week

This column provides information on how often the load is used per week. For example, a cleaner or a washing machine does not run every day but once per week in many cases.

Column 8: Average Watt-hours per Day

To build this column the Quantity of consumers of a certain type (Column 2) × Watts (Column 5) × hours (Column 6) × days (Column 7) divided by 7 days a week provides the average watt-hours per day for this appliance, on a weekly basis.

The total at the bottom of this column provides the total power (energy in Wh) on an average day.

Column 9: Starting Surge in Watts

This column gives the surge power which is useful information about any consumer. Any appliance with a motor has a starting surge. This means that before the motor is up to the operating speed, it draws more current than its rated operating power. This is true especially when the motor is starting under load.

Refrigerators, as well as pumps, and most power tools have starting surges. Motors surge between three and seven times their rated wattage. To reduce such a surge power electronic devices are connected to the motor.

Other appliances that may have starting surges are TVs, computer monitors, and any appliance with an internal power supply. They can surge up to three times their rated wattage.

They are relatively short in the surge effect which lasts for some milliseconds. Hence, this behavior does not make much of a difference in the amount of energy an appliance consumes.

Starting surges are important, however. Inverters must be sized to handle the starting surge of AC appliances; see the Fronius inverter (550 Watts nominal power) efficiency as shown in Figure 10.

Battery banks must also be sized to handle the voltage depression caused by a high current surge.

Voltage depression can cause an inverter to shut down even if the inverter itself is large enough to handle the surge.

Column 10: Phantom Load

Some appliances consume power even when turned off. e.g. loads that are in the stand by mode in order to be ready for use like chargers, inverters and diesel generator. During this stage the power they consume is much lower.

Reminder:

For a more effective load management, the daily profile of the load must be studied, see Figure 9.

Column 11: Types of loads

In this column the energy engineer has to specify after some consideration and the advice of the user, if the loads are **critical** or **non critical** ones. The concept of the critical load implies that it should be off power not more than 80 hours per year.

Column 12: Time period each load operates

This is a very useful column as the time period each load is assumed to be on is recorded in it. Therefore, it provides the necessary information to build the daily profile of the loads. That leads to the estimation of the inverter's size, the discharge current etc.

3.3   Sizing Procedures-an Introduction

The photovoltaic systems applications have developed rapidly in many fields [1-14, 15-18]. There have been developed and are under monitoring medium and large scale PV-power plants. In addition to that, a more systematic analysis than in [1, 2, 26] to reach optimum PV-configurations is needed [3, 4].

Various PV-configurations have been proposed and tried in order to reach cost effective solutions, according to the requirements set by the specific applications. The attempt to reach the most cost effective PV configuration depends on many factors and parameters which are affected by the following factors:

- The configuration type, i.e. stand alone, grid connected, hybrid, the PV-system circuitry e.g. direct feed and/or indirect feed to loads; also, PV+thermal (combi) systems, storage etc.
- The components of the PV system itself, i.e. types of PV panels, batteries, the inverter, the MPPT, electronics, the cabling layout, the charge controller etc.
- The management of the loads; priorities set to cover specific loads when non-critical [2, 11, 12, 13].
- The management of the energy source, that is the PV generator size, wiring and setup.
- The back-up system type, size, circuitry.
- The overall system sizing for a given f-value [14].
- The life cycle of each one of the components.

A combined and systematic approach is therefore required in order to determine the most cost effective solution from out of the many possible configurations and (energy) solutions, which are, in general terms, acceptable and effective.

To reach the state that someone is able to determine the most cost-effective solution, a methodology has been developed, the structure of which is outlined in the following paragraphs. The methodology is divided in two parts.

In part A, the package proceeds to the sizing of the overall proposed or possible PV configurations. The user may decide or choose one of the available configurations as built-into the package and draw the proper power transmission lines. All possible corrections, mainly to the PV generator, the battery bank, the transmission and the charger/inverter are introduced.

In part B the package proceeds to a Life Cycle Cost Analysis (LCCA) [16, 19, and 21] of the promising effective PV configurations, which were the results of the part A analysis.

There, a search for possible effective solutions takes place, following a detailed simulation process.

A series of loops is then repeated choosing different types of PV-system elements such as, PV-panels, battery types and inverters. At the end, the package compares the cost effective solutions according to the f value i.e. the % coverage of the loads by the PV-generator, set at the beginning along with the V_s that is the voltage power that is transferred.

These two parameters, V_s and f are also varied in order that LCCA provides the most cost effective PV configuration.

A review gives numerous optimal sizing methodologies including analytical solutions and numerical method approaches, for many scientists have developed stand-alone systems.

There has been proposed an analytical solution to the sizing problem which as it is claimed can bring the price of stand-alone photovoltaic systems to economic viability at today's hardware and fuel prices, and can also enable local designers in developing countries to design these systems economically. Also, there has been proposed a simulation approach that uses hourly meteorological data and hourly load data to simulate the energy flow in a PV system and predict the system reliability for various array and battery sizes.

Nevertheless, a detailed evaluation of the sensitivity of a numerical sizing method, as developed, has shown that the influence of some parameters on the sizing, i.e., simulation time step, input and output power profile is very important. It is therefore important to have knowledge of the daily profile at least on an hourly basis.

For the sizing optimization of hybrid systems a very friendly method has been developed in [4], as outlined below. Specifically two approaches are to be followed for the sizing optimization of stand alone or hybrid or grid connected systems

1. The **Wh method** and
2. The **Ah method**

For a better understanding both of these methods will be approached by sequential steps till a final result is reached. Both methods can be used for any PV system. In autonomous systems, for example, one may determine exactly the PV configuration and the battery bank needed, so that the load is covered with on the most optimum way. The same method can, also, be used for hybrid PV systems, where the back up system can be a Diesel generator, or Fuel Cells or a Wind generator.

The Wh and Ah methods used for PV systems sizing take into consideration the impact of the climatic data into the efficiency of the PV modules, the batteries and the rest components of the PV systems. Also, they take into account the charge-discharge rate of the batteries.

4.   THE SIMULATION APPROACH TO THE COST EFFECTIVE CONFIGURATION

An outline of the two parts of the simulation methodology and the process considered to optimize the overall PV-system follows.

4.1   Part A

In the first part, the steps to be followed are the following:

a) Determination of the daily load profile; DC and AC loads; priorities as critical or non-critical ones.

b) Downloading the weather data. Data are retrieved by a databank. The user may also introduce other available and reliable weather data or download online from some addresses provided by the software [5,6].

One may also use a model developed for the prediction of the solar intensity, $I(h)$, for any day, hour(h) and site [25]. That means the PV sizing software may become self-adequate for the $I(h)$ data input concerned.

c) Choice of possible PV configurations, as one of the built-in patterns; see Figure 11.

d) Decision for the inclination of the PV array, the geometry of the PV generator, the site of the installation, the type of the tracking mode of the PV array etc.

e) Decision about the power transmission voltage value, Vs. In the package, Vs functions as a parameter, i.e. the software will run several loops changing the voltage value for the power transmission, towards spotting a cost effective solution.

f) Then, the user has to estimate the total length of the cables according to the layout, so that to find the power loss due to cabling.

g) In this step, the user decides about the type of the PV panels. All the info concerning the characteristic values of the PV panels such as Voc, isc, im, Vm, Pm, NOCT, FF, temperature coefficients are loaded on the software.

h) The battery type is to be chosen in this step. The same holds for the inverter model, the charge controller and the MPPT model from several types loaded into the data bank of this simulation package.

In addition, similar data such as efficiency, voltage drops etc as it concerns the other components (battery, inverter and controller) are loaded in the package. If the values of the above parameters are not available by the manufacturer there is a built-in methodology to estimate them as e.g. the estimation of the PV panel temperature, T_C, the temperature coefficients for i_{sc} and Voc, P_m, which are used to estimate the correction to the peak power, delivered by the PV generator.

After those preliminary steps have been taken successfully, the program brings in all the necessary corrections for the components of the system; specifically the PV panels, the battery bank, the inverter, the charger and the MPPT.

The first correction concerns the power output of each panel, the open circuit voltage and the short circuit current of the PV panels. The correction coefficients depend directly on the temperature and the irradiance. The equations which provide for the corrections for i_{sc}, V_{oc}, the efficiency, the power and the loads are the following:

$$i_{sc} = 0.034 \left(\frac{A}{cm^2} \right) \times A_c \times C \times [1 + 3 \times 10^{-4} (T - 300)],$$

(1)

where C is the beam irradiance concentration on the PV cell, A the surface of the PV cells, in cm^2.

This leads to:

$$\frac{1}{i_{sc}} \frac{di_{sc}}{dT_c} = 3 \times 10^{-4} K^{-1}.$$

(2)

If i_{sc} is given a value about 3.2A, then,

$$\frac{di_{sc}}{dT_c} = 3 \times 10^{-4} \, {}^o K^{-1} \times 3.2 A = 0.096\%/K.$$

Also, the current output is related to the solar intensity by:

$$i_{sc}(T_C, I_T) = i_{sc}(T_{C,0}, I_{T,0})[1 + h_t(T_C - T_{C,0})] \times I_T / I_{T,0}.$$

(3)

where $h_t = 6.4 \times 10^{-4} K^{-1}$.

The PV cell temperature is given in [5]:

$$T_c = T_a + h_w I_T.$$

(4)

where $h_w = 0.03 m^2 \times K/W$.

T_c may be determined, also, by the **NOCT** (Normal Operating Cell Temperature) as given by the following equation

$$T_c = \text{Monthly average temperature} + ((\text{NOCT-20 }^o C)/0.8).$$

(5)

This is derived by the linear relationship between the temperature and the solar intensity.

The V_{oc} temperature coefficient is given for Si PV-cells by:

$$\frac{dV_{oc}}{dT_c} = -2.3 \times 10^{-3} (V/{}^o C) \times n_s,$$

(6)

where n_s is the number of PV cells in series.

The temperature coefficient of V_{oc} to be inserted into the package, as requested, is about 0,08V/°C per panel. This is due to 0.0023 V/°C per cell multiplied by the number n_s of PV cells in series.

The temperature coefficient for the efficiency is given by:

$$\frac{\partial \eta}{\partial T_c} = \eta_c \left[\frac{1}{V_{oc}} \times \frac{dV_{oc}}{dT_c} + \frac{1}{i_{sc}} \times \frac{di_{sc}}{dT_c} + \frac{1}{FF} \times \frac{dFF}{dT_c} \right]. \tag{7}$$

Considering that FF does not change significantly with the temperature and that $\frac{1}{i_{sc}} \times \frac{di_{sc}}{dT}$ is negligible compared to $\frac{dV_{oc}}{dT}$, solving over $\frac{1}{\eta} \times \frac{\partial \eta}{\partial T}$ and

substituting $V_{oc}=0.58$Volts and $\frac{dV_{oc}}{dT}=-2.3\times10^{-3}(V/°C)$ which holds, generally,

for Si-cells, one finds that $\frac{1}{\eta} \times \frac{\partial \eta}{\partial T} = -0,4\%/°C$.

It can be derived by the definition of the PV cell efficiency that,

$$\frac{1}{\eta} \times \frac{\partial \eta}{\partial T} = \frac{1}{P} \times \frac{\partial P}{\partial T}.$$

The temperature coefficient of P_m is now determined equal to: -0,4%/°C.

Notice: The temperature coefficient of P_m may be obtained also by the manufacturer.

The next step concerns the loads correction according to:
- their nature, i.e. if they are AC or DC;
- the power losses through the different components and the routes followed along the PV configuration - see Figure 11.

The investigation for the effective PV configuration studies various possible routes-circuitries - see Figure 11.

The approach used is the **Wh** method as elaborated in detail in [1, 2, 4, 22, 26].

The correction to daily loads takes the form:

$$DC_{L,cor}=DC_L \times (1+((100-\eta_{c.c.})/100)+((100-\eta_b)/100)+((100-\eta_C)/100), \tag{8}$$

$$AC_{L,cor}=AC_L \times (1+((100-\eta_{c.c.}/100)+((100-\eta_b)/100)+((100-\eta_C)/100)+((100-\eta_{inv})/100), \tag{9}$$

where:

$\eta_{c.c.}$: charge controller efficiency,

η_b: battery efficiency,

η_C: cable efficiency,

η_{inv}: inverter efficiency.

The total daily load is given by: $\mathbf{DC_{Lcor} + AC_{Lcor}}$.   (10)

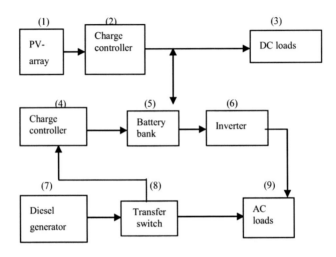

Figure 11. Different routes one can follow for the supply of a load, upon the determination of the PV system configuration.

The methodology, as presented above, may be repeated for several levels of power coverage, **f** values. Hence, the sizing of the PV generator for each level i.e. for each assumed **f** value is obtained. The parameter **f** is defined by:

$$f = \frac{E_{PV}(Wh/day)\,by\;the\;PV}{Q_L(Wh/day)\;daily\;load}.$$ (11)

After the first corrections are made, the package provides the number of the PV panels required to meet the loads.

The equation used for the peak power of the PV generator is

$\mathbf{P_{pv}=f \times total\;daily\;corrected\;load/PSH}$ (12)

P.S.H.: the Peak Solar Hours is defined as,

$$PSH = \int_{\omega_{sr}}^{\omega_{ss}} I(h)dh / 10^3 \; \frac{W}{m^2}.$$ (13)

The number of PV panels required is given by:

$$N_{pv} = P_{pv} / P_{m(cor)}.$$ (14)

The number of PV panels in series and parallel are:

$$N_{p,s} = V_s / V_{oc(cor)},$$ (15)

$$N_{p,p} = N_{pv} / N_{p,s}.$$ (16)

Hence, a corrected value is finally obtained:

$$N_{pv,cor} = N_{p,s} * N_{p,p}.$$ (16a)

The next step to be followed in this sizing methodology, concerns the decision on the type and the size of the inverter. The inverter size depends on the loads profile the PV-generator is to meet, as well as the circuitry the inverter follows.

The inverter should be carefully determined for its type i.e. PWM, square wave inverter, multi-step sine inverter, sine wave inverter, etc [27] so that it matches efficiently the loads and is sized properly.

A reliable and efficient operation is considered throughout the day. About its efficiency an approach was attempted in the description of column 4 in par. 3.3.

A complementary and very important next step to it is the size of the back up, i.e. Diesel generator, system.

For the sizing of the back up system the equations used are:

$$\text{Diesel generator} - \text{set size} = \frac{\text{max energy demand}}{(\text{max operating hours/day}) * \text{max load factor}}, (17)$$

which is based on the Wh method or

$$\text{Diesel generator} - \text{set size} = \frac{i \times V_s}{\eta_{inv} \times \eta_b \times \eta_C},$$ (18)

where **i** is the charging current which has to be less than **C/10** with C - the Capacity of the battery storage system. The value of i is equal to i_{pv} i.e. the current the PV generator provides. This i_{pv} equals to N_{pp} x i_m, which is based on the **Ah** value of the energy storage system.

In the next step, the method proceeds to the battery system sizing and determines the capacity of the batteries introducing all possible corrections plus the load type (DC/AC), the load characteristics i.e. critical or non critical type and the load demand, too. It considers, also, the charge/discharge rates, the days of autonomy, unless the PV system is grid connected and the configuration type, too.

The capacity of the battery storage system changes due to ageing, charge/discharge rate, cycles, temperature and flow of electrolyte [22, 23], see Eq. (26).

All those parameters have to be taken into account to provide for the right and reliable battery system capacity.

The number of batteries required to provide the preferable autonomy and meet the above criteria, are outlined later in this text. The equations used in this step are given below:

$$d_{crit} = -1.9 \times PSH_{min} + 18.3 \text{ (see [2])},\tag{19}$$

$$d_{non-crit} = -0.48 \times PSH_{min} + 4.58,\tag{20}$$

$$Q_L = (L_{crit} \times d_{crit} + L_{non-crit} \times d_{non-crit}) / V_s.\tag{21}$$

This load is in Ah/day. If V_S is not included in the formula above, then Q_L is expressed in Wh/day. The following relationships hold in this case:

$$DOD \times C = Q_L \times d,\tag{22}$$

where DOD is the depth of charge, C is the total capacity, Q_L is charge per day and d is the number of days for energy independence.

Remark: The load should be expressed in (Ah). If the load Q_L is in Wh/day then $DOD \times C = Q_L \times d / V_s$ where V_s is the voltage the charge is delivered to the load.

The battery capacity, $\mathbf{C_N}$, to meet the load $\mathbf{Q_L}$, with an energy independence of **d** days, is estimated by:

$$C_N = \frac{C_L}{1 - t_b \times (C_c + C_a)},\tag{23}$$

where

$$C_L = \frac{Q_L \times d \times f_c}{V \times DOD},\tag{24}$$

which is a more generalized expression than the equation given before.

t_b: no. of years that the battery will run effectively (according to specifications)

Q_L: daily load (Wh/day). It depends on the external consumers or loads

$C_c \approx 0.007\text{-}0.01$. it is a correction factor due to cycles/ recycling

C_a: correction due to the ageing of the battery. i.e. charging-discharging. More specifically:

$C_a \approx 0.015$ (for a battery with flow of electrolyte) and 0.020 (for conventional electrolyte)

f_c: correction due to Joule effect in battery : $f_c \approx 1.1$

V_s: voltage across battery, or the voltage, the PV-charge is transferred.

Another set of corrections is related to temperature effect and the charge/discharge rate effect. The corrected capacity is then equal to:

$$C_{cor} = (C_N \times f)/(f_{b,T} \times f_{b,ch} * DOD), \qquad (25)$$

where C_N, is the nominal capacity of the battery bank as given above.

f is a recovery correction coefficient, given below, which multiplies f_c and takes into account the effect of discharge rate. This coefficient is given below; see eq.(26).

The number of battery banks, the system consists of, is determined by:

$$N_{bp} = C_{cor} \times \text{days of autonomy}/DOD \times \text{Battery capacity}. \qquad (26)$$

The batteries in series, in each bank, are determined by:

$$N_{bs} = V_s/V_b \text{ (battery voltage)}. \qquad (27)$$

Correction coefficient due to charge discharge rates:

$$f_{b,cd} = \frac{i_{ch/disch \ manufacturer}}{i_{ch/disch \ appl}}. \qquad (28)$$

where $i_{ch/disch \ manufacturer}$ is taken as recommended by the manufacturer and $i_{ch/disch \ appl}$ - according to the application data.

The correction coefficient due to temperature variation is

$$f_{b,T} = \frac{C}{C_o} = \frac{C \text{ at } T°C}{C_o \text{ at } 25-27°C} = 0.01035 \times T°C + 0.724 \cdot \qquad (29)$$

Correction coefficient due to discharge rate: if t_d is large a recover from the battery's degradation is achieved. This is provided by the correction coefficient below which is bigger than 1.

$$f = 1 + \frac{4.3 \times t_d^{0.355}}{100},$$ (30)

$$t_d = C_N / i_d.$$ (31)

After all those parametric corrections the program can loop again for several other parameters selecting new power transmission voltage, V_s, then, new PV panel and other battery type. In total, 3 loops are followed for a given configuration and a given **f** value to cover the loads by a PV-system. Hence the loops are 4, in total.

At the end of the part A, the program verifies if the selected batteries meet the criteria i.e. if the type chosen and its specifications meet the requirement to manage the load.

A control is made in such order as to satisfy the criteria
1. $i_{ch,max} < C/10$,
2. DOD for the number of days of autonomy, d, to be smaller that what the manufacturer suggests. The value of the DOD reached affects the life cycle deviation which depends on the cycles and the depth of discharge.

The final value of the installed peak power of the PV generator may be given, also, by the relationship [22]:

$$P_{PV} = \frac{(d + N_R) \times E_L \times F}{N_R \times \eta_{ch} \times [1 - t_b(C_\alpha + C_c)]} \times \frac{W}{H_D \times \eta_{pv} \times [1 - n(T) \times (T_c - 25^0 C)]},$$ (32)

where:
E_L: daily load in Wh/day, as in (21);
d: days of autonomy;
DOD: depth of discharge;
$N_{b,p}$: number of batteries in parallel;
$N_{b,s}$: number of batteries in series;
N_R: days needed for recharging the battery bank after a deep discharge. This number depends on the design operation of the stand alone system.
F: correction coefficient due to Joule effect;
t_b: life in years of the battery bank , set by the manufacturer;
C_α: % decrease of the capacity of batteries due to ageing.

A typical value for the reduction of the capacity due to ageing is 0,015 for batteries with flowing electrolyte and 0,020 for conventional batteries. [22 - 23]
C_c: % decrease of the capacity of batteries due charge discharge

W: the power provided by $1m^2$ of PV panels in W/m^2, as given by the manufacturer in Standard Test Conditions (S.T.C);

H_D: daily irradiance in $Wh/m^2/day$ on the PV array plane;

η_{pv}: the efficiency of the PV panel;

T_c: temperature under which the PV panels operate;

n(T): correction coefficient of the size of PV panels due to temperature.

4.2 Part B

In part B the steps that the user has to follow are:

a) To determine the cost of the fuel (€/l) for the back up system [24] or the cost of the electric energy (€/kWh) required if the system is grid connected.

b) To determine the life period for each component of the PV-system, the inflation rate, i, and other economical data.

c) Estimate the maintenance cost per year and the repair or replacements in the system (€)

d) In this part, the LCCA (Life Cycle Cost Analysis) takes as inputs the different configuration results as obtained from part A, in order to determine the most cost effective solution.

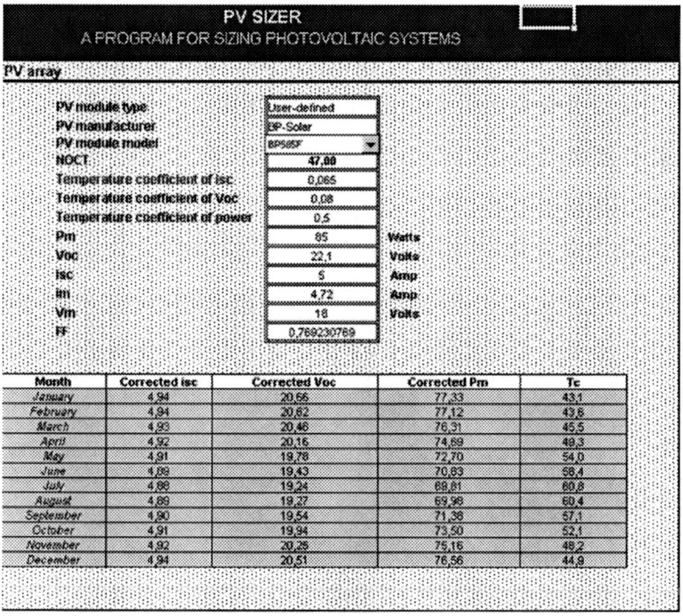
PV SIZER
A PROGRAM FOR SIZING PHOTOVOLTAIC SYSTEMS

PV array

PV module type	User-defined	
PV manufacturer	BP-Solar	
PV module model	BP585F	
NOCT	47,00	
Temperature coefficient of isc	0,065	
Temperature coefficient of Voc	0,08	
Temperature coefficient of power	0,5	
Pm	85	Watts
Voc	22,1	Volts
isc	5	Amp
im	4,72	Amp
Vm	18	Volts
FF	0,769230769	

Month	Corrected isc	Corrected Voc	Corrected Pm	Tc
January	4,94	20,66	77,33	43,1
February	4,94	20,62	77,12	43,6
March	4,93	20,46	76,31	45,5
April	4,92	20,16	74,69	49,3
May	4,91	19,78	72,70	54,0
June	4,88	19,43	70,83	58,4
July	4,88	19,24	69,81	60,8
August	4,89	19,27	69,98	60,4
September	4,90	19,54	71,38	57,1
October	4,91	19,94	73,50	52,1
November	4,92	20,25	75,16	48,2
December	4,94	20,51	76,56	44,9

Figure 12. Spreadsheet for determining the most cost effective solution

The equations used in these step are:

$$\text{LCC Energy cost} = \sum \text{Ann. Energy cost} \times [1+i]^{n-1}, \tag{33}$$

$$\text{LCC Maint. Cost} = \text{Ann. Maint. cost} \times [1+i]^{n-1}, \tag{34}$$

$$\text{LCC Repl. cost} = \sum \left[\text{Item Cost} \times \{1+i\}^{n-1} \right], \tag{35}$$

in which **n** is the life period of the PV system considered and **i** is the inflation rate.

The package then makes a final loop for various **f** values. In fact, the **f** value is given an initial estimation e.g. 10% of the peak power load. Hence, it depends on the daily load profile. Usually steps of 10% increase are followed.

Some screenshots of the developed program are shown below. Results of the cost effectiveness vs the **f** value keeping the PV panels type and the batteries type as parameters, are shown in Figures 12, 13 and 14.

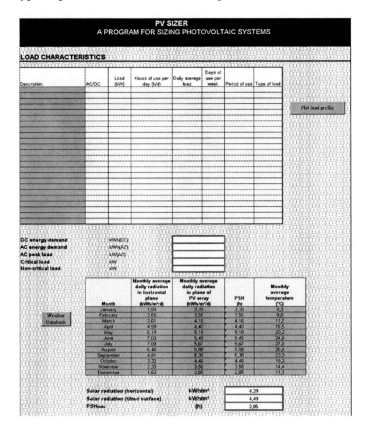

Figure 13. Spreadsheet for the determination of the load and the solar insolation.

LIFE CYCLE COST ANALYSIS

Plot

YEAR	CAPITAL €	ENERGY €	MAINTENANCE €	REPAIR/REPLACE €	TOTAL COST €
1	€ 0,00	€ 0,00	€ 0,00	€ 0,00	€ 0,00
2	€ 0,00	€ 0,00	€ 0,00	€ 0,00	€ 0,00
3	€ 0,00	€ 0,00	€ 0,00	€ 0,00	€ 0,00
4	€ 0,00	€ 0,00	€ 0,00	€ 0,00	€ 0,00
5	€ 0,00	€ 0,00	€ 0,00	€ 0,00	€ 0,00
6	€ 0,00	€ 0,00	€ 0,00	€ 0,00	€ 0,00
7	€ 0,00	€ 0,00	€ 0,00	€ 0,00	€ 0,00
8	€ 0,00	€ 0,00	€ 0,00	€ 0,00	€ 0,00
9	€ 0,00	€ 0,00	€ 0,00	€ 0,00	€ 0,00
10	€ 0,00	€ 0,00	€ 0,00	€ 0,00	€ 0,00
11	€ 0,00	€ 0,00	€ 0,00	€ 0,00	€ 0,00
12	€ 0,00	€ 0,00	€ 0,00	€ 0,00	€ 0,00
13	€ 0,00	€ 0,00	€ 0,00	€ 0,00	€ 0,00
14	€ 0,00	€ 0,00	€ 0,00	€ 0,00	€ 0,00
15	€ 0,00	€ 0,00	€ 0,00	€ 0,00	€ 0,00
16	€ 0,00	€ 0,00	€ 0,00	€ 0,00	€ 0,00
17	€ 0,00	€ 0,00	€ 0,00	€ 0,00	€ 0,00
18	€ 0,00	€ 0,00	€ 0,00	€ 0,00	€ 0,00
19	€ 0,00	€ 0,00	€ 0,00	€ 0,00	€ 0,00
20	€ 0,00	€ 0,00	€ 0,00	€ 0,00	€ 0,00
21	€ 0,00	€ 0,00	€ 0,00	€ 0,00	€ 0,00
22	€ 0,00	€ 0,00	€ 0,00	€ 0,00	€ 0,00
23	€ 0,00	€ 0,00	€ 0,00	€ 0,00	€ 0,00
24	€ 0,00	€ 0,00	€ 0,00	€ 0,00	€ 0,00
25	€ 0,00	€ 0,00	€ 0,00	€ 0,00	€ 0,00
26	€ 0,00	€ 0,00	€ 0,00	€ 0,00	€ 0,00
27	€ 0,00	€ 0,00	€ 0,00	€ 0,00	€ 0,00
28	€ 0,00	€ 0,00	€ 0,00	€ 0,00	€ 0,00
29	€ 0,00	€ 0,00	€ 0,00	€ 0,00	€ 0,00
30	€ 0,00	€ 0,00	€ 0,00	€ 0,00	€ 0,00
31	€ 0,00	€ 0,00	€ 0,00	€ 0,00	€ 0,00
32	€ 0,00	€ 0,00	€ 0,00	€ 0,00	€ 0,00
33	€ 0,00	€ 0,00	€ 0,00	€ 0,00	€ 0,00
34	€ 0,00	€ 0,00	€ 0,00	€ 0,00	€ 0,00
35	€ 0,00	€ 0,00	€ 0,00	€ 0,00	€ 0,00
TOT.	€ 0,00	€ 0,00	€ 0,00	€ 0,00	
TOTAL LIFE CYCLE COST FOR ALTERNATIVE NO.			0		€ 0,00 €

Figure 14. Spreadsheet for the selection of a PV module with all the necessary data provided and all the corrections as introduced in this sizing model

5. A CASE STUDY: THE PV SIZING OF A SOLAR HOUSE IN BUCHAREST

The owners of a house in Bucharest decided to cover the energy needs of a house with R.E.S. technologies. For hot water and space heating the solution was to use solar collector systems; see http://solar-net.teipat.gr , the IP file, while all the electric appliances are to be supplied from a stand-alone PV-system.

The sizing of this PV-system will be approached using two methods:

- the method of **Wh,**
- the method of **Ah.**

5.1 PART A: Sizing the PV generator by the Wh method

Steps:

Table 2. Typical loads for a household

Appliance	Qty	AC DC	P Y/N	Run Watts	Hours /Day	Days /Week	W-Hours /Day	Percent of Total	Surge Watts	Ph-L Y/N
Fluorescent Lights	4	AC	Y	15	5.0	7	300.0	7.7%	0	N
Fridge Sun Frost 16 cu. ft.	1	DC	N	48	11.3	7	540.0	13.9%	1300	N
Blender	1	DC	N	350	0.1	2	10.0	0.3%	1050	N
Microwave Oven	1	AC	N	900	0.3	7	225.0	5.8%	1200	Y
Food Processor	1	AC	N	400	0.1	5	28.6	0.7%	1200	N
Espresso Maker	1	AC	N	1350	0.1	7	135.0	3.5%	1350	N
Coffee Grinder	1	AC	N	150	0.1	7	7.5	0.2%	200	N
21" Color TV	1	AC	Y	125	5.0	7	625.0	16.0%	570	Y
Video Cassette Recorder	1	AC	Y	40	2.5	7	100.0	2.6%	80	Y
Satellite TV System	1	AC	Y	60	2.5	7	150.0	3.8%	1600	Y
Stereo System	1	AC	Y	30	8.0	7	240.0	6.2%	60	Y
Computer	1	AC	Y	45	6.0	3	115.7	3.0%	135	Y
Computer Printer	1	AC	N	120	0.3	3	12.9	0.3%	360	Y
Power Tool	1	AC	N	750	0.5	3	160.7	4.1%	2250	N
Radio Telephone (receive)	1	DC	N	6	24.0	7	144.0	3.7%	0	N
Radio Telephone (transmit)	1	DC	N	20	1.0	7	20.0	0.5%	0	N
Phone Answering Machine	1	AC	Y	6	24.0	7	144.0	3.7%	0	N
Washing Machine	1	AC	N	800	0.5	4	228.6	5.9%	100	Y
Clothes Dryer (motor only)	1	AC	N	500	1.0	4	285.7	7.3%	1500	Y
Sewing Machine	1	AC	N	80	2.0	1	22.9	0.6%	400	N
Vacuum Cleaner	1	AC	N	650	0.5	4	185.7	4.8%	1950	N
Hair Dryer	1	AC	N	1000	0.2	7	200.0	5.1%	1500	N
Ni-Cd Battery Charger	1	AC	Y	4	15.0	2	17.1	0.4%	25	Y

Total Daily Average Watt-hrs 3898.4

| Inverter Priority Wattage 325 | Max. AC Wattage 1350 | Max. AC Surge Wattage 2250 |

1. Determine the loads per day. For the case of the solar house in Bucharest the considered load is shown in Table 2.The total load is equal to 2500 Wh/day i.e.: 1000 Wh/day in DC; that is: 40% DC; 1500 Wh/day in AC; that is: 60% AC

2. Site's details:

   The inclination (β) to horizontal chosen as $\beta \approx \varphi = 45^0$, the METEO data of this site, are given in Table 3. Such an inclination was decided in order to achieve an optimum annual performance.

   PSH per month and its mean annual value are given in the same table; Latitude: 44.4536°, Longitude:-26.0978°, Altitude: 88 m.

3. Elaboration for the daily load profile:

   DC Load: We consider the **DC** load split in a DC day load and DC night load with 40% during the day and 60% during the off operation hours for the PV-panels:

   a) 40% during the day when PV is on operation means 0.4×1000Wh=400Wh/day.

   b) 60% during the time when PV-generator is off, at night; that is: 0.6×1000Wh=600Wh/day.

   AC Load:

   Similarly, as above:

   a) 40% during the day directly PV to load via a DC/AC inverter: 0.4×1500Wh=600Wh/day

   b) 60% during the night PV through batteries: 0.6×1500Wh=900Wh/day and then DC/AC.

Table 3

Month	$\overline{H}$		$\overline{H}_b$		$\overline{H}_d$		PSH
	kWh/m²	MJ/m²	kWh/m²	MJ/m²	kWh/m²	MJ/m²	H
Jan	41	147.6	18	64.8	23	82.8	1.32
Feb	55	198.0	25	90.0	30	108.0	1.96
Mar	89	320.4	41	147.6	48	172.8	2.87
Apr	133	478.8	71	255.6	63	226.8	4.43
May	168	604.8	91	327.6	76	273.6	5.41
Jun	192	691.2	115	414.0	77	277.2	6.40
Jul	196	705.6	118	424.8	78	280.8	6.32
Aug	176	633.6	108	388.8	68	244.8	5.68
Sep	122	439.2	69	248.4	54	194.4	4.06
Oct	84	302.4	44	158.4	40	144.0	2.71
Nov	42	151.2	18	64.8	24	86.4	1.40
Dec	28	100.8	11	39.6	18	64.8	0.90
Year	1322	4773.6	726	2613.6	597	2149.2	3.63

Table 4

Month	RH Relative Humidity	WS (m/s) Wind Speed	WD (degrees) Wind Direction	RR (mm) Rain Precipi-tation	$H_d / \overline{H}$ Diffuse/ Global solar radiation	Ta (^{0}C) Ambient tempera-ture
Jan	88	2.4	225	40	0.56	-2.4
Feb	85	2.7	36	36	0.55	-0.1
Mar	78	2.8	36	38	0.54	4.8
Apr	75	2.6	36	46	0.47	11.3
May	74	2.1	36	70	0.45	16.7
Jun	76	1.7	36	77	0.40	20.2
Jul	74	1.6	36	64	0.40	22.0
Aug	73	1.4	36	58	0.39	21.2
Sep	73	1.5	36	42	0.44	16.9
Oct	78	1.7	36	32	0.48	10.8
Nov	87	2.2	225	49	0.57	5.2
Dec	90	2.2	225	43	0.64	0.2
Year	79	2.1	31	595	5.89	10.6

4. Rough / preliminary determination of the PV-configuration.

The PV configuration to be studied according to the description made may have the following lay-out:

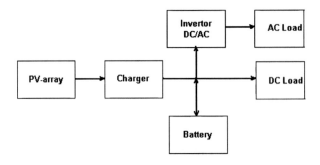

Figure 15. A possible PV-lay-out (stand alone) to meet the loads of a Solar House

5. Inclination to horizontal

Decide on PV-array: inclination, rotation axes etc, ground area required etc.

An investigation on various PV-inclination/rotation configurations has to be carried out for any inclination, in order to determine the most effective solution for the values of the parameters, e.g. when $\beta = \varphi$.

6. Days of autonomy

Decide on days of energy autonomy of the system d.

Discuss on the Critical and non-Critical loads to determine d. Use the formulae below to estimate d.

Re-discuss the PV-system-configuration to be adopted (Figure 16).

$d_{n-cr} = -1.9 \times (PSH)_{min} + 18.3 (days)$

$d_{n-cr} = -0.48 \times (PSH)_{min} + 4.58 (days)$

For Bucharest (PSH) average annual value is 3.63 while minimum is 1. So, d is to be 4. However, as seasonal storage or a supplement source may be used we keep **d=3**, to decrease costs in batteries.

7. Correction in the loads due to losses

Table 5

DC LOADS			AC LOADS	
Losses	%		Losses	%
Cabling PV-directly to loads	5			5%
Charger/cables (when via battery)	5			5%
Battery efficiency 80%	20	ch / disch in the Wh method		0%
DC/AC invertor	15	invertor efficiency 85%	DC/AC inverter	15%

The application of the above values of the losses to loads in Step 3 produces correction of loads as in table 6.

Table 6. Correction of Loads

Load	Route (see Figure 9)	Watt	Correction Factor	Final Load (correction value)
DC	1.2.3	400	1.05	400×1.05=420Wh
DC	1.2.4.3	600	1.25	600×1.25=750Wh
AC	1.2.5.6.7	600	1.20	600×1.20=720Wh
AC	1.2.4.5.6.7	900	1.40	900×1.40=1260Wh
Total				**3150Wh = 3.15kWh**
Total Final Load				**2500Wh = 2.50kWh**
Total Initial Load				

After the analysis made till this stage the PV-system configuration may change to the lay-out in Figure 16:

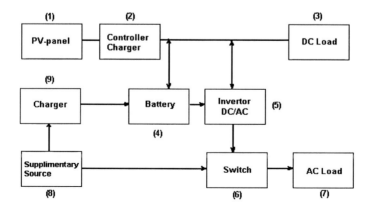

Figure 16. **A** PV-System configuration: a Hybrid Solution

8. Initial / Rough W_p determination

P_m = Load (corrected to losses): $(PSH)_{ann}$ = 3150 Wh / 5.68h = 554.6W_p

Notice:

The designer decided to size the system on summer conditions.

9. Types of PV-panels to be installed

Decide on the PV-panels to be installed:

Let a PV type chosen whose characteristics are:

i_{sc} = 3.45A   Vsc = 21.7Volts

i_m= 3.15A V_m = 17.4 Volts P_m =$i_m \times V_m$= 54.8 ≈ 55W_p

10. Correct P_m, V_m, i_m for the field values of the parameter, T_c

Lets take a PV-panel whose **NOCT** is equal to 46^0C.

Then, the operating temperature, T_c, of the PV-panels is determined as follows:

$$\frac{T_c - T_a}{I_T} = \frac{NOCT - 20^0}{0.8kW/m^2}$$

I_T should be 1 kW/m^2 using **S.T.C.** Then,

$$T_c = T_a + \frac{46^0 C - 20^0 C}{0.8kW/m^2} \times I_T = T + \frac{26^0 C}{0.8kW/m^2} \times 1,0kW/m^2 = T_a + 32,5^0 C$$

The ambient temperature for Bucharest in August is given in Table 4.

Assuming that for August the mean ambient temperature is $\overline{T}_{a,Au}$ = 21.2^{0}C for Bucharest:

T_c=21.2^{0}C+32.5^{0}C=53.7^{0}C,

we estimate the i_{sc} and V_{oc} while **FF** is considered constant. From these new conditions we get:

a) i_{sc} = 3.45A; i_{sc} is assumed non-dependent on temperature,

b) V_{oc} = 21.7Volts – 36 × 0.0023 Volts/^{0}C × (53.7 – 25)^{0}C = 19.32Volts,

$$FF = \frac{55W}{3.15A \times 21.7Volts} = 0.735.$$

Notice: It is assumed that **FF** does not change substantially with T_c.

Finally,

P_m (10^{3}W/m^2, T_c=53^{0}C) = $i_{sc} \times V_{oc} \times$FF = 3.45 × 19.32Volts × 0.735 = 49W,

instead of 55 W_p under S.T.C.

11. Determine the number of PV-panels, N_{pv}

$$N_{pv} = \frac{P_w}{P_m} = \frac{554{,}6W_p}{49W_p} = 11.32 PV - panels \approx 12 PV - panels \cdot$$

Notice: If we used P_m from the specifications (**S.T.C.**), then we would have:

$$N_{pv} = \frac{P_w}{P_m} = \frac{554{,}6W_p}{55W_p} = 10.1 PV - panels \approx 10 PV - panels \cdot$$

12. Decide on the voltage value V_s, for Power transfer i.e. 24, 48, 120 Volts
    The decision affects the PV-system elements and PV-panels electrical connections.
    Consider 2 cases: Vs=48 Volts and 120 Volts.
    If, **V_s=48 Volts**, then:

$$(N_{p,s})_{48V} = \frac{V_s}{V_m} = \frac{48 Volts}{17{,}4 Volts} = 3PV ,$$

so we have panels in series, that is $N_{p,p}$ =12: 3= 4 strings of PV-panels in parallel; each string has 3 PV-panels in series.
    If, **V_s=120 Volts**, then:

$$(N_{p,s})_{120V} = \frac{V_s}{V_m} = \frac{120 Volts}{17{,}4 Volts} = 8PV ,$$

so again panels in series, $N_{p,p} = 2 \Rightarrow N_p = 16$ in total.
Notice: V_m is 17.4 V at S.T.C. conditions. Hence, in field conditions one must use the corrected value as to be discussed later in this problem.
13. Confirmation
    In step 11, we estimated Np = 12 PV-panels.
    Hence, 12 × 49 W = 588 Wp.
    This has to be compared with the 554.6 W_p, estimated in step 8.

5.2 PART B: Approach to the same sizing problem via the Ah methodology

Steps **1 – 6** are the same as in the Wh method.
7. Determination of the charge [Q(Ah)] delivered daily by the PV-generator

Assume that the power from the PV-generator is transferred at 48 Volts or 120 Volts. So, then:

$$\frac{2500Wh}{48Volts} = \frac{2500A \times V \times h}{48V} = 52.08Ah, \text{ for 48 Volts, or}$$

$$\frac{2500Wh}{120Volts} = 20.83Ah, \text{ for 120 Volts.}$$

Let's follow both scenarios to get analytic results.
A) **DC Loads** – directly met by the PV-generator:

1. $\frac{400Wh}{48Volts} = 8.33Ah/day$;   2. $\frac{400Wh}{120Volts} = 3.33Ah/day \cdot$

Indirect coverage via batteries:

1. $\frac{600Wh}{48Volts} = 12.50Ah/day$;   2. $\frac{600Wh}{120Volts} = 5.00Ah/day \cdot$

B) **AC** Loads – directly met by the PV-generator through the DC/AC charger:

1. $\frac{600Wh}{48Volts} = 12.50Ah/day$;   2. $\frac{600Wh}{120Volts} = 5.00Ah/day \cdot$

Indirect coverage via batteries and the **DC/AC** charger:

1. $\frac{900Wh}{48Volts} = 18.75Ah/day$;   2. $\frac{900Wh}{120Volts} = 7.50Ah/day \cdot$

So, the total Ah per day is: 52.08 Ah/day for DC voltage; 48 Volts.

Remark: The same value would be obtained if we divided the load of 2500 Wh by the voltage of 48 Volts:

$Q(Ah) = E: V_s = 2500Wh: 48Volts = 52.08 Ah$

8. Correction to Ah due to losses in various PV-system elements

The correction is similar as in the Wh method. The only difference is in the battery efficiency, which in the case based on Ah is assumed much higher e.g. close to 100%.

– **DC Loads** directly met by the PV-generator: 8.33Ah × 1.05 = 8.75Ah
– **DC Loads** via batteries: 12.50Ah × 1.05 = 13.13Ah

(Notice: in the Wh method the correction factor was 1.25)

- AC Loads via inverter: 12.50Ah × 1.20 = 15Ah
- AC Loads via batteries and DC/AC: 18.75Ah × 1.20 = 22.5Ah. Total: 59.38Ah

9. Determination of the mean annual current from the PV-generator. Since, total daily load is 59.38Ah and $(PSH)_{ann}$ is 3.63h, the annual mean current $\bar{i}_{pv}$ = 59.38Ah / 3.63h = 16.358A

10. Determination of the PV-panels; $N_{p,p}$, $N_{p,s}$ in parallel, in series

The string in parallel (N_p): N_p =16.358A / 3.15A = 5.19. For N_p=6, we look for the value of V_m in field conditions, i.e. $V_m = P_m / I_m$ = 49W / 3.15A = 15.56 Volts, while in the Wh method, V_m was used equal to S.T.C. value: V_m=17.4 Volts. As a remark, in the Wh method, in step 10, we estimated P_m=49 Watts and V_m=15.56 Volts. This leads to: N_s= 48Volts / 15.56Volts = 3.08 $\Rightarrow$ N_s=3. Total: N_{pv}= N_p × N_s = 6× 3 = 18.

So, it results to a higher number for N_{pv} as N_p was well oversized. This approach will provide a PV-generator which generates much more energy than required. It is recommended to keep N_s=3 so that the system has N_{pv} = 6× 3= 18 PV-panels and not oversize N_s=3.08 → 4, as such a decision would drastically oversize the PV-generator resulting to very high costs and unused energy production.

5.3   PART C: Sizing the battery banks

1. Determine the days of autonomy, d(see Wh and Ah method)
   There is no difference either method is used.
   Decide on d=3 based on the formulae in step 6 in Wh method.
2. Determination of the load storage for the days of autonomy
a) Wh method:The load as said is 2.5kWh/day to be transferred at 48Volts.

$$Q(Ah) = \frac{2500kWh / day \times 3days}{48Volts} = 156.25Ah.$$

b) Ah method: The loads per day to be delivered by batteries are 52.08Ah, so, for 3 days there must be stored: 52.08Ah × 3days = 156.25Ah.
3. Correction in the Ah value of the batteries due to temperature
   Temperature of the batteries affects their efficiency. The capacity **C** decreases as **T** decreases below 25 – 27 °C. For high charge – discharge rates, **C**, changes as in figure 17. When **T** changes, **C** has to be corrected:

$$f_{b,T} = \frac{C}{C_0} = \frac{CatT^0C}{C_0at25-27^0C} = 0.01035 \cdot T^0C + 0.724 . \tag{36}$$

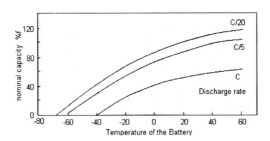

Figure 17. Impact of temperature and of discharge rate to the energy delivered (the case of a PV-acid battery).

Lets take $f_{b,T} = 1$, for the case of Bucharest. Taking into account that for North Germany it was estimated $f_{b,T} = 0.8275$, if **T** is expressed in ^{0}**F**, then:

$$\frac{C}{C_0} = 0.00575 \times T + 0.5 A \quad (\text{T in } ^0\text{F}). \tag{37}$$

4. The correction coefficient due to the charge/discharge rate
   A correction factor due to charge/discharge rate, $f_{b,cd}$, has to be studied as

$$f_{b,cd} = \frac{i_{ch/dish}(as \quad recommend)}{i_{ch/dish}(as \quad in \quad the \quad case)}. \tag{38}$$

So, the corrected capacity C_{cor} is given by the formula

$$C_{cor} = \frac{C(Ah/day)}{f_{b,T} \times f_{b,ch} \times DOD} \tag{39}$$

and for **d** days of autonomy:

$$C_{cor} = \frac{C(Ah/day) \times d(days)}{f_{b,T} \times f_{b,ch} \times DOD}. \tag{40}$$

Notice: if i_{ch} multiplied by **10h** i.e. $(i_{ch} \times 10h)Ah > C_{cor}$ - from equation (40) then, Ccor = $(i_{ch} \times 10)Ah$. In our case i = $i_m \times 6$ strings = 3.15Ah $\times$ 6 = 18.9Ah, $(i_{ch} \times 10)Ah$ = 18.9A $\times$ 10h = 189Ah. Compare $(i_{ch} \times 10)Ah$ to C_{cor} where:

$$C_{cor} = \frac{52.08 \dfrac{Ah}{day} \times 3 days}{1 \times 1 \times 0.8} = 195.3 Ah$$

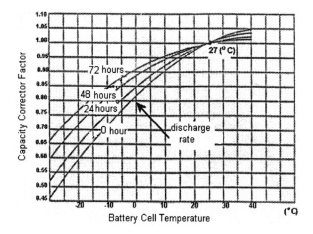

Figure 18. Capacity correction factor of a Pb battery versus battery cell temperature (^{0}C) and the discharge rate.

As $C_{cor} > (i_{ch} \times 10)Ah = 189Ah \Rightarrow$ then we accept that the battery storage system capacity should be 195.3Ah.

5. Determine the type of batteries to be used

One should choose the type of the battery to meet the requirements and the prerequisites of the problem as in the following considerations:

a) Total capacity 195.3Ah, i.e. about 200Ah
b) The voltage across batteries bank to be 48 Volts
c) The DOD value to be higher than 20%. In fact, DOD is related to **d**:

$$\frac{d}{d+1} = DOD \Rightarrow \frac{3}{4} = DOD \Rightarrow \frac{d}{d+1} = DOD_{max},$$

$$DOD_{max} = \frac{3}{7+1} = 0,75.$$

d) The decision of the battery type is complex and depends not only in the above characteristics, but also on the unit price, the life cycles, duration, etc. From the Table 7 let's choose, at first, the battery type: GNB Absolyte: C=59A, V=12Volts, DOD=0.8. Hence, 4 batteries of this type in series are required to provide: 4 ×12Volts = 48Volts. The batteries in parallel are determined by the formula:

$$N_{b,p} = \frac{Q_L \times d}{DOD \times C} = \frac{(2500Wh / 48Volts) \times 3days}{0.8 \times 59Ah} = 3.13.$$

Therefore, we assume 4 strings of batteries, in parallel.

6. Confirm that during the discharge process DOD < DOD_{specs}

We decided before, in step 5, to use 4 batteries of 59Ah with DOD=0.8.

Then, in step 7 of the Ah method the daily total charge Load is equal to 52.08Ah, while total capacity is $4 \times 59Ah = 236Ah$.

Therefore, the daily discharge is:

52.08Ah / 236Ah = 0.22 or 22% < 80% ,

as specified of the type of the battery chosen.

As total C=236Ah, DOD=0.8, the total available capacity is:

0.8×236Ah= 188.8Ah.

This is higher than the 156.23Ah required for the autonomy of the 3 days.

Finally, even if batteries would operate for all 3 days, the discharge level would be:

156.25Ah / 236Ah = 0.662 or 66.2% < 80%.

7. The decision on the battery type choice, provided that this type would meat the technical requirements and the pre-requisites as presented above, should be the outcome of a financial analysis.

Table 7. Types of Batteries

Manufacturer and Type	Model	Nominal Capacity (Ah)	Nominal Voltage (V)	DOD (%)	Life Cycles	Total to be delivered Energy (kWh)
GNB	638	42	6	50	1000	126
Absolyte	1260	59	12	50	1000	359
	6 – 35A09	202	12	50	3000	3636
	3 – 75A25	1300	6	50	3000	1700
Exide	6E120 – 5	192	12	15	4100	1417
Tubular Modular				20	3900	1797
	6E120 – 9	538	12	15	4100	3970
				20	3900	5036
	3E120 – 21	1346	6	15	4100	4967
				20	3900	6299
Delco – Remy	2000	105	12	10	1800	227
Photovoltaic				15	1250	236
				20	850	214
Global Solar	3SSSSRC –125G	125	6	10	2000	150
Reserve gel Cell	SRC – 250C	250	2	10	2000	100
	SRC – 375G	375	2	10	2000	150
Globe	GC12 – 800-38	80	12	20	1500	288
		80	12	80	250	240
GNB Absolyte	638	40	6	80	500	96
	1260	56	12	80	500	269
	6 – 35A09	185	12	80	1500	2664
	3 – 75A25	1190	6	80	1500	8568

6.　CONCLUSIONS

There are many software tools in the market for PV systems sizing like the Retscreen and the PV f-chart [14, 26].

In most of the existing software the sizing and the LCCA are combined. This methodology is quite familiar as it helps to determine and built the PV-configuration which suits the needs best.

In the systematic approach to determine the most cost effective PV-configuration, most of the foreseen corrections to i, V_m, P_m, power losses in cables, efficiency factors in the various elements (such as charge controllers, inverters and batteries) are introduced.

For this reason even the hourly load profile is taken into account especially as it concerns inverter's efficiency as well as the batteries life which is dependent on the cycles depth discharge occurred, charge/discharge rate, electrolyte etc.

Finally, another correction to the loads is introduced due to the fact that batteries rate of recharging is considered as an additional load - see [25].

Various solutions or PV-systems configurations are examined in sequential loops where parameters examined are:

a)　the f factor, i.e. the % coverage of the loads by the PV-generator,
b)　the voltage power is transferred,
c)　the type of PV panels,
d)　the type of batteries,
e)　the type of controllers and inverters,
f)　the monthly based PV-sizing is examined i.e. oversizing when based on a summer month or undersized when PV sizing is based on a winter month.

The software loaded on Excel is quite simple and friendly to run but it appears that there is a need for systematic approach. The development of flexible, easy to use software was the target of this attempt in order to respond to the above requirements. Also, the whole approach attempts to develop software with a potential integration of cost effectiveness tasks in the same tool.

REFERENCES

1.　T. MARKVART, *Solar Electricity* (John Wiley &sons 1996, ISBN 0-471-94161-1).
2.　R. MESSENGER, J. VENTRE, *Photovoltaic systems engineering* (CRC Press ISBN 0-8493-2017-8).
3.　solar-net.teipat.gr *Lectures on the I.P. course on Photovoltaics* (Sponsored by the E.C. Socrates programme. Coordinator S.Kaplanis).
4.　S. Kaplanis, *Lectures on PV-Systems engineering* (M-Sc course ''Energy Systems'' FH Aachen, June, 2004).

5. METEONORM 4.0 *Global meteorological database for solar energy and applied climatology.*
6. http://eosweb.larc.nasa.gov/
7. S. H. EL-HEFNAWI, *Photovoltaic Diesel generator hybrid power system sizing Renewable Energy*, Vol 13, no1, 1998, pp. 33-40
8. H. MABCHOUR et all, *Modeling and performance evaluation of a single crystalline silicon PV-array in Rabat* (World Renewable Energy Congress 1-7 July 2001, Brighton).
9. A. HADJ ARAB et al., *Performance of PV water pumping systems.*(Renewable Energy 18, 1999, pp.191-204).
10. J.G. MCGOWAN, J.F. MANWELL, *Hybrid Wind/PV/Diesel system experiences* (Renewable Energy 16, 1999, pp 928-933).
11. A. ZAHEDI, *Photovoltaic hybrid power systems for off-grid rural electrification, design, cost and performance prediction.* (World Renewable Energy Congress VI (WREC 2000) pp 2065-2068).
12. K KOUZAM et al., *Optimum matching of Ohmic loads to the photovoltaic generator.* (Solar Energy, Vol. 46, 1991, pp101-108).
13. O.E. IBRAHIM, *Load matching to PV generator* (Renewable Energy, 6, 1995, pp 29-34).
14. PV F-CHART software, Photovoltaic system analysis, Wisconsin University.
15. T. HOVE, *A method for predicting long-term average performance of photovoltaic systems.* (Renewable Energy Vol.21, 2000, pp.207-229).
16. M.M.H. BHUIYAN et al., *Economic evaluation of a stand alone residential photovoltaic power system in Bangladesh.* (Renewable Energy Vol.21, 2000 pp.403-410).
17. C.MUERER et al., *PHOEBUS, An autonomous supply system with renewable energy: six years of operational experience and advanced concepts.* (Solar Energy Vol.67, 1999, pp.131-138).
18. A. CHERIF, A. DHOUIB, *Dynamic modeling and simulation of a photovoltaic refrigeration plant* (Renewable Energy Vol.26, 2002, pp 143-153).
19. A. OFFIONG, *Assessing the Economic and Environmental Prospects of Stand-By Solar Powered Systems in Nigeria.* (J. Appl. Sci. Environ. Mgt. Vol.7 no.1, 2003 pp 37-42).
20. D. BERGERON, *Solar Powered Refrigeration for Transport Applications. A Feasibility Study.* (Final Report SOLUS, 2001).
21. *Life Cycle Cost Analysis Handbook*, State of Alaska, Department of Education & Early Developments, 1999
22. C. ARMENTA DEU, *Improving photovoltaic system sizing by using electrolyte circulation in the lead acid batteries.* (Renewable Energy Vol.13, 1998, pp 215-225).
23. C. PROTOGEROPOULOS et al., *Battery state of voltage modeling and an algorithm describing dynamic conditions for long term storage simulation in a renewable system.* (Solar Energy Vol.53, 1994, pp 517-528).
24. M. KOLHE, S. KOLHE, J.C. JOSHI, *Economic viability of stand alone solar photovoltaic system in comparison with diesel-powered system for India*
25. S. KAPLANIS, *A model to predict global hourly solar insolation in a site.* (Research report, T.E.I. Patra, Greece, 2003).
26. Retscreen software, PV2000, www.retscreen.net
27. *"Postgraduate Distance Learning Series in Renewable Energy Systems Technology"*, Solar Power, Unit 7. © CREST 2000

THE SPANISH ELECTRIC POWER GENERATION SECTOR AND ITS POSITIONING BEFORE KYOTO PROTOCOL

S. Cabezudo Maeso and C. Ochoa Laburu
University of the Basque Country, San Sebastián, Spain

Abstract: The most important result of The European Climate Change Program is the Directive 2003/87 that establishes a scheme for greenhouse gas emission allowance trading within the Community. This Directive in force since October 2003 is already moving the entire energy sector in each member country in order to get a favorable National Allocation Plan. This scheme covers 46% of the current EU carbon dioxide, percentage that will be change with the EU enlargement. In the present scenario, some industries are taking proactive steps by reducing or sequestering their greenhouse gas emissions, as others face this shift as an important penalty. This paper analyzes different strategies shown in the electricity generation sector and its move to renewable energies.

Key words: electric power, Kyoto protocol.

1.   INTRODUCTION

Nowadays climate change is without doubt the main environmental challenge that humanity faces, due to its vast effects as well as the global character of its possible solutions and, most of all, to the economic consequences of them. The Kyoto Protocol was at its time a political consequence of this environmental problem. Currently it has been approved by 119 countries that together represent 44.2% of the emissions in the reference year of 1990. While awaiting the answer from Russia whose approval would mean addressing more than 55% of the world emissions (Russia contributes with the 17.4% of the total emissions), the European Union has provided its economies with the tools needed to cope with its

509

D. Talabă and T. Roche (eds.), Product Engineering, 509–518.

compromise. So far, independently of the Russian position, the gas emissions reduction is legally compulsory for the EU member states.

In this sense, the entrance into force of the Directive 2003/87 in 2004 makes feasible the beginning of a European emissions market from the very start of January 2005. The aims at this first stage are certain components and activity sectors. There are six main gases that contribute to the greenhouse effect but only CO_2 whose contribution to the global warming means 55% will be controlled.

Table 1. Common Gases with Greenhouse Effect

Gas	Main source	Warming contribution
Carbon Dioxide (CO_2)	Fossil fuel combustion (77%)	55
Chlorine fluoride	Deforestation (23%)	
Carbon (CFC; HFC; HCFC)	Industrial use: refrigeration, aerosol	24
	Intensive agriculture	
	Carbon mining	
Methane (CH_4)	Gas escape	
	Deforestation	15
Nitrous oxide	XXX Fermentation	
	Intensive agriculture and forest exploitation	6
	Biomass combustion	
	Fertilizers	
	Fossil fuel combustion	

(Source: Spanish Office for Climate Change)

Similarly, of all the implicated activity sectors only six will have limitations in its emissions to the atmosphere: i.e. electric power generation, petrol refining, iron and steel production, concrete production, glass and ceramics and paper production. At this first stage, chemicals, dangerous or municipal (domestic) waste installations and research laboratories are excluded. From 2006 on, the gases to be controlled as well as activity sectors lists will be extended.

The European Union accepted a compromise to reduce its CO_2 emissions by an 8% taking 1990, the year in which the Protocol was approved, as a reference year. From the beginning there were quotas allocated to each Member State but nothing was established at a sectorial level. Anyway it is now 15 years since those reference values were taken and before the definite start up of the economic tools some important changes on the reduction aims have to happen. For example, Spain was allocated a 15% maximal increment on the emissions measured in 1990 but taking into account the economic growth that obviously has implied the emissions increase, so the actual aim is to reduce about a 20% of these emissions. That means that by 2012 Spain will be able to emit to the atmosphere a maximum of 333 million tones CO_2.

This paper treats only the electric power generation sector that is the main contributor to the global warming. In it we discussed briefly the tools that are considered by the Kyoto Protocol as well as the European Union and the diverse reactions from the leader companies in that activity in Spain.

2.   KYOTO PROTOCOL

The global warming results from the increase of CO_2 emissions that stay in the atmosphere so the solutions are either to limit these emissions or increase the absorption capacity of the drains. So that and with the aim of smoothing the change towards cleaner economies, the Protocol foresees three kinds of tools that would allow the initial exchange between countries:

a) Joint Implementation (JI): The Protocol 6th article allows the investment from a developed country in another developed country in projects aiming at the emissions limitation or the carbon fixation. This mechanism recognizes the possibility of transfer or acquisition from other country emission reductions on a project basis.

b) Clean Development Mechanism (CDM): Article 12th foresees investments from a developed country in another developing country in projects aiming at the emissions limitation or the carbon fixation. This path allows emissions credits obtained from projects undertaken in developing countries.

c) International Emissions Trading (IET) Market: Pollution Permission exchange between countries that have acquired the compromise of fulfill in the Kyoto Protocol.

Thanks to these mechanisms, the total reduction costs would be optimized by promoting the change to alternatives with less emissions in those companies that show a minor marginal cost, providing the emissions concentration in other companies with bigger reduction costs. These mechanisms have been studied over the last forty years by authors like [9, 13. 21, 4]. They demonstrate the socioeconomic advantages of these tools instead of command and control solutions.

"The inclusion of the Kyoto Mechanisms will further and significantly reduce costs. Whereas price estimates for a purely internal market range between 18 and 33 euros/t CO_2, forecasts for world market prices range from 2 to 8 euros/t CO_2. The World Bank's Prototype Carbon Fund pays about 3.5 euros/t for millions of tons of CO_2." [12].

It should be added to these three tools the chapter of measures and policies like the promotion of renewable energies use, the drains (mainly forest and agriculture) protection and improvement and the progressive reduction of the market inefficiencies like fiscal incentives and subsidies that were opposite to the Convention's aims. The purpose of it all being to optimize the energy use on the consumer side while companies are encouraged to improve their technologies to reduce the emissions. In this way the market we will find a new Pareto optimum where Society as a whole will benefit.

3.   EUROPEAN EMISSIONS MARKET

Nowadays, the European Union has developed the legislative framework that will allow the home market of pollution rights to start by January 2005. It is by no means the first world initiative in this area and some authors have revised and compared up to ten pollution rights system currently at work and that apply either to CO_2 markets as well as to some initiatives in the field of NOx and SO_2 and its implications in acid rain [6, 19, 10].

In the European Union's case, some of the variables that have a decisive influence in the results that these systems accomplish as geographical and spatial cover are guaranteed. The start up of the Directive in the fifteen member countries an its application later in the new associated countries means the creation of a market big enough to avoid rights prices manipulation problems as is the case in an oligopolistic situation [7].

In a model like this another variable is introduced refers to permits receivers. From a theoretical point of view it could be end consumers as well as companies the receiving emissions rights ands some work such as [1] show the advantages and disadvantages of both. In Europe's case it will be the companies that should manage its pollution leaving consumers in second plane.

While this item has not been questioned at any moment, the allocation of rights between the participant companies has been. This item is crucial for the efficiency the good work of the whole system. The models suggest three ways for its distribution between the agents: auction, free distribution based on historical pollution records or purchase based on the actual production. The first one being the most in favour of costs saving as it implies the higher pollution permit prices, so that the way to reduce the CO_2 before the break-even point. This hypothesis is the favorite one with the ecology/environmental groups for this system accelerates the accomplishment of the Kyoto aims. However the business organizations argue that this alternative will affect negatively the industrial competitiveness. First of all because of the high cost that it implies in the short term. And most of all it is the uncertainty increase that this system brings. The industry have to integrate this uncertainty in its decision making process which will influence negatively the investments done (Business Europe, 2002).

Moreover in the case of auction of permits based on production indicators, the most favored business will be those that had introduced improvements in its production systems in the past and they will harvest the fruit of their investments straight after the permits markets are in force. On the contrary if the rights were to be distributed based on historical emissions records those companies will be punished because reducing the first pollution units is much cheaper than doing the same in the subsequent stages.

4. APLICATION OF HOFFMAN MODEL TO SPANISH ELECTRIC POWER GENERATION SECTOR

Some big multinational companies (Du Pont, Shell, Alean Aluminum, BP, Suncor Energy, Pechiney and Notario Power Generation) affirm that the introduction of some measures directed to fight the climate change through energy savings have contributed to improve its economic performance. This same argument is presented in Spain by the second company in terms of invoiced power, Iberdrola. This company is acting as a leader of a group of companies from its own industry, basically the younger ones, to promote the accomplishment of Kyoto Protocol aims in a shorter period of time. Supporting this idea, some relevant authors have published arguments that relate almost inevitable environmental investments and competitive position improvement [15, 8, 14, 20].

However this way of thinking is not mainstream in Europe now, it is not shared by many big companies that oppose strongly the new emissions market. Returning to the Spanish situation, the bigger company in the Spanish electric power generation sector, Endesa, together with other important companies such as Union Fenosa, Hidrocantabrico, Eléctrica del Viesgo, are trying to delay the start up of the Directive. Some authors [11] try to balance the very optimistic vision of some companies with another more realistic one that mean important changes in the very actual conception of the economic paradigm and that would explain the extreme confrontation between the different positions around environmental regulation. In its work Hoffman analyses four variables to understand the positioning, in favour or against, that show the companies: capital investment, market competitive position, international competence, institutional change.

4.1   Capital investment

Recovering the capital investment needs to have a guaranteed minimum lifespan. A premature substitution prevents the necessary cash flow generation that would allow new investments. In the case of the electric power generation sector, responsible for the 50% of CO_2 emissions, the more difficult investments to replace right now are the carbon related one. (Table 2). It is true that this sector position is quite comfortable because it could reduce its emissions in the short and medium term introducing renewable energies and the use of natural gas without reducing production. In contrast with this situation, the rest of the sectors would need to reduce production in order to reduce emissions.

Table 2. Impact and emissions estimation per industry sector in Spain

Sector	Year 1990	Year 2000	Year 2010	Increment 1990-2010	Déficit
Electric power gener	63,7	87,7	85,5	34%	19%
Petrol refining	12,2	14,5	16,6	36%	21%
Concrete	21,5	25,5	34,5	60%	45%
Glass and ceramics	8,8	12,4	15	70%	55%
Paper	2,7	3,5	3,7	36%	21%
Iron and Steel	14,2	11	13,4	-6%	0
Total	123,2	154,6	168,6	37%	

(Source: PricewaterhouseCoopers)

However for those carbon intensive companies, though technically could substitute their electricity generation sources with some other can not afford it financially. This is the case of the main agents in the electric power generation sector, e.g. the Endesa group have 12 carbon generation stations while the ones in favor of Kyoto Protocol are less dependent on it.

4.2   Market competitive position

This theoretical model suggests that the second variable that will explain the companies behavior will be the possibility of generating profits through environmental investments. It should be understood that the climate change control regulation is another strategic variable to be considered. So for electric power generation companies nuclear or hydraulic sources could be favored by this Directive but should consider the effect of the regulation on nuclear waste or the environmental impact of dams. In this case the bet for CO_2 emissions market is a logical solution that can provide some profits from permits sold or bought in the market.

Obviously the companies that see themselves as sellers are interested in the creation of this type of market [12], while the potential buyers try to hinder it. So that, if at a global level the market mechanisms reduce the total costs, they do not favor equally all of the companies in a certain industry.

Moreover this comparative cost advantage is not the one that generates the climate change. The bet for green energies turning to be a product with a market segment is becoming more and more important [17]. The immediate outcome is the offer from some companies to their customers to choose the supply of electric power from green energy sources. In the US there exist labels and certifies that inform the consumer about the energy source and the environmental impact they have [5]. In Europe case there is an electric power certification system that takes account of the source used (RECS: Renewable Energy Certificate System).

The Iberdrola competitive strategy is then doubled. While in the debate forum advocates for the strict accomplishment of Kyoto Protocol, has thrown a new product to the market that in coincidence with the liberalization of the sector and the possibility for consumers of choosing their supplier give them the opportunity to subscribe to the green energy offered by them.

Again it appears the competitive advantage is a central argument in support of climate change regulation. Moreover, as this green energy is more expensive than the regular one, Iberdrola see how their profit grows.

4.3   International competence

Russia refusal of Kyoto Protocol leaves it without legitimate value, what makes impossible the creation of an emissions permits world market. All the sources agree that the bigger the permits market the lesser the price of an emission right. So European companies feel discriminated as long as they will be the ones that should fulfill with agreements that will benefit the whole world but whose cost they should assume alone.

Danish companies in a joint manifesto before the European Commission show its disagreement with the directive on emissions control following the previous arguments. It has to be considered also the rejection of the US administration to the Protocol would introduce a new distortion on the energy comparative prices between Europe and the USA.

A way to try to minimize those environmental costs are the flexibility mechanisms as the Joint Implementation that would allow the investment in emissions reduction projects in third world countries were the reduction of a CO_2 tone is cheaper than in the member countries.

From a global point of view there is no doubt about the dissemination effect that will have these efforts from European companies in the rest of developing countries. The development of new technologies, the economic acceleration of clean energies, the investments in B ANEX countries are some examples of these multiplicative and beneficial effects. As some authors affirm all of it will return multiplied to the promoting countries.

4.4   Institutional change

The business role on this process and the negotiating power of each of the agents involved is the last variable introduced in the Hoffman model. In the European Union's case this role is important most of all when designing the national allocation plans for each member state.

The Government has two alternatives to fix the emissions rights to distribution between their companies. The first one is to distribute them proportionally to the contribution to greenhouse effect. In this way the electric power generation sector would receive an allocation proportional to its emissions, about two thirds of the total. But this is not the only alternative, not even the most rational from an economic point of view. A second criteria is to minimize costs and business competitiveness, so that it has to be considered the cost that takes to reduce by a ton the CO_2 emissions. In this second scenario things change a lot for the electric power generation sector has a lower marginal reduction cost and so they would receive much less rights than other industries with much higher costs.

The British Government has been the first one to publish its allocation plan has followed the second criteria.

Also the European Union is giving the first steps into unifying the CO_2 emissions measurement system. These measurements have to have two main characteristics. First of all, data reliability. Second the on line input of the emissions reduction so that business could trade fast the permits surplus. A possibility is to integrate them in the measurement systems considered in ISO 14000 [18]. From this point of view the business that had incorporated environmental management systems in the past were more prepared and should have a better learning curve.

5.   CONCLUSIONS

The Kyoto Protocol changes at large the "game law" in the markets. While there are a lot of economists and businessmen that say that it is a too strict and very expensive process, most of the scientists and environmentalists foresee that the real effect on the planet temperature will be weak. Businesses take part on the debate and put pressure on their Governments. This debate can be explained to some extent through the theoretical work developed by Hoffman in which four explicative variables are used: capital investment, market competitive position, international competence, institutional change.

The creation of an European permits market would improve the expectations of some companies but would affect negatively others.

However it is necessary that it happens like this. There would not exist a market if there were not sellers and buyers, that are winners and losers. The reason why one particular company stand in one position or another could be explained by the Hoffman model. So business that were not prisoners of past investments, that could create a competitive advantage in this new frame, that were not affected in its international position and that would make sure that the allocation criteria would favour them will be sellers while the rest should be obliged to buy emissions rights.

However this position is not universal. It should be enough to modify the environmental aims of the Protocol and focus the environmental regulation on nuclear energy or the big hydraulic dams to change the panorama.

The principle of "It pays to be green" represent once again only half of the reality: it will benefit the companies that had changed towards environmental management before it was compulsory to do that [8]. It will also benefit the whole of the European Union if we start on our own the way towards an economy with less CO_2 emissions, that is a way that the rest of the countries should follow in a near future.

REFERENCES

1. M. AHLHEIM, F. SCHNEIDER, *Allowing for Household Preferences in Emission Trading* (Environmental and Resource Economics, April 2002, pp 317-342).
2. Anonymous, *Emissiones trading at a crossroads* (Business Europe, oct 30, 2002; 42, 21, pp 6).
3. E. BARCLAY, E. COOK, *Safe Climate, Sound Business* (Corporate Environmental Strategy, vol 9, issue 4, December 2002, pp 338-344).
4. W. J. BAUMOL, W. E. OATES, *The Theory of Environmental Policy* (Cambridge University Press, Cambridge, UK., 1988).
5. L. BIRD, *Understanding the Environmental Impacts of electricity: Product Labeling and Certification* (Corporate Environmental Strategy, vol. 9, Issue 2, May, 2002, pp. 129-136).
6. C. BOEMARE, P. QUIRION, *Implementing greenhouse gas trading in Europe: lessons from economic literature and international experiences* (Ecological Economics, vol 43, issue 2-3, December, 2002, pp. 213-230).
7. T. CASON, L. GANGADHARAN, C. LUKE, *Market power in tradable emission markets: a laboratory testeb for emission trading in Port Phillip Bay, Victoria* (Ecological Economics, vol 46, issue 3, October, 2003, pp. 469-491).
8. F. CAIRNCROSS, *Green.Inc* (Earthscan Publications Ltd, London, 1995).
9. J. DALES, *Pollution, Property and Prices* (University of Toronto Press, Toronto, 1968).
10. D. HARRISON, D. B. RADOV, *Evaluation of alternative initial allocation mechanisms in a European Union Greenhouse gas emissions allowance trading scheme* (National Economic Research Associates for the European Commission (DG Environment, 2002).
11. A. J. HOFFMAN, *Examining the rhetoric: the Strategic Implications of Climate Change Policy* (Corporate Environmental Strategy, vol 9, Issue 4, December, 2002, pp. 329-337).
12. A. MICHAELOWA, S. BUTZENGEIGER, *The Eu proposal for Emissions Trading: A reasonable Approach?* (CESifo Forum, Spring 2003).
13. D. MONTGOMERY, *Markets in licenses and efficient pollution control programs* (Journal

of Economic Theory n 5, 1972, pp. 395-418).

14. K. NORTH, *Environmental Business Management* (International Labor Office Geneva, Management Development Series, n° 30, 1992).

15. M. PORTER, C. LINDE, *Green and Competitive. Ending the Stalemate* (Harvard Business Review, Sep-Oct, 1995, pp. 120-134).

16. A. RAHMAN, *Market System to curtail emissions magnitudes evolved from electricity generation* (2002).

17. I. ROWLANDS, D. SCOTT, P. PARKER, *Consumers and Green electricity: profiling potential purchasers* (Business Strategy and the Environment, Jan/Feb, vol 12 n°1, 2003, pp. 36-48).

18. R. THIRNTON, H. SANGEM, *Environmental Management Systems and Climate Change* (Environmental Quality Management, Autumn 2001; 11, 1, 2001, pp. 93-100).

19. R. SCHWARZE, P. ZAPFEL, *Sulphur allowance trading and the regional clean air incentives market: a comparative design analysis of two mayor cap-and-trade permit programs* (Environmental and Resource Economics number 17, 2000, pp. 279-298).

20. J. SKEA, *Environmental Technology* (in Folmer, Landis and Spschoor (ed), "Principles of environmental and resource economics: a guide for students and decision-makers", Edwards Elgar Publishing Limited, UK, 1995).

21. T. H. TIETENBERG, *Emissions Trading: an Exercise in Reforming Pollution Policy* (Washington, D.C., Resources for the future, 1985).

22. J. WACKERBAUER, *Emissions trading with greenhouse gases in the European Union* (CESifo Forum, Spring 2003, 4).

HELICAL TURBINE FOR AEOLIAN SYSTEMS AND MICRO-HYDROSTATION

I. Bostan, V. Dulgheru and R. Ciupercă
Technical University of Moldova, Republic of Moldova

Abstract: According to the wind cadastre existing in the Republic of Moldova and the development priorities of the wind energy conversion systems, which recommend the development of small power electric-energetic plants (approximately 3 – 5kW/h) for private consumers, the utilization of efficient working parts is crucial. According these rigors a helical wind and water micro turbine is proposed.

Key words: helical turbine, wing profile.

1. ELABORATION OF MATHEMATICAL MODEL FOR A HELICAL TURBINE

Knowing the aerodynamic characteristics is highly important for the design of the optimal wing profile of helical rotor and setting some constructive parameter values which can make efficient its functioning at usage conditions' variation and various wind speed. The design of the helical turbine (Figure 1) includes axis *1* where blades *2* are stiffly fixed on the constant pace helical line. The wing profile (Figure 2) is characterized by its blunted foreside and sharp backside. Its central line is the geometric place of circuit centres inscribed in the profile.

The main geometric parameters of the profile are:
a) relative thickness of the profile $\vec{c}$, which was determined as the relation of the peak thickness of the profile c towards the chord length b, $\vec{c} = c/b$;
b) relative hollow $\vec{f}$ which was determined as the relation of peak bending-deflection of the axial curve f towards the chord length b, $\vec{f} = f/b$;
c) the camber, which was determined through the bending angle of the central line ε, that is the angle between the tangent lines at the central line

519

D. Talabă and T. Roche (eds.), Product Engineering, 519–528.
© 2004 *Springer. Printed in the Netherlands.*

of the profile in its foreside and backside.

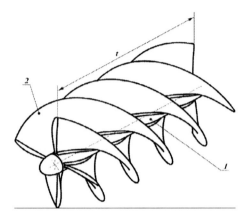

Figure 1. Helical turbine

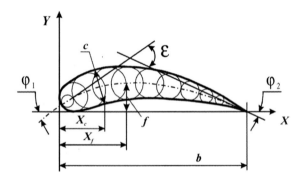

Figure 2. Wing profile

Positions $\vec{c}$ and $\vec{f}$ were determined through the relative abscissas: $\vec{x}_c = x_c / b$ and $\vec{x}_f = x_f / b$. The converse position of the profile in the reticule is characterized by pace t, position angle Θ (the angle between the chord of the profile and the flank of the reticule), and angles φ_1 și φ_2 between the tangents at central line of the profile in its points and the flank of the reticule. The relative pace of the reticule was determined by the relation of pace t towards the chord length b, $t = t/b$. The reticular density, which is the inverse value of the relative pace, was determined from the relation $\tau = 1 / \vec{t} = b / t$.

The elaborated mathematical design allows determining the basic kinetic energy parameters of the helical turbine. In order to do this, the motion equation system of the perfect incompressible and isoenthropic fluid was considered, which describes the air movement round the Aeolian rotor with a rather high accuracy:

$$\rho \operatorname{div} \bar{V} = 0 \text{ - continuity equation,}$$

$$\rho \frac{\mathrm{d}V}{\mathrm{d}t} = -\operatorname{grad} p + \rho f \text{ - pulse equation,} \tag{1}$$

$$\rho \frac{\mathrm{d}e}{\mathrm{d}t} = -p \operatorname{div} V \text{ energy equation.}$$

The solutions to these equations comply with the limit conditions on the turbine rotor propeller and at big distances in undisturbed limits of the fluid. Generally the setting of these conditions presents certain difficulties related to the constructive form and the operating conditions of the Aeolian turbine. That is why it was resort to their determination for certain optimal working conditions, where the speed values in the outflow of the rotor and of the induced speed in the propeller blade were known. Thus, through the given integral equation system (meant for the examined Aeolian turbine) the immediate calculus of the aerodynamic characteristics was possible.

This research has been carried out in order to determine the air-mass speed and the corresponding generated forces. The effect force of the current over a unitary profile in cross direction to the figure plan was determined at careening of a infinite reticule profile by a continuous parallel air flow. Sections *1* and *2* (Figure 3), parallel to the figure flank, were pointed out in the current and spaced from the reticule at a distance that permitted the acceptance of a constant speed and pressure in each section, that is where the current will not be disturbed. The current lines *AB* and *CD* have been set out at the distance of reticule pace *t.*

Motion quantity equation was applied to *ABCD* pointed out space:

$$F \Delta T = m \vec{w}_2 - m \vec{w}_1. \tag{2}$$

The resultant projections of all forces which act in this space on *Z* axis and *U* reticule flank are:

$$P'_U = M \left(-w_2 \cos \beta_2 + w_1 \cos \beta_2 \right) = \\ = M \left[-w_{2U} + w_{1U} \right], \tag{3}$$

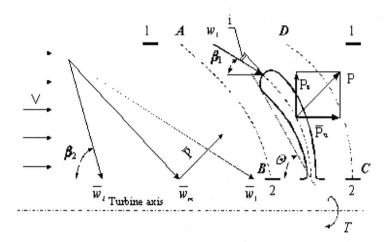

*Figure 3.*Diagram of forces generated by air currents

where P_U' is the resultant force projection on U axis;

$M = m / \Delta T$ – air-mass which passes per second through the reticle with pace t and unitary length (in cross direction to the figure plan).

From the continuity equation:

$$M = \rho_1 w_{1Z} \cdot t \cdot 1 = \rho_2 w_{2Z} \cdot t \cdot 1. \tag{4}$$

For an incompressible gas:

$$\rho_1 = \rho_2 = \rho \text{ and } w_{1Z} = w_{2Z} = w_Z, \tag{5}$$

and the resultant force projection on Z reticle axis is determined by the relation:

$$P_Z' + (\rho_1 - \rho_2) \cdot t \cdot 1 = M(w_{2Z} - w_{1Z}) = 0. \tag{6}$$

$\vec{P}$ force projection, which acts over the profile – wing of unitary length:

$$P_U = -P_U' = -M(w_{1U} - w_{2U}) = -\rho w_Z t(w_{1U} - w_{2U});$$
$$P_Z = -P_Z' = (\rho_1 - \rho_2)t. \tag{7}$$

Thus, $\vec{P}$ is the resultant force that acts over the profile, but $\vec{P}'$ – the force applied to the computational load.

According to Bernoulli equation:

$$p_1 + \rho w_1^2 / 2 = p_2 + \rho w_2^2 / 2, \tag{8}$$

where p_1 and p_2 are static pressures in sections *1* and *2*; $\rho w_1^2 / 2$ and $\rho w_2^2 / 2$ – dynamic pressures in sections *1* and *2* accordingly. Thus:

$$p_1 - p_2 = \rho/2\left(w_2^2 - w_1^2\right) =$$
$$= \rho/2\left(w_{2U}^2 + w_{2Z}^2\right) - \rho/2\left(w_{1U}^2 + w_{1Z}^2\right) = \tag{9}$$
$$= \rho/2\left(w_{2U}^2 - w_{1U}^2\right)$$

Let us determine now the speed circulation on *ABCD* contour accepting as positive the counter-clockwise direction:

$$G_{ABCD} = G_{AB} + G_{BC} + G_{CD} + G_{DA}. \tag{10}$$

As *AB* and *CD* current lines are congruent, and the speed distribution on them is the same, then

$$G_{AB} = -G_{CD}; \quad \begin{aligned} G = G_{ABCD} = \oint_{ABCD} c \cdot \cos(\vec{c}, \vec{s}) ds = \\ = -w_{2U} \cdot t + w_{1U} \cdot t = t(w_{1U} - w_{2U}). \end{aligned} \tag{11}$$

2. DETERMINATION OF THE AERODYNAMIC FORCES AND PERFORMANCES FOR THE HELICOID ROTOR

The average geometric vector of $\overrightarrow{W}m$ speed is determined by the relation:

$$\overline{w}_m = (\overline{w}_1 + \overline{w}_2)/2 \tag{12}$$

The projection of this vector on *U* axis is equal to $\left(w_{1U} + w_{2U}\right)/2$, and on *Z* axis respectively $\left(w_{1Z} + w_{2Z}\right) = 2w_Z/2 = w_Z$.

The direction of the average geometric speed is determined:

$$ctg\,\beta_m = \frac{w_{mU}}{w_{mZ}} = \frac{w_{1U} + w_{2U}}{2w_Z} =$$

$$= \frac{1}{2}\left(\frac{w_{1U}}{w_Z} + \frac{w_{2U}}{w_Z}\right) = \frac{1}{2}(ctg\,\beta_1 + ctg\,\beta_2).$$

$$(13)$$

Thus the resultant of all velocities, which reacts on the reticulum from the part of the incompressible gas current, is equal to the product between the density, medium geometric speed and speed circulation around the profile. Its agency direction is perpendicular to the medium geometric speed vector. In order to determine the direction of *P* force we rotate vector W_m with the angle of 90^0 counter-clockwise.

As previously shown, over a unitary profile, which moves with a peripheral speed *U*, there is a lifting acting force $\vec{P}$. The projections of this force on the reticulum axis and on the frontal line are equal to P_Z and P_U respectively.

Applying the quantity of motion equation towards an elementary annular section with unitary thickness the following relations are obtained:

$$P_z = -\frac{\rho G}{2}(w_{1U} + w_{2U}),$$
$$P_U = -\rho G w_z$$

$$(14)$$

$$P = \sqrt{P_z^2 + P_U^2} = \rho G \sqrt{\frac{(w_{1U} + w_{2U})^2}{r^2} + w_z^2} = \rho G w_m.$$

$$(15)$$

To analyze how the viscosity of an incompressible gas reacts on the resultant of all velocities which reacts on the profile in the reticulum, the quantity of motion equation will lead to the relations for axial and frontal components of the current reaction on a unitary profile of the reticulum (Figure 4):

$$P_z = (P_1 - P_2)$$
$$P_U = -\rho w_z t (w_{1U} - w_{2U})$$

$$(16)$$

Taking into account the viscosity, the Bernoulli equation for the sections *1* and *2* becomes:

$$p_1 - p_2 = \frac{\rho}{2}(w_{2U}^2 - w_{1U}^2) + \Delta p,$$

$$(17)$$

where Δp are summary losses of the overall pressure which occur due to the viscosity.

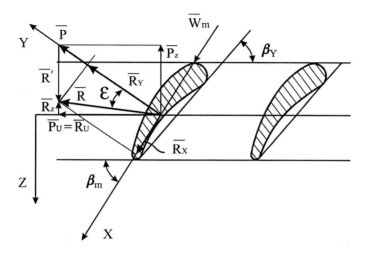

Figure 4. Calculus schema of the radial and frontal components of the force generated by the real gas

Thus,

$$\Delta p = p_1 - p_2 + \frac{\rho}{2}(w_{1U}^2 - w_{2U}^2)$$

$$R_z = (p_1 - p_2)t = -\frac{\rho t}{2}(w_{1U}^2 - w_{2U}^2) + t\Delta P$$

(18)

Comparing the relations of the lifting force projections at the motion of a perfect gas with the similar relations at the motion of a semi fluid gas, for a perfect gas we have:

$$P_U = -\rho G w_z ,$$

$$P_z = -\rho G \frac{w_{1U} + w_{2U}}{2}$$

(19)

and for a semi fluid gas:

$$P_U = -\rho G w_z ,$$

$$P_z = -\rho G \frac{w_{1U} + w_{2U}}{2} + t\Delta P.$$

(20)

The additional member $t \cdot \Delta P$ expresses the projection of the resistance force $\vec{R}'$ on the reticulum axis.

The projection of this force on the axis is equal to zero, i.e. parallel to the reticulum axis. We introduce the average geometric speed $\vec{w}_m = \left(\vec{w}_1 + \vec{w}_2 \right)/2$ and we obtain the relation:

$$R = \rho \cdot Gw_m, \tag{21}$$

which formally does not differ from the one for the perfect gas. But here both w_m and G are determined according to the real velocities $\vec{w}_1$ and $\vec{w}_2$.

The interaction resultant force (total force) $\vec{R}$ of the viscous gas draught with the net outline is equal:

$$\vec{R} = \vec{P} + \vec{R}'. \tag{22}$$

Thus, as $P_U = R_U$, the resultant force does not influence the torque moment of the outlines net.

We decompose the force $\vec{R}$ into components:

$$\vec{R} = \vec{R}_X + \vec{R}_Y, \tag{23}$$

where R_X is the frontal resistance force and R_Y – the lifting power.

The frontal component of the resultant force R_X characterises the power action of the draught upon the working wheel, and the axial component R_Y determines the loading force of the turbine bearing. The relation between the lifting power of the profile and the frontal resistance force is the quality of the outline:

$$K = R_Y / R_X = ctg\varepsilon \tag{24}$$

These non-dimensional coefficients of the integrated outline forces or of the net depend on the outline and the net geometry, on the angle of the attack i, on the dynamical pressure $\rho \cdot W_m^1 / 2$ and on other auxiliary factors.

The resultant force for the net composed of n outlines with integrated height is determined as:

$$R = C_R nb\rho w_m^2 / 2 \tag{25}$$

and the components for the integrated outline as:

$$R_y = C_y b \rho \cdot w_m^2 / 2,$$
$$R_x = C_x b \rho \cdot w_m^2 / 2.$$

(26)

Here, C_R, C_Y, C_X are the coefficients of the aerodynamic force, of the lifting power and of the frontal resistance force.

The coefficient $C_Y = \dfrac{2t}{b} \cdot (\mathrm{ctg}\beta_1 - \mathrm{ctg}\beta_2) \cdot \sin \beta_m$ allows the determination of the lifting power according to the known specific features of the outlines net.

3. COMPUTER ASSISTED OPTIMIZATION FOR HELICAL TURBINE GEOMETRIC PARAMETERS

The model of the "helical turbine–fluid" system consists of two elements (one is fluid with the flow in the outflow and the other one is solid and fix on the direction of the fluid flow) with implicit links between them. This system has an essential particularity, that is, the impossibility to set strict interaction conditions of the fluid element with the rotor on one side, and of the modification laws of the upstream, downstream fluid and horizontally to it depending on the turbine usage conditions and of the global status of the undisturbed area. To simplify the research methodology and to reduce the measuring points in the inside of the air flux field it was agreed upon the concept of combining the experimental method with the theoretical one. The computerized model of the helical turbine with four beginnings is represented in Figure 5.

Figure 5. Computerised model of the helicoid turbine

REFERENCES

1. I. BOSTAN, M. TOPA, V. DULGHERU, R. CIUPERCĂ, *Helicoidal turbine.* (Patent no. 2126MD, 2003).
2. I. BOSTAN , M. ŢOPA, V. DULGHERU, A. OPREA, R. CIUPERCĂ, *Hydraulic plant.* (Patent no. 2288MD, 2003).

DYNAMIC MODELLING OF WIND FARMS: A COMPARATIVE STUDY BETWEEN TWO MODELLING APPROACHES

I. Zubia[1], S.K. Salman[2], X. Ostolaza[3], G. Tapia[3] and A. Tapia[3]
[1]Department of Electrical Engineering, University of the Basque Country, [2]School of Engineering, The Robert Gordon University, [3]Department of Systems Engineering and Automation, University of the Basque Country

Abstract: World's energy needs have encouraged the development of wind power sources. An increasing number of wind farms are being connected to the power grid and accurate the development of wind farm models need to be developed to study the dynamic behavior of them. This paper reports an investigation to compare two different modeling approaches of a real wind farm. This study is applied to the dynamic model of a real wind farm located at El Perdón, in Navarre, (Northern Spain). It consists of 30 Squirrel Cage Induction Machine (SCIM) wind generators of 660 kW and a nominal voltage of 690V. The wind farm feeds three local loads at 66 kV and is connected through two transformers to the network of 220KV.

Key words: wind-energy, modeling of electrical systems, simulation of power plants.

1.   INTRODUCTION

The development of dynamical models for wind farms could be made from different approximations. The first one, could be the consideration of the whole set of N parallel connected generators of the farm as a unique equivalent N-machine. This approach is the most simple, and hence, the most common in different references. In a previous work the development of the **N-machine model** has been done [1]. Where the model of the N-machine generator has been developed as a particular case of the electric machine general model: "The Quadrature-Phase Slip-Ring Model" [2]. This includes the mechanical power characteristic of the wind hitting the blades of the wind turbine.

D. Talabă and T. Roche (eds.), Product Engineering, 529–539.
© 2004 *Springer. Printed in the Netherlands.*

This model allows the analysis of the effects between the global farm and its surrounding electrical distribution network: contribution to fault current, interaction with power system protection system [3]. The N-machine model also brings interesting information about transient behavior and stability of wind farms under electrical disturbances. This problem has been studied for the case of three phase faults and for islanding operation [1]. The important results obtained via this approximation lead to a second approximation which concerns the study of wind farm internals.

A second approximation could be a deeper modelling of the wind farm in order to asses the interactions between particular generators, under different wind speeds and, hence, different mechanical generating powers [4]. The development of this Multi-machine model is more complex than the N-machine, not only due to the increasing number of non-linear state-equations of machines, but also because of the Dyn connection of step-up transformers used in the Spanish electrical system.

The purpose of this work is to present a comparative study between these two modelling approaches. This study is applied to the dynamic model of a real wind farm located at El Perdón, in Navarre, (Northern Spain). The wind farm consists of 30 wind generators of 660 kW. The line parameter values, and load and capacitor values have been provided by IBERDROLA S.A. electric generating company.

2.   CASE STUDY

Both models have been implemented as a C-MEX S-Function for Simulink, a dynamic system simulation package for MATLAB. The model inputs are the voltage at infinite busbar and the wind speed of each generator. Outputs are: induction machine stator currents, stator voltages, rotor mechanical speed, electromagnetic torque and active and reactive power; wind farm busbar voltages, currents in the distribution lines and surrounding loads; fault currents and currents in the connecting point with the infinite busbar [4]. A comparison of model implementation characteristics is shown in Table 1.

The N-machine model gives a set of 14 non-linear state equations for the wind farm. Furthermore, a set of 12 state-variables describes the electrical system that connects the bus at 20 kV with the network at 220 kV. The Multi-machine model gives a set of $(14*N+12)$ nonlinear state-equations. (In this case 432). Both mathematical models of the wind farm have been directed to the possibility of checking their performance under different conditions of the electrical network.

Table 1. Model implementation comparison

	N-machine	Multi-machine
Inputs	(3+N)	(3+N)
Outputs	10+15	(10*N+15)
Non-linear state equations	26	(14*N+12)

2.1   Normal operation

Several cases of the normal operation of the wind farm are studied. First, the case in which the wind speed of all wind turbines is vw= 13 m/s and suddenly decreases to vw= 8 m/s in 15 of 30 wind turbines. Figure 1 shows the evolution of the RMS value of the voltage at 20 kV busbar (the common busbar of the wind farm, where all the induction machines are connected). In this figure both responses are not identical. This is due to the fact that the N-machine model is a scaled machine that works with full wind power,

$$P_w = \sum_{i=1}^{N} P_w \,. \tag{1}$$

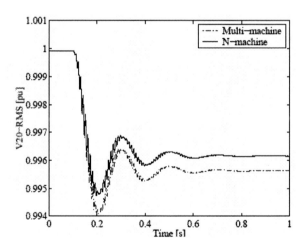

Figure 1. Voltage at 20 KV busbar. Wind speed decreases

The approximation of a total unique generating power for a unique N-machine is equivalent to suppose that all machines work at the same operation point. This operation point corresponds to the continuous curve in figure 2. It can also be observed in Figure 2 that, in fact, there are two operation points in the wind farm, related to the different wind speeds. The dotted upper curve shows the evolution of slip of the induction machines with vw= 13 m/s. The dotted lower curve is the evolution of slip with vw=8 m/s. And the equivalent wind farm operates with an intermediate slip.

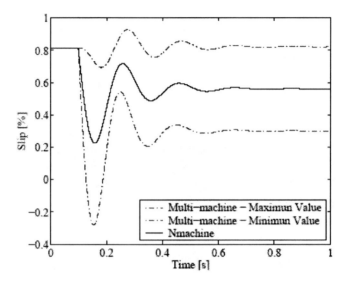

Figure 2. Comparative slip. (wind speed decreases)

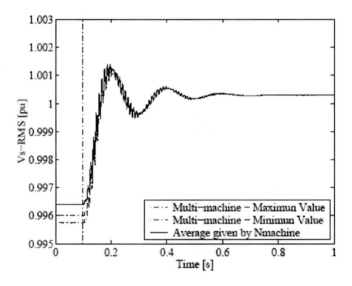

Figure 3. Wind generators stator voltage (wind speed increases)

Next, the effect of wind speed increase is analyzed. In this case 15 wind generators have vw= 13 m/s. The other 15, have initially a wind speed of vw= 8 m/s and suddenly increases to vw= 13 m/s. Figures 3 and 4 show the evolution of the RMS value of wind turbines stator voltage and the RMS value of subterranean lines currents, respectively. In this case, at initial conditions, two different operation points are observed in the Multi-machine model simulation and the N-machine works in some average operation point. When all the induction machines have the same wind power, both models responses are identical.

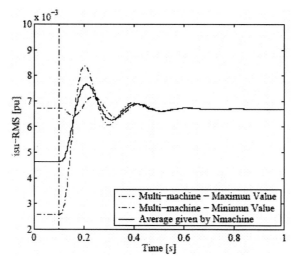

Figure 4. Subterranean lines currents (wind speed increases)

2.2   Short Circuits

In the study of short-circuits, all induction machines have different wind-speeds. Figure 5 shows the evolution of currents in subterranean lines that connect each machine with the common busbar of the wind farm in normal operation.

The impact of different short-circuits has been studied: one-phase, two-phase and three-phase resistive short-circuits. The resistance value is variable: from direct ground fault (R=0 pu) to a very resistive fault (R=1 pu). The fault can be placed at different positions into the intermediate network at 66 kV. In all cases the fault length is set to 100 msec.

Figures 6 and 7 show simulation results of single-phase fault located at the connection point of the wind farm with the distribution network. Figure 6 is the evolution of the RMS value of the voltage at the wind farm busbar (20kV).

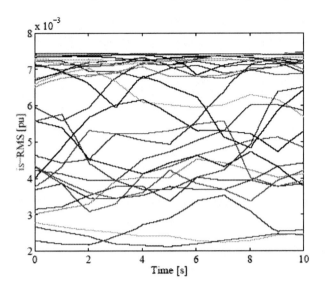

Figure 5. Subterranean lines currents (normal operation)

Figure 7 presents the evolution of the RMS value of currents at the connection point with the transport network at 220 kV (infinite busbar).

In this case, both models lead to practically indistinguishable results. This means that the transient evolution of external magnitudes could be evaluated with the N-machine model.

Figures 8 and 9 show simulation results of two-phase fault located at the connection point of the wind farm with the distribution network. Figure 8 is the evolution of the RMS value of the voltage at the wind farm busbar (20kV). Figure 9 is the evolution of the RMS value of currents at the connection point with the transport network at 220 kV (infinite busbar).

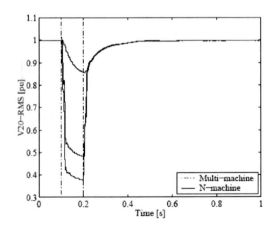

Figure 6. Voltage at 20 KV busbar (one-phase fault)

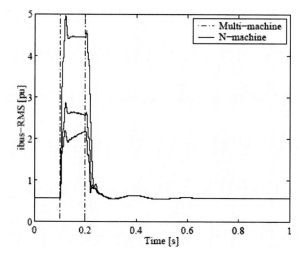

Figure 7. Currents in the infinite busbar (single-phase fault)

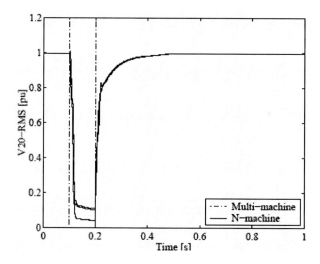

Figure 8. Voltage at 20 KV busbar (two-phase fault)

In this case, both models leads to similar results. It can be observed that the two faulted phases are coupled in voltages and in currents values.

Figures 10 and 11 show simulation results of three-phase fault located at the connection point of the wind farm with distribution network.

Figure 10 presents the evolution of the RMS value of the voltage at the wind farm busbar (20 kV). Figure 11 is the evolution of the RMS value of currents at the connection point with the transport network at 220 kV (infinite busbar).

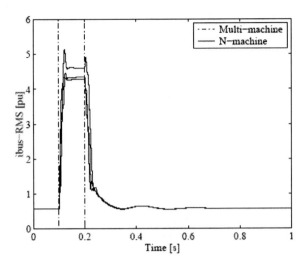

Figure 9. Currents in the infinite busbar (two-phase fault)

For the symmetric three-phase fault both models leads to similar results. It can be observed that all three phases have the same voltage drop and the same overcurrent behavior.

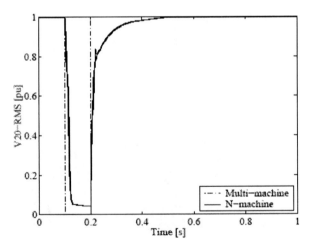

Figure 10. Voltage at 20 KV bus (three-phase fault)

Furthermore, the dynamic response of generators can be observed with the multi-machine model. Figures 12, 13 and 14 show the evolution of the stator currents of induction machines with different wind speeds when single-phase, two-phase and three-phase faults are located at the connection point of the wind farm with the distribution network.

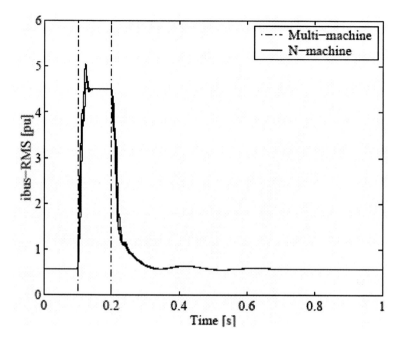

Figure 11. Currents in the infinite busbar (three-phase fault)

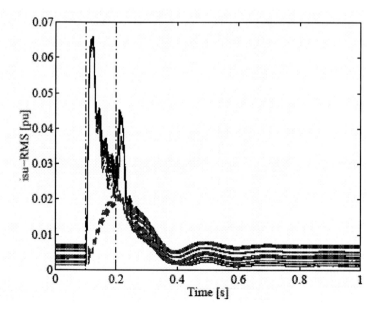

Figure 12. Induction machines stator currents (single-phase fault)

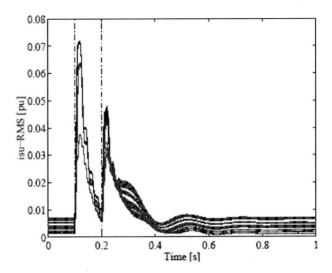

Figure 13. Subterranean lines currents (two-phase fault)

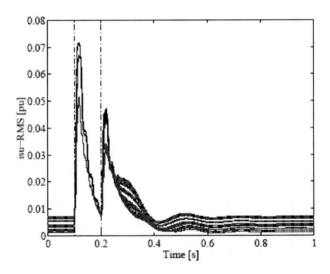

Figure 14. Subterranean lines currents (three-phase fault)

3.   CONCLUSIONS

The systematic modeling of the electrical network and general wind farm is the main contribution of the study carried out.

The N-machine models purpose is an in depth study of the fundamental magnitudes of the wind farm and the surrounding network. The simulation results show that the equivalent N-machine works at somewhat representative (but not real) slip. Nevertheless, the study of different short-circuits has revealed that this difference is not significant and the evolution of the main wind farm external electrical magnitudes can be simulated with the N-machine. Therefore, this modeling could support the design of the electrical protections and operational strategies for the wind farm taken as a whole.

On the other hand, the development of a complex Multimachine model explains the particular behavior of each generator of the wind farm. This model gives the power exchange between generators. With this model, different short–circuits have been simulated.

These experiments allow the evaluation of current flows and stabilization times. Both values are valuable information for the coordination of the system protection. The choice of the Multi-machine or N-machine model depends on the application.

ACKNOWLEDGEMENTS

We would like to thank the electric generating company IBERDROLA S.A. and especially R. Criado, J. L. Berasategi and M. Irizar for their helpful comments and for their contribution in providing the electrical line parameters. The authors are also grateful to J. Garde, from the induction generator manufacturing company, INDAR for his contribution providing the electric machine characteristics.

REFERENCES

1. X. ZUBIA, G. OSTOLAZA, G. TAPIA, A. TAPIA., *Dynamic behavior of a real wind farm in front of short-circuits and islanding operation.* (in *Proc. ICEM2002*, Vol. 1, pp. 122, and Conference CD).
2. P. VAS, *Vector Control of AC Machines.* (Oxford University Press, New York, 1990).
3. S. K. SALMAN, I. M. RIDA, *Investigating the impact of embedded generation on relay settings of utilities electrical feeders.* (IEEE Transactions on Power Delivery, Vol 16, N° 2, pp 246-251, 2001).
4. I. ZUBIA, X. OSTOLAZA, G. TAPIA, A. TAPIA, *Dynamic behavior of a real wind farm: Analysis of interactions between generators.* (in Proc. UPEC 2002).

Printed in the United Kingdom
by Lightning Source UK Ltd.
126901UK00006BE/2/A

9 781402 029325